Plumbing

A Practical Guide for Level 2

F

Plumbing

A Practical Guide for Level 2

Robert Boyce
Arnold Masterman

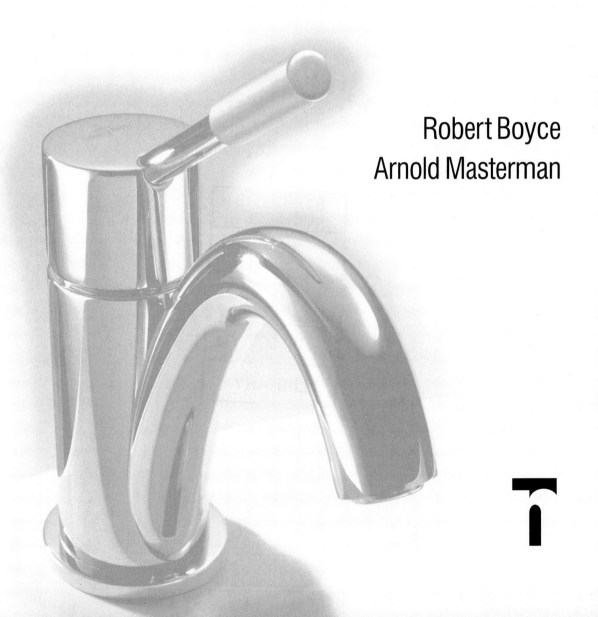

Published in 2005 by:
Nelson Thornes Ltd
Delta Place
27 Bath Road
CHELTENHAM
GL53 7TH
United Kingdom

05 06 07 08 09 / 10 9 8 7 6 5 4 3 2 1

A catalogue record for this book is available from the British Library

ISBN 0 7487 9275 9

Page make-up by Florence Production Ltd

Printed in Croatia by Zrinski

Contents

Preface

The introduction of a new City and Guilds specification, the updating of *Regulations* and the entry into the European Community has brought about many changes in the teaching and of the requirements of the modern-day Plumber.

This book is written to include all these changes, the text follows closely the requirements of the new syllabus in Craft Technology, working processes and related associated subjects, the aim being to assist the student towards that qualification.

Although mainly written for those commencing a career in the *Mechanical Services Industry* it is also essential reading for the mature person or anyone wishing to keep abreast with the changing techniques and technology in this section of the *Construction Industry*. In addition to students in the *Mechanical Services Industry* this book should prove invaluable for *Technical Students*, and for students attending *Links, Foundation*, or *Youth Training Scheme* programmes.

THE BOOK AIMS TO

- Provide you with the information relating to mechanical services studies enabling you to be a competent crafts person and to obtain the necessary qualification.

- Provide you with support material and increase your knowledge and competence in this subject.

- Provide you with evidence of competence that may be included in your Assessment Portfolio and used to provide evidence of the underpinning knowledge specified in the various units.

Acknowledgements

The authors and publisher are grateful to the following for permission to reproduce textual material and illustrations:

The Swiss Society of Engineers for Figs 1.10, 1.12, 1.13, 1.14, 1.24, 1.25, 1.26 reproduced from *Berufskunde fur Spengler*; Tubela Engineering Co. Ltd for Figs 2.55, 2.56; HMSO for Table 2.6; Record and Ridgid for Fig. 2.71; Rothenberger for Figs 2.72, 2.85; BCIRA for Figs 2.97 and 2.98; Barking-Grohe for Figs 4.39–4.41; Potterton for Fig. 5.19; Landis and Gyr for Fig. 5.38; Armitage Shanks for Figs 6.1, 6.2, 6.3, 6.5–10, 6.13, 6.16, 6.17, 6.29; Twyfords for Fig. 6.31; The Lead Development Association for Figs 7.13–15, 7.17, 7.18, 7.21–23.

Every effort has been made to reach copyright holders, but the publisher would be grateful to hear from any source whose copyright they may unwittingly have infringed.

1 Safety

After reading this chapter you should be able to:

1. State the basic requirements of the Health and Safety at Work etc. Act 1974.
2. List the main hazards involved in handling materials and equipment, including scaffolding.
3. State the correct use of and maintenance procedure for tools.
4. State the important points to observe when lifting.
5. Explain the possible hazards of falling items, fragile roofs and welding.
6. State the requirements to observe when working from ladders.
7. State the safe working procedures for electrical tools.
8. List the problems associated with untidy sites.
9. Demonstrate general first aid including artificial respiration.
10. Name the correct equipment for fire fighting.
11. State the correct procedure in the event of an accident.

Introduction: safety is everybody's business

There are far too many accidents in the construction industry, many of which could be avoided with thought and common sense. The Government publishes a guide entitled Health and Safety At Work etc. Act 1974. If the procedures in this were adhered to by both employers and employees, then the number of accidents would be very greatly reduced.

Accidents are generally caused by people disregarding the recommended procedures. They may feel that accidents only happen to other people and that, in any case, they have done that operation many times before without any problem.

Regardless of how good an Act is, it will only succeed if all participants, both employers and workers, are awake to their obligations and respond accordingly. Failure to comply to this Safety at Work Act could lead to a criminal conviction. Employers are responsible to maintain the work place (i.e. site) in a safe environment and to instruct the employees of their responsibilities. In the case of large building sites there is generally a safety officer employed.

The two basic principles of accident prevention are:

1. Implement safe methods of working to reduce the chance of a mistake.
2. Implement precautions to reduce the chance of injury even if somebody does make a mistake.

This chapter looks briefly at matters of general safety and highlights many of the common causes of accidents in the construction industry. The procedure and the correct type of extinguishers to be used in the case of fire are listed. Basic first aid treatment and the methods of procedure of accidents is dealt with.

Health and Safety at Work etc. Act 1974

The Health and Safety at Work etc. Act 1974 (HASAWA) has wide implications. Its purpose is to provide the legislative framework to promote, stimulate and encourage high standards of health and safety and to ensure the welfare of all personnel at work, as well as the health and safety of the public as affected by work activities. It concentrates on the promotion of safety awareness and effective safety organization and performance channelled through schemes designed to suit the particular industry or organization. The Act is an enabling measure superimposed over existing health and safety legislation and consists of four main parts: Part I relates to health, safety and welfare at work, Part II relates to the Employment Medical Advisory Service, Part III relates to building regulations, and Part IV relates to a number of miscellaneous and general provisions.

The scope of the Act includes all 'persons at work', whether employers, employees or self-employed persons. It also covers the keeping and use of dangerous substances and their unlawful acquisition, possession and use. Requirements can be made imposing controls over dangerous substances in all circumstances, including all airborne emissions of obnoxious or offensive substances that are not a danger to health but would cause a nuisance or damage the environment.

All of the existing health and safety requirements operate in parallel with the HASAWA until they are gradually replaced by new regulations and codes of practice etc. made under the Act.

HASAWA objectives

The four main objectives of the HASAWA are as follows:

1 To secure the health, safety and welfare of all persons at work.

2 To protect the general public from risks to health and safety arising out of work activities.

3 To control the use, handling, storage and transportation of explosives and highly flammable substances.

4 To control the release of noxious or offensive substances into the atmosphere.

These objectives can be achieved only by involving everyone in health and safety matters. This includes:

■ employers and management

■ employees (and those undergoing training)

■ self-employed persons

■ designers, manufacturers and suppliers of equipment and materials.

Employers' and management duties

Employers have a general duty to ensure the health and safety of their employees, visitors and the general public. This means that the employer must:

1 Provide and maintain a safe working environment

2 Ensure safe access to and from the workplace.

3 Provide and maintain safe machinery, equipment and methods of work.

4 Ensure the safe handling, transport and storage of all machinery, equipment and materials.

5 Provide their employees with the necessary information, instruction, training and supervision to ensure safe working.

6 Prepare, issue to employees and update as required, a written statement of the firm's safety policy.

7 Involve trade union safety representatives (where appointed) with all matters concerning the development, promotion and maintenance of health and safety requirements.

Safety signs

Safety signs communicate information such as warning of a hazard, showing the way to a fire exit, or instructing employees to wear personal protective equipment. Regulations specify certain types to be used. There is a standard range of safety signs including visual, audio (spoken and acoustic), hand signals and pipework markings. Safety signs must be displayed in appropriate places and kept in good condition (colour coded).

Figure 1.1 *Safety Signs*

There are four main categories of safety signs which are:

1 *Prohibition* Informing people where they must not go or no smoking.

2 *Warning* Informing you of hazards ahead such as flammable materials.

3 *Mandatory* Telling people they must obey, i.e. wear protective clothing.

4 *Safe conditions* These show which way to go, such as fire exits.

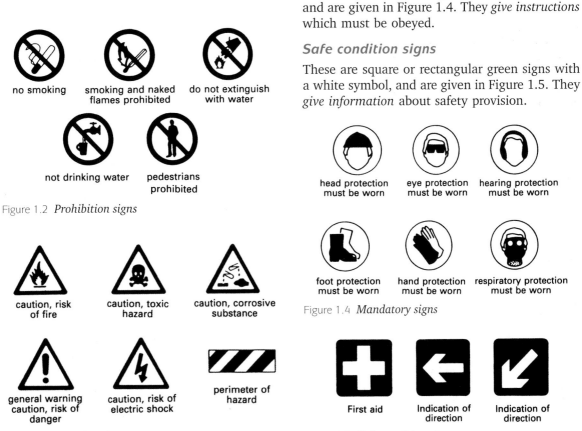

Figure 1.2 *Prohibition signs*

Figure 1.3 *Warning signs*

Prohibition signs

These are circular signs with a red border and white interior. Examples are shown in Figure 1.2. They are signs prohibiting certain behaviour.

Warning signs

These are triangular yellow signs with a black border and symbol, and are given in Figure 1.3. They *give warning* of a hazard or danger.

Mandatory signs

These are circular blue signs with a white symbol, and are given in Figure 1.4. They *give instructions* which must be obeyed.

Safe condition signs

These are square or rectangular green signs with a white symbol, and are given in Figure 1.5. They *give information* about safety provision.

Figure 1.4 *Mandatory signs*

Figure 1.5 *Safe condition signs*

Tools

Most accidents with tools are caused by workers striking themselves or other workers; through the use of defective tools, or through misuse of tools.

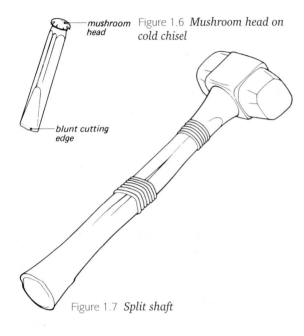

mushroom head

blunt cutting edge

Figure 1.6 *Mushroom head on cold chisel*

Figure 1.7 *Split shaft*

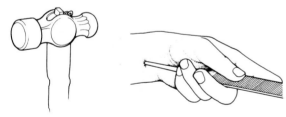

Figure 1.8 *Loose head*   Figure 1.9 *No handle on file*

Figure 1.10

Figures 1.6–1.9 show some typical examples of defective tools. They must be repaired immediately or taken out of service. Figures 1.10–1.12 show a few typical examples of the correct application of tools. The basic principles to follow are:

■ Make sure you use the right tool for the job – do not make do with the wrong one.

■ Wear unbreakable goggles when:

chipping welds;

de-scaling boilers;

cutting concrete, brick, etc.

■ When using Stillsons and wrenches make sure the pull forces the jaws together, otherwise the tool might slip away.

■ Never leave a defective tool about for others to use.

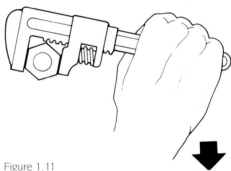

Figure 1.11

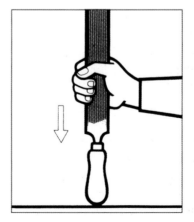

Figure 1.12

Lifting

The six major points to remember when lifting are:

1 Back straight.
2 Chin in.
3 Arms close to body.
4 Feet slightly apart.
5 Bend knees and lift by straightening the legs.
6 Grip with palm of hands, not just fingers.

The correct and incorrect methods of lifting are illustrated in Figure 1.13. The following checklist details the procedure to be observed:

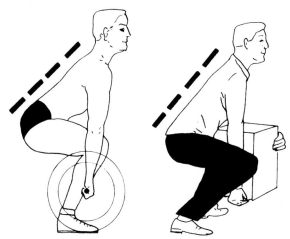

Figure 1.13 *Correct method of lifting*

- Size the job up: look out for splinters and jagged edges on the object to be lifted.
- If you are going to carry an object make sure you have an unobstructed path.
- Beware of slippery surfaces.
- Feet 200–300 mm apart, one foot in advance of the other, pointing in the direction you intend to go – feet together can cause a rupture.
- Chin in – avoid dropping head forwards or backwards.
- Bend the knees to a crouch position – back straight but not necessarily vertical.
- Arms as close to the body as possible so that the body takes the weight (instead of the fingers, wrists, arm and shoulder muscles).

- Get a firm grip with the palms of the hands and the roots of the fingers – using just the finger tips means more effort and more chance of dropping the object.
- Lift with the thigh muscles by straightening the legs.
- Lift by easy stages – from floor to knee, from knee to carrying position.
- Make sure you can see over your load when carrying.
- Do not change grip while carrying – rest the load on some firm support, then change.
- Reverse the lifting procedure to set the object down.
- Wear gloves when handling sharp or slippery objects.

Team lifting

The same basic principles apply when two or more men are lifting the same object (see Figures 1.14 and 1.15). In addition, remember the following procedure:

- Lifting gangs must work as a team.
- Everyone in a lifting team should be roughly the same height.

Figure 1.14

- The operation should be planned from 'lifting' to 'setting down' – route to be taken, signals, etc.
- Make sure everyone in the team knows precisely what to do.
- Appoint one person as team leader.

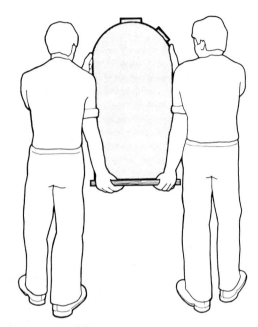

Figure 1.15 *Team lifting*

Lifting gas cylinders

Special care should be taken when moving gas cylinders. Use a trolley wherever possible. To lift a cylinder on to a trolley, first lift to the vertical position with a straight back and bent knees and then use your thighs as shown in Figure 1.16.

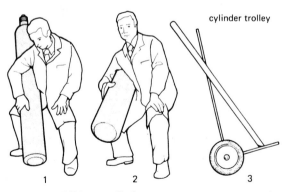

Figure 1.16 *Lifting a cylinder*

Falling materials

It is well to remember that an object gathers tremendous energy when falling. For example, a 19 mm nut falling and striking a person's unprotected head from 20 m high can kill.

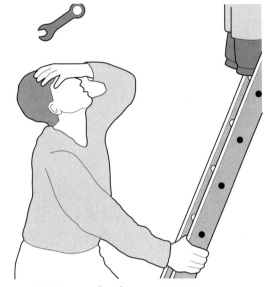

Figure 1.17 *Lacerated scalp*

It is therefore essential that great care must be taken not to place tools, tins of jointing paste, etc. on pipes or ledges, where they could topple off. Always observe the following safety rules:

- Do not place materials where they are likely to fall or get knocked off.
- If possible position yourself and other workers where there is no danger from falling objects.
- When working under scaffold make sure it has a toe-board.
- Lower materials – never throw things down from scaffolds or ladders or throw things to other workers on the ground.
- Make sure materials are securely stacked and withdrawn in the correct order.
- Always wear safety helmets and safety boots.
- Provide tool boxes for tools.

Stacking materials

Location

Materials should be readily accessible and as close as possible to the point of use, but:

- Not in quantities so great as to limit working space unnecessarily.
- Not where they will cause obstruction.
- Not close to edges of excavations.
- Not close to moving machinery or overhead lines.
- Not where they will interfere with new deliveries.

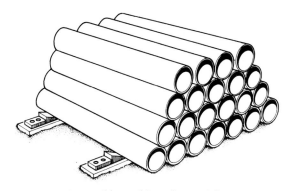

Figure 1.18 *Acceptable stacking of materials*

Foundations and supports

A firm, even base is essential. The foundation – for example, the floor – must be strong enough to support the total weight, which may be considerable. Stacking materials against a wall may be dangerous, as the wall may not be designed to take the sideways thrust.

Size

Stacks should not normally be higher than a man, to permit easy withdrawal. The shorter base should be about one-third the height.

Structure

Batter (i.e. step back) every few tiers. Chock or stake rolling objects (for example, drums, pipes) with sound material. Bond to prevent the stack collapsing. Avoid unnecessary protrusions – protrusions which cannot be avoided should have a distinctive marker tied to them. Oxygen cylin-

ders may be stacked horizontally. Acetylene cylinders *must* be stored and used vertically.

Use

Withdraw materials from top of stack – never from bottom or sides. Do not climb on to stacks – use a ladder.

Excavations

Accidents can arise when people slip into trenches, sometimes while trying to jump across them or while climbing out of them or when supports give way. Other accidents are due to people falling into uncovered manholes while walking across a site. Always observe the following safety rules (see Figures 1.19 and 1.20):

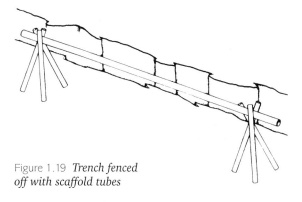

Figure 1.19 *Trench fenced off with scaffold tubes*

Figure 1.20 *Mark covers over holes*

- Warn people to look where they are going.
- Provide proper walkways across trenches.
- Erect barriers round excavations where necessary.
- Do not leave manholes uncovered or unfenced.

- Take care when doing manhole work.
- Fence openings or cover them with heavy material appropriately marked.
- Never walk along pipes.

Ladders

Used incorrectly or in bad condition a ladder becomes a hazard. Working from a ladder is inherently dangerous. Where possible always provide a working platform. Most falls from ladders are the result of a person simply slipping or falling from the ladder, but movement of the ladder also causes a considerable number of accidents – the ladder either slips outwards at the bottom or sideways at the top. Accidents are also caused by missing or broken rungs or by the ladder itself breaking. Always observe the following points:

Use of ladders

- Stand on a firm, even base. Trussed side underneath if reinforced.
- Never wedge one side up if ground is uneven. Either level ground or bury foot of ladder.
- Set at correct angle of repose – four up to one out. The ladder should project at least 1 m above any landing place.
- Beware of wet, icy or greasy rungs. Clean any mud or grease from boots before climbing.
- Watch out for live overhead cables, particularly when using metal ladders.

Typical examples of the incorrect use of ladder which results in accidents

Figures 1.21–1.26 show six examples of the dangerous uses of ladders.

- Never use a ladder which is too short or stand it on something, for example a drum or dustbin, to get extra height (Figure 1.21).
- Do not over-reach sideways from a ladder – move it (Figure 1.22).
- Never support a ladder on its rungs (Figure 1.23).
- Never work on a ladder set at the incorrect angle (75° is recommended) (Figures 1.24 and 1.25).

Figure 1.21 *Asking for trouble*   Figure 1.22 *Never over-reach*

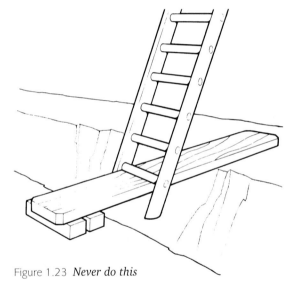

Figure 1.23 *Never do this*

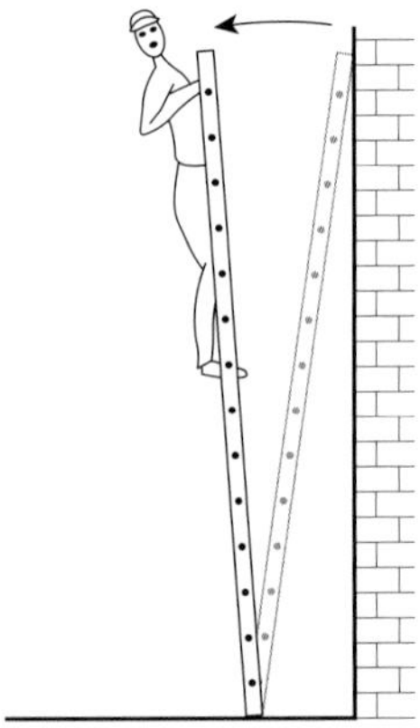

Figure 1.24 *Too steep*    Figure 1.25 *Not steep enough*

▪ Never overload a ladder (use a hoist) (Figure 1.26).

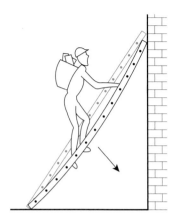

Figure 1.26 *Never overload a ladder – use a hoist*

Correct method of climbing and descending a ladder

1 Be aware of your limitations.
2 Check ladder for security
3 Clean mud, etc. from your footwear.
4 Face ladder squarely. Using both hands either
    (a) grasp stiles, or
    (b) grasp rungs (fireman fashion).
5 Keep feet placed well into the rungs.
6 Eyes should be directed at working level above – do not look down.
7 Don't carry anything in your hands. Tools and materials may fall when carried up ladders, even if carried in pockets. Where possible provide a hoist line. Alternatively, use a shoulder bag.

Securing ladders

Too much importance cannot be placed on this part of your work. In addition to the obvious lashing of the top of the ladder, the foot of the ladder should also be suitably anchored. When it is not possible to tie the top then side guys should be used. These should be secured to the stiles (never the rungs) and should form an angle with the horizontal of approximately 45° (see Figures 1.27 and 1.28).

Figure 1.27 *Ladder staked and guyed*

Figure 1.28

Platforms

A hook-on foot is useful in providing something more comfortable than a rung to stand on (see Figure 1.29). Foot platforms should be capable of being easily fitted and removed and should give a level surface with the ladder placed at 75°. Platforms are designed to protect either in front of or behind the ladder. One disadvantage is they can be difficult to climb past. Various hook-on tray attachments are available to enable tools and components to be readily accessible.

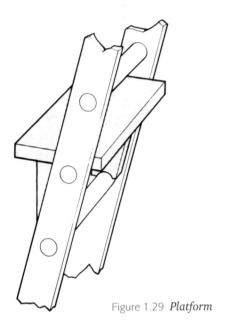

Figure 1.29 *Platform*

Standoff

These fitments hold the top of the ladder off from the wall (see Figure 1.31). They are particularly useful when working on gutters, etc. as they overcome the need to lean outwards. It is advisable to secure them to the ladder to prevent side-to-side movement.

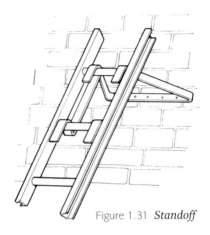

Figure 1.31 *Standoff*

Correct positioning of a ladder

The recommended ideal working angle of a ladder is 75° to the horizontal or one unit out to four units up (see Figure 1.32). It is also recommended that the top of the ladder should extend 1 m past the working platform. The lift of a single ladder should not be more than 8–10 m.

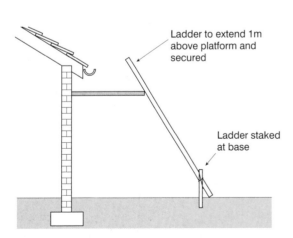

Ladder to extend 1m above platform and secured

Ladder staked at base

Figure 1.30

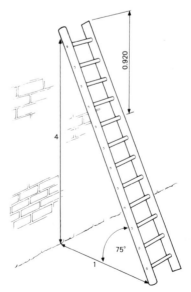

Figure 1.32

Raising and carrying a ladder

Figure 1.33 illustrates the correct method of raising a ladder.

Ladders should be carried vertically or with the front end elevated (see Figure 1.34). It takes two people to move a tall ladder.

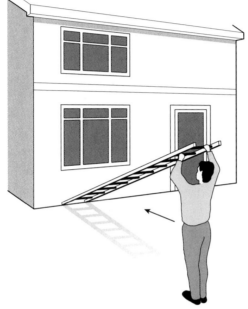

Figure 1.33 *Correct method of raising a ladder*

Figure 1.34

Storage of ladders

Do not leave ladders on wet ground or leave them exposed to the weather. Store at normal temperatures under cover to prevent warping. Support at intermediate points and not just at the ends (see Figure 1.35).

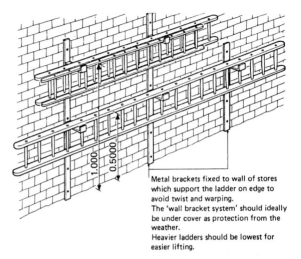

Metal brackets fixed to wall of stores which support the ladder on edge to avoid twist and warping.
The 'wall bracket system' should ideally be under cover as protection from the weather.
Heavier ladders should be lowest for easier lifting.

Figure 1.35 *Correct method of storage*

Inspection and maintenance

Always inspect a ladder before use. Ladders should *not* be subjected to a severe test.

- The rungs, particularly at the point where they enter the stiles.
- The wedges, which should be properly in position.
- The stiles for warping, cracking or splintering.
- The condition of the feet.
- Any ropes or metal attachments.
- To test the rungs tap each rung with a mallet – a dull sound indicates a defective rung.

Take defective ladders out of service immediately, mark them defective and do not use them again until they are repaired. Destroy unfit ladders which cannot be repaired.

Treat new ladders with clear wood preservative, particularly round the end grain of the rungs and coat with clear varnish. Painting a ladder is not recommended as the paint may hide defects.

Electrical work

The following are the most common causes of faults in electrical apparatus:

1 Improvised junction boxes with, for example, wires jammed in sockets with matchsticks or nails.

2 Insulation damage through flexing – cables should be protected by heavy rubber sleeves at the point where they enter the tool plug.

3 Earth wires pulled out of terminals and touching live conductors. (When making connections the earth wire should have some slack so that, if the cable tends to pull out, the earth wire will be the last to fail.)

4 Powered tools run off lamp sockets so that they cannot be earthed. There is also the danger that the earth wire, if tucked into the lamp adaptor, may make contact with a live conductor.

5 Wrong connections in plugs or joints, usually caused by unauthorized persons tampering with electrical apparatus or confusion of British and Continental colour codes.

6 Plugs forced into the wrong socket.

7 Use of incorrect fuses. Fuses should be rated as closely as possible to the normal working current of the tool. Using a 30-A fuse for a 3-A load will result in greater injury should there be an accident.

8 Cables lying around where they can get damaged or wet.

9 Cables hung up on nails, is not a secure method of fastening and can damage the insulation. Proper cleats should be used.

Remember

Be especially careful of overhead power lines when using mobile scaffolding. Do not work near trailing cables – get them suspended. Inexperienced people should not tamper with electrical connections. In case of failure in breathing owing to electrical shock, artificial respiration must be started immediately.

Always use 110v supply, centre tapped to earth for portable equipment.

Powered tools

In inexperienced hands, powered tools can become a source of danger, but are much safer now due to:

1 Double insulated tools.

2 Power charged tools.

Electrically operated tools

■ Look our for:
faulty leads;
trailing leads;
faulty plugs;
unearthed equipment.

■ Check that all tools are properly earthed and insulated before use. Make sure that leads are suspended and not trailing in oil or water.

■ Be especially careful with unguarded threading and cutting machines.

■ If a machine has a guard, make sure it is fitted.

Cartridge operated tools

■ Anyone using these tools must be fully instructed in their use.

■ Unbreakable goggles must be worn.

■ *Always* check the material into which the fixing stud is to be fired – make sure there is no danger of the stud going right through the material.

■ Make sure protective guard is fitted.

■ Never leave unexploded cartridges lying about.

■ Always unload when not in use.

Goggles

■ Make sure goggles are of the unbreakable sort.

■ *Always* wear goggles when:
grinding;
cutting concrete;
using a cartridge operated tool;
drilling metal.

■ Wear goggles over your eyes – they are no good on top of your head (see Figure 1.36).

Figure 1.36
Goggles – wear them over your eyes

Scaffolding

Falls from scaffolds are usually serious, sometimes fatal. The main causes are:

1 Badly constructed scaffolds.
2 Alterations.
3 Obstructed walkways.
4 Absence of guard-rails and toe-boards.

Defects to look for

Something removed and not replaced.
No toe-boards.
No guard-rails.
Split or knotted boards.
Loose boards.
Overlapping or protruding boards.
Gaps between boards.
Obstructed walkways.
Uneven foundations.
Bent or rusty poles.
No bracing.
No tie-ins.
No base plates.
Wrong couplers used.
Worn or rusty couplers.
Ledgers protruding.

Use of scaffolds

■ Erection, dismantling and alteration of scaffolding should be done by experienced workers.
■ *Always* inspect scaffolding before starting work on it and also after wet or frosty weather.
■ If you have to alter scaffolding to get a particular job done get it altered by a scaffolder and make sure the alteration is made good after the work has been done.
■ Never overload a scaffold – if in doubt, find out what the safe load is.
■ Make sure stacked materials cannot fall off – get wire mesh frames between the guard-rail and toe-board.
■ Keep scaffolds tidy.
■ Provide unobstructed passageways for workers and materials.
■ Use the access ladders – never jump from or climb up and down a scaffold.

■ Hoist materials up and down – don't throw them.
■ When using suspended scaffolds or cradles, make sure that the rope, pulley and hoisting gear are in good condition and that the cradle or scaffold is adequately counterbalanced for the weight it is taking.
■ When using a trestle scaffold make sure it is firm and that the platform is properly supported.

Fragile roof coverings

The most common roof covering in this category is either plain or corrugated. It can look and feel deceptively strong and safe but is liable to break suddenly under a concentrated load, i.e. a person standing or walking on it.

Safety rules

1 Warning notices should be clearly displayed on all asbestos and other fragile roofs (see Figure 1.37).
2 No person to work directly off the roof.
3 Loads to be distributed over as large an area as possible by means of battens and/or crawling boards (roof ladders) (see Figures 1.38 and 1.39).

Figure 1.37

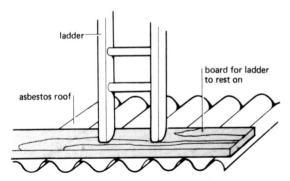

Figure 1.38 *Method of support when working on a fragile roof*

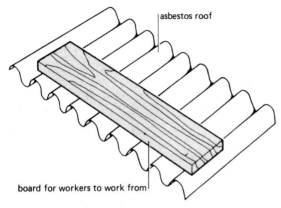

Figure 1.39

Welding

Eye protection and burns

Great care must be exercised by welders to avoid the possibility of workers being burned or fires or explosions.

Gas cylinders

- Keep all gas cylinders, especially the regulators, clear of oil, grease and dirt; do not handle them with greasy hands; spontaneous combustion may occur if gas and oily vapours mix.
- Do not expose gas cylinders to excessive heat.
- Gas cylinders should not be loaded loosely on lorries.

- Store oxygen cylinders separate from cylinders containing combustible gases – acetylene, propane, butane, hydrogen, coal gas.
- When storing or stacking oxygen cylinders make sure they are secured – do not lean them up against a wall. They must not be stacked more that four high.
- Acetylene cylinders and liquefied petroleum gas cylinders must be kept upright, in storage and in use.
- Propane should be stored above ground in a place with adequate ventilation and away from excavations. Propane is heavier than air, and if stored below ground level or near excavations any leakage will collect in the lower level.

Precautions to be taken during gas welding

- Wear protective clothing – goggles and gloves.
- Wear a protective apron when sparks are flying.
- Make sure the surrounding area is free of combustible materials and that the cylinders are clear of falling sparks.
- Use protective blankets to cover materials which cannot be moved.
- Keep hoses clear of walkways.
- Purge hoses before using equipment – the explosion of mixed gases in hoses causes the majority of welding accidents.
- Check all equipment before use, especially hoses and regulators for leaks.
- Use soapy water to check for leaks *not* a naked flame.
- See that the nozzle of the blow pipe is free from obstruction.
- Mark completed work '*hot*'.
- If welding in a confined space, see that adequate ventilation is provided.

Welding or cutting tanks or vessels

- Tanks or vessels which have contained inflammable or explosive materials should always be cleaned out thoroughly before welding or cutting.

- Clean the container by steaming or with boiling water. Resteam daily if work is to continue on following days.
- *Never* blow out the container with oxygen.

Tidy site and workshop

It may surprise you to learn that untidiness is a major contributor to accidents involving people falling. Tidiness is everybody's business.

Safety rules

1 A basic safety rule is 'keep the site tidy' (not like Figure 1.40).
2 Keep walkways free of obstructions.
3 Watch where you are going.
4 Remove any hazard you come across, even those left by others.
5 Nails should be removed or knocked down flat in discarded timber (not left as in Figure 1.41).

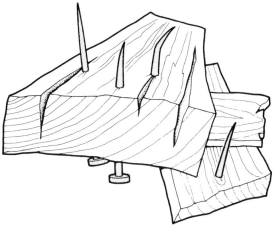

Figure 1.41 *Discarded timbers*

Fire

Section 51 of the Factories Act requires that in every factory and building site there shall be provided an appropriately maintained means of

Figure 1.40 *Unsafe and inefficient*

fighting fire, which shall be so placed as to be readily available for use.

For fire fighting in rooms where there are no exceptional risks to life, portable fire extinguishers are normally sufficient, provided that the right type is used. For fire fighting in certain buildings, such as factories and large commercial premises, it is necessary to install a fixed system of sprinklers because of the risk of fire breaking out after working hours. Premises which are deemed to have a fire certificate fall within: Section 40 of the Factories Act 1961; Section 29 of the Offices, Shops and Railway Premises Act 1963; Hotels and Boarding Houses (Fire Precautions) Act 1971; and all premises specified under Section 1 (2) of the Fire Precautions Act 1971 as and when designated by a Commercial Order.

Sources of fire danger

- Heating devices and stoves.
- Electric wiring and equipment.
- Flammable liquids, materials and pastes such as fuels, lubricants, paint, timber, and oily rags.
- Welding operations, particularly below overhead welding.
- Beneath and near hot roofing.

- All areas exposed to sparks and heat if refuse burning takes place.
- Compressors, engine generators and all internal combustion engines and their fuel supplies.

Fire precautions

- Maintain free access from street and hydrant.
- Provide hosing where possible near water taps.
- Provide and maintain fire extinguishers.

Fire extinguishers

- Provide the proper kind of extinguisher for risk (see Table 1.1 and Figure 1.42).
- Extinguishers should be inspected regularly and recharged immediately after use.
- Protect from freezing in cold weather by enclosing them.
- Instruct workers in the use of extinguishers.

General first aid

This advice is concerned only with first aid. It is not a substitute for attention by a doctor or a trained nurse. If medical aid is going to be

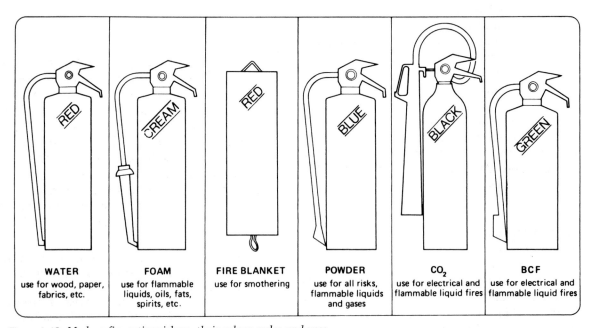

WATER	FOAM	FIRE BLANKET	POWDER	CO$_2$	BCF
use for wood, paper, fabrics, etc.	use for flammable liquids, oils, fats, spirits, etc.	use for smothering	use for all risks, flammable liquids and gases	use for electrical and flammable liquid fires	use for electrical and flammable liquid fires

Figure 1.42 *Modern fire extinguishers, their colour codes and uses*

Table 1.1 *Choice of portable extinguisher and extinguishing agents*

Class of fire risk	Remarks	Water	Dry powder e.g. sodium bicarbonate	Carbon dioxide	Foam	Chlorobro-momethane
Wood, cloth, paper or similar combustible materials	Water best agent	Most suitable	Not recommended	Not recommended except for small surface fires	Not recommended except for small surface fires	Unsuitable – dangerous fumes given off
Flammable liquids – petrols, oils, greases, fats	Smothering effect required	Unsuitable	Most suitable for general use	Most suitable where contamination by deposits must be avoided	Most suitable where reignition risk is high*	Effective for small fires but dangerous fumes given off
Live electrical plant but not including electric wiring or individual electric motors	Smothering required by non-conductor of electricity	Unsuitable – dangerous	Suitable	Suitable	Unsuitable – dangerous	Effective but dangerous fumes given off

*Special foams required for liquids which mix with water.

needed urgently, send for a doctor or ambulance immediately.

General

If the casualty has stopped breathing from whatever cause, artificial respiration must be started at once before any other treatment is given and should be continued until breathing is restored (see page 19). Where there is shock, keep the casualty lying down and comfortable. Cover with a light blanket or clothing but do not apply hot water bottles. Do not give drink or anything by mouth if there seems to be an internal injury. Wash your hands before treating wounds, burns or eye injuries. Always wear surgical gloves when treating injuries.

Minor wounds and scratches

All wounds and scratches, even minor ones, should receive attention immediately. Delay increases the risk of infection. Cover the wound as soon as possible with a sterilized dressing* or adhesive wound dressing.§ If it is necessary to clean the skin round the wound, avoid washing the actual wound because this can wash germs

into it. Warn the casualty that this is the first dressing and that further attention may be needed; if an injury becomes inflamed, hurts or festers, then get medical attention.

Serious injuries

Bleeding

Stop the bleeding at once and send promptly for a doctor or an ambulance. To control bleeding by direct pressure, apply a pad of sterilized dressing or bandage firmly, adding, if need be, sterilized cotton wool; finally apply a triangular bandage. It will sometimes be possible to stop arterial bleeding by pressing the artery with the finger or thumb against the underlying bone.

If bleeding cannot be controlled by direct pressure, a rubber bandage or pressure bandage may be applied to a limb between the wound and the

---

* *Sterilized dressing:* an unmedicated complete dressing with bandage, sterilized and put up in an individual sealed packet.
§ *Adhesive wound dressing:* special type of dressing approved by certificate of HM Chief Inspector of Factories.

heart for no longer than 15 minutes at a time, pending medical attention. It is most important that this time limit is not exceeded.

Fractures

Do not attempt to move a casualty with broken bones or injured joints until the injured parts have been secured with triangular bandages so that they cannot move. An injured leg may be tied to the uninjured one, and the injured arm tied to the body, padding between with cotton wool.

Electric shock

Switch off the current. If this is impossible, free the person using something made of rubber, cloth or wood or a folded newspaper; use the casualty's own clothing *if dry*. Do not touch the skin before the current is switched off. If breathing is failing or has stopped, give artificial respiration and continue for some hours if necessary. Get help and send for a doctor.

Gassing

Carry the casualty into fresh air; do not let them walk. If breathing has stopped, give artificial respiration, get help and send for a doctor or an ambulance.

Mild cases should be kept resting and, after recovery, sent home by car.

Special injuries

Burns and scalds

If serious, send promptly for a doctor or ambulance.

Put a sterilized dressing on the burn or scald.

Never use an adhesive wound dressing. If extensive, cover with clean cold water towels and secure loosely. Do not burst or remove clothing sticking to the burn or scald.

Chemical burns

Remove all contaminated clothing and flush the burn with plenty of cold water. Apply a sterilized dressing.

Eye injuries

Something in the eye

If the object cannot be removed readily with sterilized cotton wool moistened with water, or if the eye hurts after removal of the object, cover the eye with an eye-pad* and bandage firmly so as to keep the eye shut and still. Send the casualty to a doctor or hospital quickly. If there is likely to be considerable delay in getting medical attention first insert eye ointment.§

Injury from a blow

Cover the eye with an eye-pad and send the casualty at once for medical attention. Do not apply an eye ointment.

Chemical in the eye or chemical burn

Flush the open eye at once with clean cold water and continue washing the eye for at least 15 minutes. (A good method is to get the casualty to place his/her face under the water and blink the eye.)

Then cover with an eye-pad. Do not apply eye ointment. Send the casualty to a doctor or hospital quickly.

Heat burns

Cover the eye with eye-pad and send the casualty immediately to a doctor or hospital. Do not apply eye ointment.

Bandaging for eye injuries

The eye-pad is kept in place by the covering bandage running under the ear next to the injured eye and above the other ear.

Arc-eye

Apply cold compresses to the eyes and bathe them with an astringent lotion, obtainable from a chemist. Do not use any eye-drops unless prescribed by a doctor.

First aid box

They should be adequately stocked and should always be available. All items used should be replaced immediately. It is also good practice to have the names, addresses and telephone numbers of qualified helpers, i.e. local first aiders,

---

* *Eye-pad:* a pack containing a sterilized pad with a long bandage attached.
§ *Eye ointment:* an ointment approved by certificate of HM Chief Inspector of Factories.

doctors and hospitals permanently displayed inside the lid of the box.

Where more than 50 people are working on a site, a qualified first-aider must be in charge of the box.

Contents of box

There is no standard list of items to put in the box, it depends on the type of work and any special risk in the workplace. The following list of items is only a guide but should be taken as the minimum requirements.

Resuscaide

Pair of gloves

Medium dressings × 2

Large dressings × 2

Small dressings (eye pad) × 2

Triangular bandage × 1

Pack of gauze × 1

Adhesive dressings approx 20

Mepore dressings × 2

Crepe bandage × 1

Ice pack

Safety pins × 6

Cleansing wipes

Clinical wastes bags

Tweezers × 1 (construction)

Paper work in box or available

First aid log

Head injury advice slip

Content list/First aid Guideline

You should not keep tablets or medicines in the first-aid box.

What is an appointed person?

This is someone chosen to:

1 Take charge when someone is injured or falls ill, including calling the ambulance if needed.

2 To look after the first-aid equipment, and to restocking the first-aid box.

Appointed persons should not attempt to give first-aid for which they have not been trained.

What is a first aider?

This is someone who has undergone a training course in administering first-aid and holds a current first-aid at work certificate. *The training has to have been approved by* **HSE**. Lists of first-aid training organizations in your area are available from HSE.

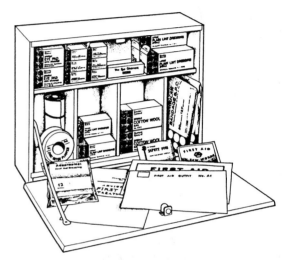

Figure 1.43 *First aid box*

Artificial respiration

Electric shock, gassing, drowning or choking may cause breathing to stop. In any of these cases *artificial respiration must be started without delay*. Do not find help if you are alone – only go for help when the patient is breathing.

Mouth-to-nose respiration is by far the most effective method of artificial respiration – the mouth-to-mouth should only be used if the mouth-to-nose is impossible.

Mouth-to-nose respiration

1 Lay the patient on his or her back, and, if on a slope, have the stomach slightly lower than the chest.

2 Make a brief inspection of the mouth and throat to ensure that they are clear of obvious obstruction.

3 Give the patient's head a backwards tilt so that the chin is prominent, the mouth closed

and the neck stretched to give a clear airway (see Figure 1.44).

4 Open your mouth wide, make an airtight seal over the nose of the patient and blow. The hand supporting the chin should be used to seal the patient's lips (see Figure 1.45).

5 After blowing, turn your head to watch for chest movement, while inhaling deeply in readiness for blowing again.

6 If the chest does not rise, check that the patient's mouth and throat are free from obstruction and that the head is tilted back far enough.

7 Blow again.

8 If air enters the patient's stomach through blowing too hard, press the stomach gently with the head of the patient turned to one side. If at any time the patient vomits, turn the head to one side so that he or she cannot inhale the vomit.

9 Commence resuscitation with four quick inflations of the patients chest to give rapid build up of oxygen in the patient's blood and then slow down to 12 to 15 respirations per minute or blow again each time the patient's chest has deflated.

Alternative method

An alternative method to the direct mouth to mouth resuscitation is for the casualty to lay on

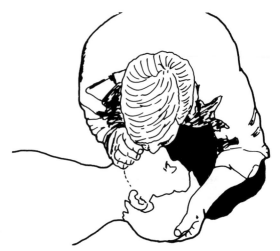

Figure 1.45 *Blow down nose*

the floor (ensure the airway is clear) place an oval shaped shield over the nose and mouth as shown in Figure 1.47 then proceed as for mouth to mouth resuscitation as described. Some types of shield have a side outlet in case the person should vomit. If no shield is available use a handkerchief over the nose and mouth to overcome direct person to person contact.

There are several purpose made aids from simple face shields to quite elaborate ones.

Figure 1.44 *Tilt head back*

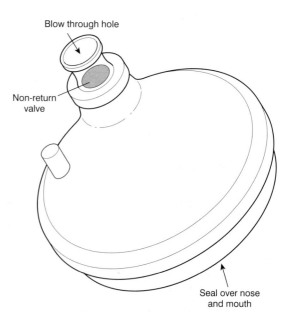

Blow through hole

Non-return valve

Seal over nose and mouth

Figure 1.46 *A shield*

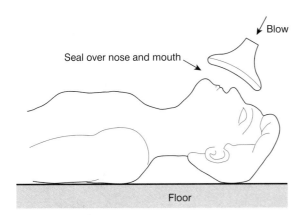

Figure 1.47 *Use of a shield*

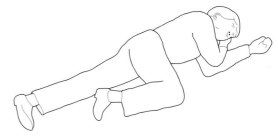

Figure 1.48 *Recovery position*

First Aid

When an accident occurs you should endeavour to follow the following procedure. Assess the situation (do not put yourself in danger).

1 Make the area safe.
2 Assess the casualties and attend to any unconscious ones first.
3 Follow the advice given as outlined.

Check for consciousness

This is done by a gentle shaking of the shoulders and shouting, should there be no response then carry out what is known as the **A. B. C.** of resuscitation.

'A' Airway
To open the airway:

1 Place one hand on the casualty's forehead and gently tilt the head back.
2 Remove any obvious obstruction from the casualty's mouth.
3 Gently lift the chin to give a clear airway.

'B' Breathing
Look along the chest, listen and feel at the mouth for normal signs of breathing, for no longer than 10 seconds.

If the casualty is breathing

1 Place in recovery position and ensure the airway remain open.
2 Send for help and monitor the casualty until help arrives.

If the casually is not breathing

1 Send for help.
2 Keep the airway open by maintaining the head tilt and the chin lift.
3 Pinch the casualty's nose closed and allow the mouth to open.
4 Take a full breath and place your mouth around the casualty's mouth making a good seal.
5 Blow slowly into the mouth until the chest rises.
6 Remove your mouth and let the chest fall fully.
7 Give a second breath then look for signs of circulation (see 'C').
8 If signs of a circulation are present continue breathing for the casualty, recheck about every 10 breaths.
9 If breathing starts but the casualty remains unconscious, put in recovery position.

'C' Circulation
Look, listen and feel for normal breathing, coughing or movement by the casualty, for not longer than 10 seconds.

If there are no signs of circulation, or you are unsure, immediately start chest compressions.

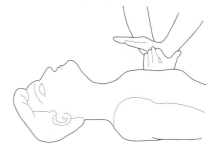

Figure 1.49 *Chest compressions*

1 Lean over the casualty and with straight arms, press vertically down 4 to 5 cm on the breastbone, then release the pressure.

2 Give rapid chest compressions (a rate of approximately 100 per minute) followed by two breaths.

3 Continue alternating 15 chest compressions with two breaths until help arrives or the casualty shows signs of recovery.

A typical accident report form is shown in Figure 1.50.

COSHH

COSHH provides a legal framework to protect people against health risks from hazardous substances used at work and means the control of substances hazardous to health; this legislation came into force in 1994.

Thousands of people are exposed to hazardous substances at their place of work. If the exposure is not prevented or properly controlled, it can result in serious illness, sometimes even death. Hazardous substances can be found in all kinds of construction, building and plumbing work environments and unless the correct precautions are taken, they can threaten the health of workers and others exposed to them. COSHH lays down a step-by-step approach to those precautions and sets out the essential measures that employers and employees have to take.

Substances or mixtures classified as dangerous to health are identified by their warning label (very toxic or toxic, harmful or irritant, corrosive) and the supplier must provide a safety data sheet for them.

It is important that the words hazard and risk are clearly understood for the correct interpretation of COSHH. The 'HAZARD' presented by a substance is its potential to cause harm. It could make your skin sore or damage your lungs. The 'RISK' from a substance is the likelihood that it will harm people in the actual circumstances of use.

Specialist literature is available from the Health and Safety Executive, which explains the role and responsibilities of employers and employees when dealing with substances, which are hazardous to health.

RIDDOR

RIDDOR requires the reporting of work-related accidents, diseases and dangerous occurrences and means The Reporting of Injuries, Diseases and Dangerous Occurrences Regulations, this legislation came into force in 1996.

RIDDOR applies to all work activities but not to all incidents. Reporting accidents and ill health at work is a legal requirement. The information obtained enables the enforcing authorities to identify where and how risks arise and to investigate serious accidents.

Specialist literature is available from The Health and Safety Executive, which explains the role of employers and employees and defines injuries, dangerous occurrences and diseases that are reportable. A sample reporting form is shown in Figure 1.51.

Risk assessment

Most of us spend a great deal of time at work, so we need to be happy and safe in our work environment. Employers and employees need to work together for the good of all, it makes sense for us all to watch out for any hazards and then to minimize the risks.

Relating to the risk assessment 'HAZARD' means anything that could cause harm, like chemicals, working on a scaffold or in a trench, working with heat or in confined spaces.

'RISK' is the chance (big or small) that someone may be harmed by it. Risk assessment does not have to be complicated but it does need to cover those risks involved in the work process. In order to assess the risk we must first look at the hazards. Begin by walking around your place of work and try to spot anything that could reasonably be expected to cause harm. Don't worry about the trivial but concentrate on the more important hazards, ask your workmates as they may know something you don't. Look at manufacturers data sheets (COSHH), check out the accident book and health records. If patterns emerge, it could indicate that there is a problem. Next you should think about the people who are most likely to be at risk, including normal workers, maintenance workers, cleaners, visitors and members of the public. At this point it is usual to

ACCIDENT REPORT FORM

Ref No: _____
(office use only)

1	Full Name of injured person (Mr. Mrs. Miss. Ms) Please Print

2 Home Address
Please Print

Postcode _____ Home Telephone Number _____ Work Telephone Number _____

3 Age _____   Sex :- Male / Female   Occupation _____

4 STATUS OF PERSON (Please tick relevant box)	Employee	Contract Staff	Student	Participant	Self Employed Person	Member of Public

5 Date and Time of Accident	6 Which Location	6a Where did the accident occur (eg: office, canteen, courtyard, workshop, project area, staircase, corridor) **GIVE ROOM NUMBER, IF APPROPRIATE**
Date _____ Time _____		

7 Injury Type	7a Body Part

8 FULL DETAILS OF ACCIDENT. Please give as much information about how the accident happened and precisely what the injured person was doing when the accident occurred, for example if a fall of person or material, plant, etc. state height of fall or type of object that fell.

9 Names, occupations and addresses of witnesses (if no witnesses, state none)

10 What do you consider was the cause of the accident? (mention defects or hazards in Machinery or Premises)	11 What action has been taken to prevent a recurrence?

12 To whom and on what date and time was the accident first reported Name _____ Date _____	13 If injured person taken to hospital say which and whether as an "in" or "out" patient Hospital _____ IN / OUT

14 Person completing this form – if different from the injured Person, otherwise put as above. Name _____ Address _____ _____ Occupation: _____ Contact Telephone No: _____	15 The information contained on this form is correct as far as I am aware. Signed _____ Designation _____ Date _____

16 Is the accident reportable under Riddor? YES ☐ NO ☐ (Office Use Only)	16a Date and time reported to Health and Safety Executive (Office Use Only) Date _____ Time _____

Figure 1.50 *Accident report form*

Health and Safety at Work etc Act 1974
The Reporting of injuries, Diseasses and Dangerous Occurrences Regulations 1995

HSE
Health & Safety
Executive

Report of an injury or dangerous occurrence

Filling in this form
This form must be filled in by an employer or other responsible person.

Part A

About you

1 What is your full name?

2 What is your job title?

3 What is your telephone number?

About your organisation

4 What is the name of your organisation?

5 What is its address and postcode?

6 What type of work does the organisation do?

Part B

About the incident

1 On What date did the incident happen?

    /   /

2 At what time did the incident happen?
(Please use the 24-hour clock eg 0600)

3 Did the incident happen at the above address?

Yes ☐ Go to question 4

No ☐ Where did the incident happen

☐ elsewhere in your organisation – give the
name, address and postcode
☐ at someone else's premises – give the name,
address and postcode
☐ in a public place – give detailes of where it
happened

If you do not know the postcode, what is
the name of the local authority

4 In which depeartment, or whre on the premises,
did the incident happen?

Part C

About the injured person

If you are reporting a dangerous occurrence, go
to Part F.
If more than one person was injured in the same incident,
please attach the details asked for in Part C and Part D for
each injured person.

1 What is their full name?

2 What is their home address and postcode?

3 What is their home phone number?

4 How old are they?

5 Are they

☐ male?

☐ female?

6 What is their job title?

7 Was the injured perso (tick only one box)

☐ one of your employees?

☐ on a training scheme? Give details

☐ on work experience?

☐ employed by someone else? Give details of the
employer:

☐ self-employed at at work?

☐ a member of the public?

Part D

About the injury

1 What was the injury? (eg fracture, laceration)

2 What part of the body was injured?

F2508 (01/96)

Continued overleaf

Figure 1.51 *Report of an injury or dangerous occurrence*

3 Was the injury (tick the one box that applies)

☐ a fatality?

☐ a major injury or condition? (see accompanying notes)

☐ an injury to an employee or self-employed person which prevented them doing their normal work for more than 3 days?

☐ an injury to a member of the public which meant they had to be taken from the scene of the accident to a hospital for treatment?

4 Did the injured person (tick all the boxes that apply)

☐ become unconscious?

☐ need resuscitation?

☐ remain in hospital for more than 24 hours?

☐ none of the above.

Part E

About the kind of accident

Please tick the one box that best describes what happened, then go to Part G.

☐ Contact with moving machinery or material being machined

☐ Hit by a moving, flying or falling object

☐ Hit by a moving vehicle

☐ Hit something fixed or stationary

☐ Injured while handling, lifting or carrying

☐ Slipped, tripped or fell on the same level

☐ Fell from a height

How high was the fall?

[   ] metres

☐ Trapped by something collapsing

☐ Drowned or asphyxiated

☐ Exposed to, or in contact with, a harmful substance

☐ Exposed to fire

☐ Exposed to an explosion

☐ Contact with electricity or an electrical discharge

☐ Injured by an animal

☐ Physically assaulted by a person

☐ Another kind of accident (describe it in Part G)

Part F

Dangerous occurrences

Enter the number of the dangerous occurrence you are reporting. (The numbers are given in the Regulations and in the notes which accompany this form.)

Part G

Describing what happened

Give as much detail as you can. For instance
- the name of any substance involved
- the name and type oof any machine involved
- the events that led to the incident
- the part played by any people.

If it was a personal injury, give details of what the person was doing. Describe any action that has since been taken to prevent a similar incident. Use a separate piece of paper if you need to.

Part H

Your signature

Signature

Date

[/   /   ]

Where to send the form
Please send it to the Enforcing Authority the place where it happened. If you do not know the Enforcing Authority, send it to the nearest HSE office.

For official use
Client number   Location number   Event number

☐ INV REP ☐ Y ☐ N

Figure 1.51 (continued)

The Management of Health and Safety at Work Regulations 1992

RISK ASSESSMENT FORM
To be completed <u>BEFORE</u> the commencement of the task

1. Job/Task Identification	2. Reference

3. Known Hazards	4. Worst Likely Injury (please tick)	
	SERIOUS	
	MEDIUM	
	MINOR	

5. Persons/Groups Involved

6. Actions Already Taken to Control/Reduce Risk

7. Overall Assessment of Risk and further Action Required

8. Authentication and Record

Assessor's Signature	Date:

Assessor's Signature

Figure 1.52 *Risk assessment form*

ask, 'Is it worth the risk?' and 'Can I get rid of the hazard altogether?' If the answer is no then it is necessary to control the risk.

The law says that we should do all that is reasonably practical to keep the workplace safe, if there could be a risk; we need to ask if it is high, medium or low. Your aim is to make all risks as low as possible. To help you to assess risks a typical assessment form is shown in Figure 1.52.

Self-assessment questions

1 (1) The Health and Safety at Work etc. Act covers all people at work.
   (2) The Health and Safety at Work etc. Act excludes domestic servants in private households.
      (a) Both statements (1) and (2) are correct
      (b) Both statements (1) and (2) are incorrect
      (c) Only statement (1) is correct
      (d) Only statement (2) is correct.

2 An employee while working in an area requiring eye protection accidentally breaks his goggles provided by his employer. If he continues working with the broken goggles could he be liable for prosecution?
   (a) Yes
   (b) No
   (c) Not if he informed his employer of the breakage before resuming work
   (d) Not if he was instructed to continue working by his employer.

3 Which fire extinguisher should *not* be used on electrical fires?
   (a) dry powder
   (b) carbon dioxide
   (c) foam
   (d) vapourizing liquid.

4 Tilting the head backwards when giving mouth-to-mouth resuscitation ensures:
   (a) an effective breathing in position for the rescuer
   (b) a clear airway into the lungs of the victim
   (c) a good supply of blood to the victim's brain
   (d) the victim's chest will rise and fall automatically.

5 When treating an unconscious person for electric shock, a number of steps need to be taken immediately:
   (1) switch off the supply
   (2) seek medical help
   (3) treat the burns
   (4) carry out artificial respiration
   (5) keep the patient warm.

   What is the correct sequence of events?
   (a) 2, 1, 3, 5,4
   (b) 1, 2, 3, 4, 5
   (c) 3, 5, 1, 2, 4
   (d) 1, 4, 2, 5, 3

6 A person lifting a load from the ground by hand should use:
   (a) bent knees and straight back
   (b) straight legs and arched back
   (c) straight legs and curved back
   (d) bent knees and curved back.

7 The fuse used in a 13 A plug is intended to:
   (a) avoid the use of an earth wire
   (b) maintain a steady voltage
   (c) allow the use of double insulated tools
   (d) fail as soon as the system is overloaded.

8 Which is the correct procedure to be adopted for severe external bleeding if no dressing is available?
   (a) wash the wound thoroughly
   (b) lie flat and keep warm
   (c) hold the sides of the wound firmly together
   (d) apply an antiseptic lotion.

9 By law an industrial accident must be reported without delay to the factory inspectorate if it involves:
   (a) hospital treatment
   (b) two or more persons
   (c) site visitors
   (d) absence from work longer than three days.

10 Mobile scaffolding should be fitted with:
   (a) locking wheels
   (b) lifting handles
   (c) wheels at one end only
   (d) an independent ladder.

11 When an accident victim is suffering from shock, the correct initial treatment would be to keep them:
   (a) sitting down in the open air
   (b) lying down in a cool place
   (c) lying down warm and quiet
   (d) mobile by walking.

12 Before mounting an abrasive wheel, the operator should:
   (a) be at least 16 years of age
   (b) be trained and certified to do so
   (c) consult the Woodworking Machines Regulations
   (d) get permission from his or her supervisor.

13 Which one of the following groups of materials is potentially the most dangerous to health?
   (a) brick, plaster, cement
   (b) sawdust, steel, PVA adhesive
   (c) timber, plastic, glass paper
   (d) asbestos, lead, mercury.

14 Before operating a machine, an employee must:
   (a) know the installation cost of the machine
   (b) be able to carry out all necessary repairs and maintenance
   (c) have training in the safe use of the machine
   (d) have written permission from the foreman.

15 The purpose of timbering trenches dug for drains is to:
   (a) provide fixings for drain pipes
   (b) assist workers in getting in and out of the trench
   (c) Support the walls of the excavation
   (d) ensure that the drains are laid to the correct gradient.

16 In order to develop an awareness of safe working practices on site, it is essential that all personnel know the:
   (a) dangers that exist
   (b) building regulations
   (c) first-aid procedures
   (d) safety officer's name.

17 Before applying a dry dressing to a simple burn, it is essential to:
   (a) apply ointment
   (b) cool the burn area with cold water
   (c) apply a cold cream
   (d) clean the burn area with antiseptic lotion.

18 If solvent cement has a flash point of 7°C, this indicates that the cement:
   (a) cannot be used when the temperature is lower than 7°C
   (b) has a constant temperature of 7°C
   (c) gives off a vapour that will ignite when the temperature is above 7°C
   (d) must not be used when the temperature is above 7°C.

19 To comply with safety regulations, a record book must be kept on site containing records of all:
   (a) hours of work
   (b) accidents
   (c) safety equipment
   (d) potential safety hazards.

20 Spontaneous combustion is caused by:
   (a) reduction
   (b) chemical action
   (c) combustion ratio
   (d) vapour pressure.

2 Common plumbing processes, tools and materials

After reading this chapter you should be able to:

1 Name the various hand tools used by a plumber.
2 Recognize different types of hand tools.
3 Select the correct tools for a particular operation.
4 Distinguish between the cutting action of various tools.
5 State the use of given tools and equipment.
6 Understand and state the purpose of different tools and items of workshop or site equipment.
7 Understand the principle of flow through pipes and its resistance.
8 Demonstrate knowledge of the various methods of bending pipes.
9 Identify and name the usual bending machines and their component parts.
10 Appreciate the movement of metal during work processes.
11 Understand the extent to which pipes should be bent, use of jigs.
12 Have knowledge of plastics pipes, their bending capabilities and plastic memory.
13 Understand the bending techniques of (a) tension, (b) compression, (c) draw, (d) push.
14 Recognize and identify faults in bends.
15 Identify and name the materials used for pipes, fittings, and components.
16 State what is meant by the properties of materials.
17 Define the terms 'ferrous' and 'non-ferrous'.
18 Appreciate the advantages and disadvantages of various metals and their application.

Tools

Good tools are indispensable to the craftsperson, and buying them can be an expensive business. Some tools can only be used for one specialized task, whilst others, like hammers and pliers, may be used for a variety of jobs.

Most apprentices and plumbers will buy their tools as experience grows, and as they need them for the job they are doing. A kit of good quality tools, built up in this manner, is a sound investment. Some employers will buy tools for their employees, which helps to reduce the cost, payment often being made by an agreed weekly deduction from the employee's wages. The cost of the tool kit will, of course, depend upon the quality and quantity of tools bought.

Manufacturers produce such a wide variety of tools that there is no limit to the possible contents of a kit. Each tool has its own particular advantages and disadvantages and everyone has their own preferences and prejudices – many plumbers also make or adapt tools to suit their own special needs. The following list has been agreed between the Joint Industry Board for Plumbing Mechanical Engineering Services in England and Wales and the Electrical, Electronic, Telecommunication and Plumbing Union to be a full kit of tools which should enable a plumber to complete a reasonable job.

Adjustable spanner (300 mm long)

Bent pin or bolt

Brace

Gas blow torch and nozzle

Boxwood dressers (large and small)

Boxwood bending dresser

Boxwood mallet (large and small)

Bradawl

Compass saw (padsaw)

Footprint wrench (225 mm long)

Flat chisel for wood (225 mm × 55 mm)

Floor board cutter

*Glass cutter and putty knife

*Hacking knife

Hacksaw

Hammers (small and large, 1 kg maximum)

Junior hacksaw

Knife, large pocket type

Lavatory basin Union key

Pliers (two holes, gas)

Rasp (250 mm long)

Rule (metric 3 m tape)

Screwdrivers (large and small)

Shave hook

Snips (250 mm Tinmans) (straight and curved)

Spirit level (225 mm long)

Springs for bending 15 mm and 22 mm light gauge copper tube

Steel chisels for brickwork (up to 500 mm long)

Stillson or similar pipe wrench (up to 300 mm long)

Tank cutter

Tool bag

Trowel (small)

Tube cutters suitable for light gauge copper tubes

Plumbers able to provide tools from the list when they are needed for a job are eligible for a tool maintenance payment for every day which plumbing work is done. This money is used to maintain the kit in good working order, and to replace items as they wear, or are lost.

The list does not include items such as pipe cutters, pipe vices, welding apparatus, stocks and dies and bending machines – these and other larger and more expensive items of equipment are provided by the employer. Neither does the list contain various tools for specialized or unusual tasks such as may be encountered when working on sheet lead, copper, aluminium or zinc roof weatherings. Many of these tools are made by the plumber, and will be described as the need for them arises.

The plumber should:

■ Select, from the tool kit, the correct tool for the job.

■ Ensure that it is in good working order.

■ Use it correctly.

■ Transport it safely.

Failure to do this can result in: Use of more physical effort than is necessary. Waste of time. Damage to fittings, appliances or client's property. Injury to the user or others.

The maintenance and care of tools is important from both the practical and safety point of view. This applies to all tools, whether they are your own or provided by your employer for your use. Damaged, blunt or worn tools will not produce good work, and they could prove dangerous to the user and others working nearby.

Metal saws

The hacksaw has a pistol grip and the frame is adjustable to take various lengths of blade, usually 250 mm or 300 mm (see Figure 2.1).

Two types of blade are commonly used, one having 22 teeth per 25 mm of blade length and

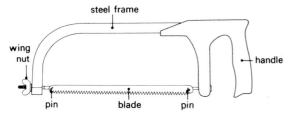

Figure 2.1 *Hacksaw*

the other 32 teeth per 25 mm. The coarser blade is used for steel pipe and the finer blade for copper tube, although many plumbers use a junior hacksaw for cutting the smaller sizes of copper tube (see Figure 2.2).

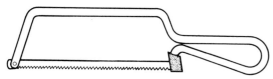

Figure 2.2 *Junior hacksaw*

The 'shetack' saw is particularly useful when cutting heavy gauge metal or corrugated materials. Its design enables the blade to cut material of an unlimited length or width, something which is not possible with an ordinary hacksaw, as the length of cut is limited by the depth of its frame.

Wood saws

There are many types of wood saw available, one of the most useful is the tenon saw. This can be used for cutting floorboards. These saws are generally 250–350 mm long and usually have about 12 teeth per 25 mm of blade. The blade teeth are 'set', that is folded outwards on alter-

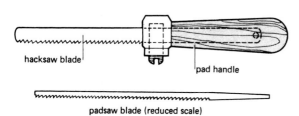

Figure 2.3 *Padsaw blade*

nate sides to provide clearance for the blade in the cut.

Padsaw

The padsaw has a plastic or metal handle into which is fitted a blade (see Figure 2.3). The blade can be a tapered one with about 10 teeth per 25 mm of blade and used for cutting wood. Alternatively a hacksaw blade can be fitted for cutting metal. The padsaw is most useful in awkward corners or when cutting floorboards in position.

Drilling tools

A ratchet brace can be used to hold a variety of drills or 'bits'. The ratchet allows the brace to be used close to a wall or corner. The chuck usually has two jaws and is most suitable for holding drills with a squared tapered shank. The jaws are called alligator jaws (see Figure 2.4).

The hand brace can be used in more confined spaces than a ratchet brace. The hand brace has a three-jaw chuck and holds round shanked drills or bits. Its gearing enables it to turn much quicker than a ratchet brace and is most suitable for drilling holes up to about 7 mm in diameter in metal, wood or masonry. The hand and breast drill is a larger version of the hand brace.

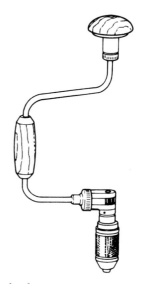

Figure 2.4 *Rachet brace*

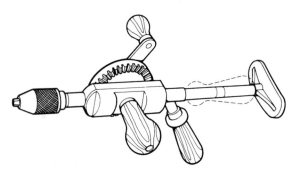

Figure 2.5 *Hand and breast drill*

Figure 2.6 *Cross pane hammer*

Figure 2.7 *Claw hammer*

Hammers

There are many different types of hammer and they are identified by their weight and head pattern. The 'pein' or 'pane' is the end of the head opposite to the face of the hammer. Hammer heads are usually made of cast steel and the handles or shafts are ash or hickory. A claw hammer is useful for removing nails from roof timbers or floorboards. The range of hammers is extensive and the choice will depend upon the particular work operation, personal opinion and experience (see Figures 2.6 and 2.7).

Chisels

Wood chisels are made in a variety of types and sizes. In the past these chisels had a steel blade and wooden handle, but they are now available made completely of steel (see Figure 2.8). These are most useful to a plumber who will usually

have a hammer in his tool kit but, not always a wooden mallet to strike the chisel. Plugging chisels are used for cutting out slots or joints between bricks (see Figure 2.9).

Cold chisels are so called because they can cut mild steel when it is cold, although a plumber will also use this tool for cutting brick and concrete (see Figure 2.10). As with wood chisels, cold chisels are available in a variety of types, shapes and sizes, and for normal domestic work a selection of chisels ranging from 150 mm to 450 mm in length and 12 mm to 20 mm in diameter will be most useful. Chisels for lifting floorboards are available (see Figure 2.11). These have a parallel blade about 75 mm wide. This blade is driven between the floorboards which are then raised by leverage via the chisel. This type of chisel is also used for cutting bricks.

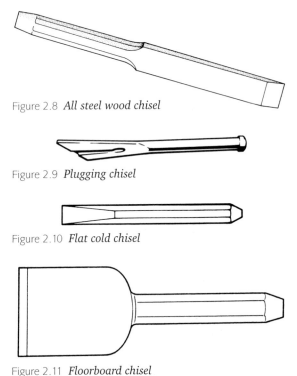

Figure 2.8 *All steel wood chisel*

Figure 2.9 *Plugging chisel*

Figure 2.10 *Flat cold chisel*

Figure 2.11 *Floorboard chisel*

Pliers

There are many types of pliers in common use, and most plumbers have several of these to enable them to perform different tasks.

Engineers' or combination pliers are available in several sizes, 150 mm to 200 mm are generally the most useful length (see Figure 2.12). Models are available with insulated grips for use on electrical circuits.

Figure 2.12 *Engineer's pliers*

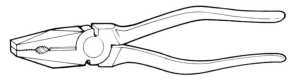

Figure 2.13 *Gas pliers*

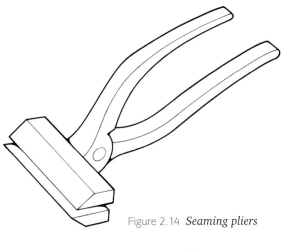

Figure 2.14 *Seaming pliers*

Figure 2.15 *Gland nut pliers*

Gas pliers are an essential item in any plumber's kit. Their circular jaws make them most useful for holding pipes or bulky components (see Figure 2.13).

Long snipe pliers are useful when riveting sheet metal and on domestic servicing work, seaming pliers are mainly used when working on sheet (aluminium, zinc or copper) weatherings to assist with folding and welting (see Figure 2.14). A variation of these have the jaws in line with the handle and formed into a 'V'. These are called dog earing pliers and are used for that operation.

Gland nut pliers can be obtained in sizes from 100 mm to 350 mm and more than one size may be included in a tool kit (see Figure 2.15). They are extremely useful for many jobs, but like most pliers with serrated jaws can damage brass or chromium surfaces or fittings.

Files and rasps

Files and rasps are made of cast steel. One end is formed into a tang on to which fits a wooden or plastics handle. (see Figures 2.16 and 2.17).

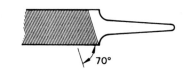

Figure 2.16 *File*

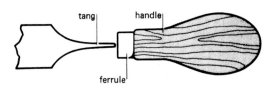

Figure 2.17 *Handle for file*

Some files and rasps are available with the tang formed into a handle. Files and rasps are identified by their:

Length 100 mm to 350 mm, in 50 mm steps.
Shape and cross-section hand, flat, half-round or round.
Cut single, double or rasp.
Grade rough, bastard, second cut or smooth.

The cut is standard for certain types of files. Flat files are double cut on the face and single cut on the edge. Hand files are similar, but have one edge uncut. Round files are usually single cut, and the half-round are double cut on the flat surface and single on the curved.

Spanners

Spanners are available in a variety of types. The most common are as follows:

Open ended

Ring

Box

Socket

Adjustable

Open ended spanners are usually double ended, with each end taking a different sized nut (see Figure 2.18). They are described by the size of the thread on which the nut screws, or by the distance across the flats of the nut.

Ring spanners fit completely round the nut to hold it very securely (see Figure 2.19). Ring spanners are safer to use than an open ended spanner as there is less risk of the spanner slipping off the nut. Also they are less likely to wear or open out. They are preferred for jobs where nuts must be tightened more securely.

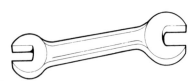

Figure 2.18 *Open ended spanner*

Figure 2.19 *Ring spanner*

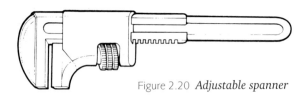

Figure 2.20 *Adjustable spanner*

Box spanners are most useful for releasing or tightening recessed nuts, or nuts in inaccessible positions such as those securing taps to wash basins, baths and sink units. Most box spanners are double ended and are turned by a steel rod called a tommy bar.

Socket spanners are a very robust type of tool. They may be used with a ring or open ended spanner or with a ratchet brace. Socket spanners are most useful for servicing work to boilers, water heaters, etc.

Adjustable spanners are available in several sizes and different designs and most plumbers include at least two different lengths in their kit (see Figure 2.20). Thin jawed adjustable spanners are most suitable for assembly and disconnection work to pipework and components.

Pipe grips or wrenches

There are four main types of wrench in use:

Stillson pipe wrench

Footprint pipe wrench

Chain pipe wrench

Self grip wrench

The Stillson wrench is a very robust tool and is most suitable for steel pipe work. These wrenches are available in a wide variety of lengths, ranging from 150 mm to 1.225 m. The most adaptable sizes for plumbers' work are 250 mm and 450 mm.

Footprint wrenches rely on hand grip pressure to secure the pipe or component. These are available in lengths ranging from 150 mm to 400 mm.

Chain pipe wrenches or chain tongs are usually associated with industrial work, but small models are available for domestic purposes. The length of lever handle may vary between 200 mm and 900 mm.

Self grip wrenches rely on hand grip pressure to secure the component although they also have a lock-on action to securely grip the wrench on to the component allowing the grip pressure to be released (see Figure 2.21). They are available with jaws of alloy steel and in lengths from 150 mm to 250 mm.

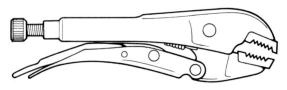

Figure 2.21 *Self-grip wrench or mole wrench*

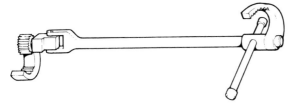

Figure 2.22 *Adjustable wrench*

Wrenches for specialist tasks are produced by various manufacturers. The shetack basin wrench is specially designed for the difficult job of fitting back nuts and union nuts behind wash basins, baths and sink units. The tool may be used in the horizontal or vertical position enabling a nut to be tightened or loosened in the most inaccessible places. The spanner is approximately 250 mm long and fits standard size backnuts. A similar model is also available with a greater distance between the jaws for waste fittings and traps.

An adjustable wrench for use in similar locations as the shetack basin wrench overcomes the difficulty of non-standard size nuts and unions (see Figure 2.22). The serrated teeth give a ratchet action which is useful when space is limited.

Screwdrivers

These are available in many types and sizes. Blades are made from high carbon steel and handles of wood or plastic. Some screwdrivers include a ratchet for ease of operation. Some manufacturers produce a set of screwdrivers of varying length with interchangeable blades having various widths and blade pattern, all fitting into a common handle. Screwdrivers are identified by their type and the length of the blade. Sometimes the width of the blade is also stated.

A cabinet screwdriver is used for bigger screws and has a large handle to provide gripping power to turn the screw. The most useful sizes are 200 mm and 400 mm.

Stubby or dumpy screwdrivers are used for larger screws which are located in awkward places. The blades are very short, usually about 25 mm long and available in a variety of widths, the most popular being 6 mm.

Electricians' screwdrivers are made for smaller screws. These screwdrivers have a plastics handle and some are available with a plastics sheathed blade.

Phillips   Posidriv

Figure 2.23 *Phillips and Posidriv screw heads*

Phillips and Posidriv screwdrivers both have cross-shaped blade points to fit cross-slotted head screws (see Figure 2.23). The Posidriv has superseded the Phillips type head, but the Phillips screwdriver is still retained because it will fit both screwheads, although the Posidriv cannot be used on Phillips screws. A Phillips screwdriver is more sharply pointed than the Posidriv, but the essential difference is in the square section between the slots of the cross which enables the Posidriv to fit closely between the blade and the screwhead. The corners of the square can be easily seen on the screwhead and are illustrated on the trade mark. Two sizes of blades are available. Cross headed screws are common on certain plumbing components such as water heaters and boilers.

Offset or cranked screwdrivers are very useful for getting at screws in awkward places. Blades may be flat or cross-slotted. These screwdrivers are usually all steel.

Gimlet and bradawl

These tools are used for forming holes in timber to enable a wood screw to start. The gimlet is used on hardwoods (see Figure 2.24) and the bradawl on softwoods (see Figure 2.25).

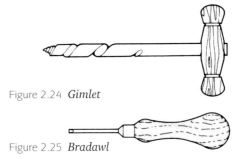

Figure 2.24 *Gimlet*

Figure 2.25 *Bradawl*

Punches

Several types are available which vary in shape according to the job to be done.

The nail punch is for punching nails below the surface of the timber which is being secured (see Figure 2.26). This is essential practice on roof boarding which is to be covered with sheet weatherings. Another function is to punch nails through floorboards so that the boards may be raised easily. The punch is usually about 100 mm long, with a 3 mm diameter head which is slightly recessed to prevent the punch from slipping off the nail head.

Figure 2.26 *Nail punch*

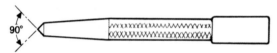

Figure 2.27 *Centre punch*

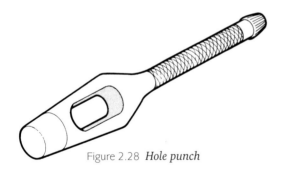

Figure 2.28 *Hole punch*

Centre punches are used to mark the centre of a hole to be drilled in metal (see Figure 2.27). The punch provides a small indentation which locates the drill in the correct position and prevents it from slipping. These punches are usually 100 mm to 200 mm long and the point is tapered at 90°.

Hole punches are used to cut a small circular hole in soft materials such as sheet lead, so enabling washers to be made (see Figure 2.28).

Taps

Taps are used for cutting internal screw threads in metal. There are three taps to a set (see Figure 2.29).

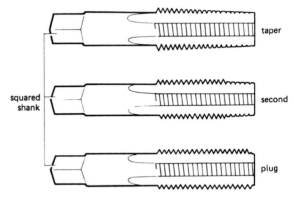

Figure 2.29 *Taps*

Taper tap

This is used to start the thread. It has no threads near its end so that it can enter the hole. This tap will cut a complete thread if it can pass right through the hole.

Second tap

This is used in a blind hole (a hole that does not pass right through the metal) to cut a thread near to the bottom of the hole, after preliminary cutting with the taper tap.

Plug tap

This is used to complete the thread started by the taper and second taps, and cuts right through to the bottom of a blind hole.

Taps are turned by means of a tap wrench (see Figure 2.30).

Figure 2.30 *Tap wrench*

Dies

Dies are used for cutting external threads on circular sections of metal, plastic rod and pipe. Dies are available in a number of different forms and are held in 'stocks', so that they can be rotated around the rod or pipe. Figure 2.31 shows circular split pattern dies which are held in the stocks shown in Figure 2.32. The split allows for adjustment to be made to the size or depth of thread being cut.

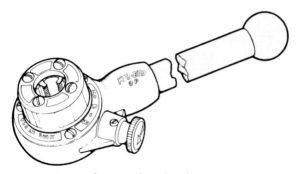

Figure 2.31 *Circular split pattern dies*

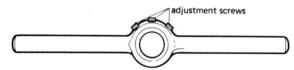

Figure 2.32 *Circular die stocks*

Figure 2.33 *Hand operated ratchet dies*

External threading of mild steel pipes is usually carried out using solid dies which have a ratchet included for ease of operation (see Figure 2.33).

Bending springs

Springs are used when bending copper tube by hand. The spring supports the tube and prevents it from kinking or losing its circular section. Springs for bending copper tube are made from square section spring steel and are available in two types: internal and external.

The internal spring is approximately 600 mm long and is most suitable for short lengths of pipe when a sharp tight bend is required (see Figure 2.34). The pipe is slightly over bent and then opened out to release the tension on the spring. The spring is then rotated in the direction required to reduce its diameter and withdrawn from the pipe.

External springs for copper tube are shorter in length than internal springs and have one or both ends opened slightly to assist tube insertion (see Figure 2.35). They are most useful for long lengths of pipe or when bending *in situ*. Generally external bending springs do not allow such sharp bends to be pulled as with the internal spring.

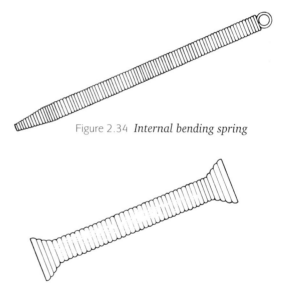

Figure 2.34 *Internal bending spring*

Figure 2.35 *External bending spring*

Pipe cutters

Pipe cutters operate by the rotation of a hardened steel cutting wheel around the outside of the pipe (see Figures 2.36 and 2.37). The wheel is gradually moved through the wall of the pipe as the adjustment is tightened until the pipe is cut. The cutting wheel is narrow and sharp, but the action of cutting produces a burr around the inside edge of the pipe which must be removed by a reamer (see Figure 2.38).

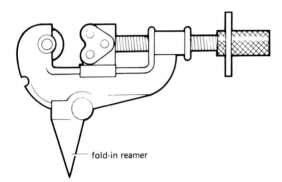

Figure 2.36 *Copper tube cutter*

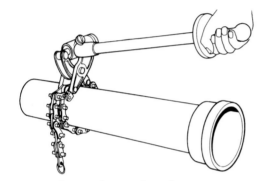

Figure 2.37 *Snap action cast iron pipe cutter*

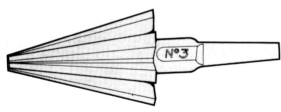

Figure 2.38 *Reamer*

Cutters are generally used for cutting copper, stainless steel, mild steel and cast iron pipes. In the case of cast iron pipes, no internal burr is produced as the pipe shears at the cutting point before the cutting wheel reaches the inside edge of the pipe.

Hole cutters

Hole cutters are used for cutting circular holes in cisterns, tanks and cylinders. The cutter shown in Figure 2.39 is used in a brace and the cutter is adjusted to the size of the hole required.

The cutter shown in Figure 2.40 is called a hole saw and resembles a circular hacksaw blade. These can be used in a brace or on an electric drill. They are available in a wide range of sizes and both the drill and blade are replaceable.

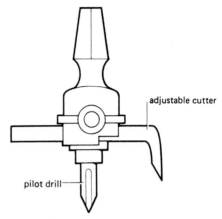

Figure 2.39 *Hole cutter*

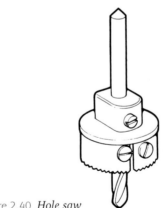

Figure 2.40 *Hole saw*

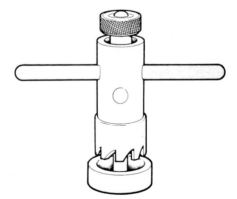

Figure 2.41 *Hand operated cutter*

Figure 2.41 shows a hand operated cutter which is most commonly used for cutting galvanized mild steel cisterns and tanks; a ratchet handle is available for easier operation in confined spaces.

Tools for expanding the diameter of copper tubes

Mandrel

This is used to open the end of a piece of tube to a 15° taper, making the tube suitable for use on certain tube fittings. The tool may be suitable for one diameter of tube, or may be a combination tool as shown in Figure 2.42, incorporating two sizes in one tool.

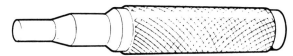

Figure 2.42 *Mandrel*

Socket forming tool

This is driven into the end of the copper tube opening it out parallel and forming a socket to receive another piece of tube of the same size (see Figure 2.43).

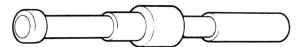

Figure 2.43 *Socket forming tool*

If the end of the tube is annealed the opening process will be simplified. The capillary joint formed is easily made by soldering or brazing.

Wooden tools

Hardwood dresser

This tool is used for dressing copper, aluminium, zinc sheet and lead sheet (see Figure 2.44). Dressers are usually made from hardwoods or high density plastics.

Bossing mallet

This is a useful tool when working on sheet roofing materials. The mallet head is made from a hardwood and the handle from malacca cane (see Figure 2.45).

Tinmans mallet

This is used on sheet copper and aluminium weatherings to assist with joining together welts and seams, and may be used to strike another sheetwork tool, for example a setting-in stick, or dresser (see Figure 2.46).

Bossing stick

This is used mainly for bossing corners on sheet lead weatherings (see Figure 2.47).

Bending dresser

This tool is also used to dress the sheet lead into its various shapes.

Setting-in stick

This is, as its name implies, a tool used to reinforce or sharpen folds or angles which have been formed in sheet weatherings (see Figure 2.49).

Chase wedge

This is a tool used for setting-in the corner of a fold or crease and is available in widths from 50 mm to 100 mm (see Figure 2.50).

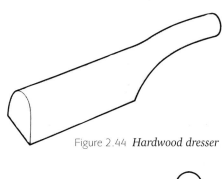

Figure 2.44 *Hardwood dresser*

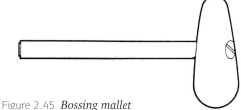

Figure 2.45 *Bossing mallet*

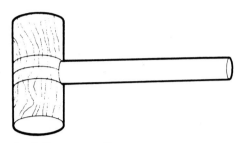

Figure 2.46 *Tinmans mallet*

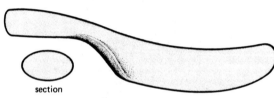

section

Figure 2.47 *Bossing stick*

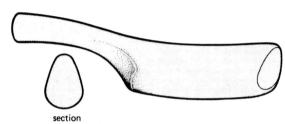

section

Figure 2.48 *Bending dresser*

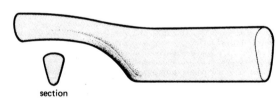

section

Figure 2.49 *Setting-in stick*

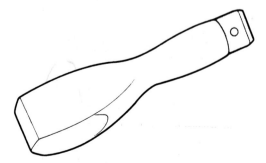

Figure 2.50 *Chase wedge*

Other tools

Shavehook

This is used to remove a thin layer of shaving from the surface of lead sheet. The blade is held at 90° to the material and drawn along it to remove the surface layer. Shavehooks are available with different shaped blades to suit a variety of tasks (see Figure 2.51).

Figure 2.51 *Shavehook*

Turnpin or tan pin

This is a cone shaped piece of steel and is used to open the end of a piece of copper tube prior to jointing.

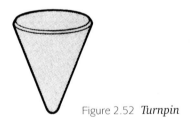

Figure 2.52 *Turnpin*

Bent bolt

This is a cranked piece of steel rod approximately 230 mm long and 14 mm diameter, tapered to 5 mm at one end (see Figure 2.53). It is used for opening up a hole in copper pipe to form a socket for a branch pipe.

Tin snips

These are used for cutting thin sheet metal and are available in several sizes, the most popular

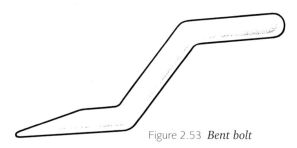

Figure 2.53 *Bent bolt*

being 150 mm and 250 mm long (see Figure 2.54). Straight and curved pattern blades are available. Universal snips are heavy duty pattern and are designed to cut either curved or straight and have open-ended handles to prevent nipping.

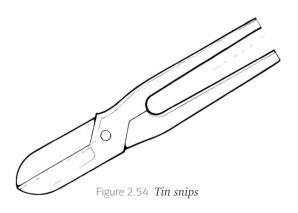

Figure 2.54 *Tin snips*

Soldering irons

These are used for soldering and tinning purposes, for example copper and brass unions or zinc and copper sheet. Soldering irons have a forged copper 'bit', steel shanks and a whitewood handle and are usually sold by weight.

Small electric soldering irons are available and are most suitable for workshop use.

Pipe vices

Pipe vices are needed to grip and secure mild steel pipes which are to be cut or threaded. They may be of the hinged type or chain type, and may be secured either to a work bench or to portable tripod stand.

Bending machines

These may be used for bending copper, stainless steel and mild steel pipes, and are necessary to bend pipes which are of large diameter or are too rigid to bend manually. They are also useful for prefabrication work, or when several bends have to be formed in a short length of pipe.

There are several types of machines: rotary (see Figure 2.55), ram (see Figure 2.56) and scissor.

Figure 2.55 *Portable rotary tube bender*

Figure 2.56 *Hydraulic ram bender for heavy-duty pipe bending. This type of bender is operated by a hydraulic pump and is commonly used for mild steel pipes*

Bending machines work by hand power or through a gear or ratchet action and employ special formers and back guides to ensure that the tube, when pulled to the required angle, maintains its true diameter and shape throughout the length of the bend. When using a machine it is advisable that the guides and formers should be lubricated and maintained in good condition.

Power tools

Although this title describes all kinds of tools driven by a variety of types of motor, so far as the plumber is concerned it is likely to include only a few tools, all of which are driven by electricity.

Electrical tools may be:

1 Mains voltage.
2 Low voltage, with a step-down transformer.
3 Double insulated.
4 All insulated.

Mains voltage

Portable tools, like drills, usually have single phase universal motors and operate on 240 V. These tools have a three core cable with the casing connected to the earth connection. If this earthing becomes faulty, particularly in wet situations, the tool can cause a lethal electric shock. When working in premises where the earthing cannot be guaranteed, mains voltage tools should not be used.

Low voltage

Transformers are used to step down the mains voltage from 240 V to 110 V for tools or 25 V for hand lead lamps. Usually both the live and neutral connections are fused on the transformer output. Low voltage tools can be fitted with special plugs so that they cannot be connected or used accidentally on full mains voltage.

Double insulated

These tools have additional insulation to eliminate risks from defective earthing. They are tested to 4000 V and may be used on 240 V supplies without an earth lead if they conform to BS 2769 and are identified by the appropriate BS symbols.

All insulated

This type of power tool is made entirely from shock-proof nylon and does not have a metal casing, therefore electricity cannot be conducted from any part, unless the casing becomes damaged. They are tested to 4000 V and may be used on 240 V without an earth like the double insulated tool.

The following notes are intended only as an introduction to power tools.

Drills

Portable electric drills generally have chucks to receive drills up to 10 mm in diameter. Two speed drills rotate at about 900 and 2400 revolutions per minute. Bench drills for larger work are usually fitted in workshops.

Percussion tools

These give the drill fast-hitting blows at the rate of about 50 per second. This hammer-action helps to penetrate hard materials such as concrete which are difficult to cut with an ordinary rotary drill. A special impact type of tungsten carbide tipped drill should be used with percussion tools to produce an accurate smooth hole.

Mechanical saws

These may be portable circular blade hand saws for cutting timber, or larger static machines for cutting mild steel pipes and rods. The item being cut must always be fixed or clamped securely and excessive pressure must not be used.

Screwing machines

There is a variety of types of machine for cutting similar to those on hand dies and many machines incorporate a pipe cutter and reamer. A suitable cutting lubricant must be used to keep the dies cool and assist with the cutting operation.

Cartridge tools

These tools act like a gun and shoot a hardened steel fixing stud into the material to which a fixing is required. Several types of fixing stud are available. Some are similar to wood nails for fixing timber, others have threaded ends to which a bracket, clip or nut can be screwed.

The foregoing tools and equipment comprise those used by the average plumber employed on work of a general nature. There are, however, several other tools and items of equipment which might be used for specific or specialist tasks.

Bending of pipes

General

Ideally for maximum flow through a pipe, the bore should be smooth, of uniform diameter and with no joints or bends. In practice, this of course, is impossible. Therefore every effort must be made to create conditions as close as possible to perfection. As stated, changes in direction are unavoidable, and the use of purpose-made bends or elbows are generally fitted to accommodate for this. The bends used wherever possible should be purpose-made and be of large radius to minimize the frictional resistance to the flow through the system. In some instances it may be necessary for one pipe to pass over another pipe or obstruction, necessitating the use of offsets (double bends) or passovers. These could be made up from purpose-made fittings or the pipe can be bent to give a very effective result.

The methods of bending vary according both to the material from which the pipe is manufactured and to the size of pipe and the thickness of the pipe wall.

When pipes are subjected to the process of bending, particularly with small radius bends, tremendous stresses are set up in the material. The stresses will be either compression or tension, depending upon where the stress reading is taken.

Figure 2.57 illustrates the possible effect of bending a pipe, showing the thinning of the material at the heel and the thickening at the throat. In the case of small-diameter pipes, if these are of a heavy-gauge material, little or no adverse effect should be experienced with bends of normal radii. As a guide this could be taken as four times the diameter of the pipe (4 × dia.).

The generally accepted methods of bending are classified as:

1 Loaded
2 Mechanical.

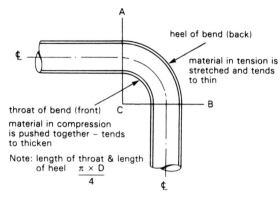

Figure 2.57 *Definitions*

heel of bend (back)

material in tension is stretched and tends to thin

throat of bend (front)
material in compression is pushed together – tends to thicken

Note: length of throat & length of heel $\dfrac{\pi \times D}{4}$

Loaded

This could be by using a steel or rubber insert or loose fill material. It is also possible to a limited extent to use air pressure on certain pipes.

Mechanical

The use of various types of bending machine is perhaps the most commonly accepted method. They are either manual (for the smaller diameter pipes) or hydraulic (for the large diameter pipes).

Note

It must be remembered that it is not always the correct procedure to bend the pipe. As in the case of thin-walled copper tube and for certain plastics pipes, the change of direction must be performed by the use of purpose-made bends or elbows.

The materials most commonly used in the industry are:

1 Steel
2 Copper
3 Plastics.

Copper tube

One of the main advantages in the use of copper tubes is the ease with which they can be bent, either by the loading method or with the aid of bending machines.

Methods of loading

1 *Springs*
   (a) internal
   (b) external

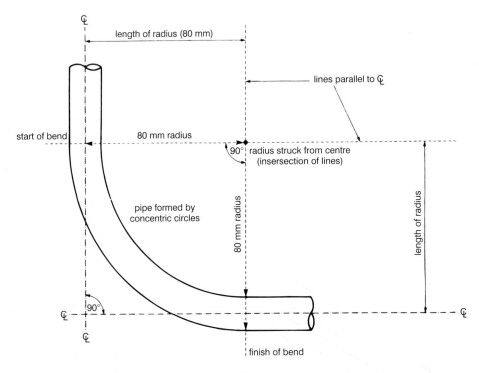

Figure 2.58 *Method of setting out 90° bend with 80 mm radius*

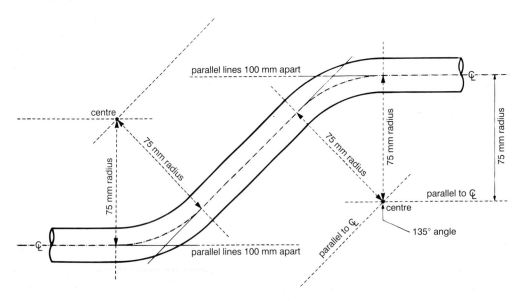

Figure 2.59 *Method of setting out 100 mm offset with 75 mm radius*

2 *Loose fill*
   (a)   dry sharp sand
   (b)   resin
   (c)   low melting point alloy
3 *Machines*
   (a)   former and guide
   (b)   internal mandrel.

It must be appreciated that when bending tubes metal is compressed at the throat and stretched and thinned at the heel. Therefore, unless the wall of the pipe is supported during the bending operation, the tube will be deformed or will completely collapse. This support is provided by the above named methods.

Template
Where bends have to be made to a given radius and accuracy and shape is important, the required bend/bends should be set out full-size in the form of a working drawing. A 4–6 mm steel *template* rod is then bent to fit the centre line, or alternatively the template could be a piece of sheet steel cut accurately to fit the inside line of the bend.

Bending springs
These are available in both internal and external types. Spring bending is a popular and quick method and is perhaps the easiest. The main advantage of spring bending is that the bend can with care be moved slightly should it be wrongly positioned.

External springs
These are used only on the smaller diameter size tubes up to 22 mm maximum. Their main advantage is that it is an easy operation to place and remove the spring in cases where the bend is required mid-way along the pipe.

Note
Small diameter copper tubes can be bent satisfactorily without being annealed.

Internal spring bending of copper tube
Care must be taken to ensure that you have selected the correct type of spring and to check whether the tightening effect is in a clockwise or anti-clockwise direction to assist in the removal of the spring after bending.

The bending of light gauge copper pipe up to 28 mm diameter can be performed fairly easily with the aid of bending springs (particularly 15 and 22 mm tube). The bends should be of an easy radius from 3 × dia. for the smaller size pipes up to 6 × dia. for the sizes up to 54 mm.

Before commencing to bend the pipe, first ensure that the bending spring passes easily down the pipe until the centre of the spring is positioned at the centre point of the bend. If satisfactory withdraw the spring.

Method
1 Set out the bend and make a template to fit the centre line of the bend.

2 Ensure uniformity of pipe and mark length of bend.

3 Anneal pipe (heat until red hot) from the *beginning* to the end of the bend. It is important not to heat either short of or outside the marks of the bend as this will affect the finished length of the bend making it either short or too long.

4 Insert the spring and pull the bend round the knee until it fits the template.

5 Bend slightly more than the required angle, then open the bend (this releases the pipe grip on the spring).

6 Turn the spring to tighten the coil and withdraw.

It is advisable to lubricate the spring.

The pipe can be bent while still hot or alternatively the pipe may be cooled, then bent. The softening is done by the heating until red hot and is equally annealed irrespective of the cooling.

Loose fill loading
A special type of sharp sand is by far the easiest and safest loose fill loading material for pipe bending (provided the sand is *dry*) and can be satisfactorily used for all the types of pipes mentioned and for all sizes (see Figure 2.60). It is possible to bend the pipes to much smaller radii and to more complex shapes. The removal of the sand is a simple operation of removing the bungs and tapping the sides of the bend, the sand being returned to a receptacle for further use.

Method
1 Set out the bend and make a template to fit the throat of the bend.

2 Seal one end of the pipe, then fill with dry sharp sand.

3 Compact the sand by tapping the side of the pipe, and gently bumping the sealed end of the pipe on the floor (this compacting is perhaps the most important point).

4 Seal the open end with a wooden bung.

5 Mark the beginning and end of the bend, heat until red hot the portion between the marks, allowing the heat to soak through the pipe into the sand (annealing process).

6 The bend can be formed while the pipe is hot or it can be allowed to cool (safer to handle).

   Note It is advisable to start the bending from the sealed end; the pipe may be held in a vice by means of purpose-made vice protectors.

7 When the bends are completed, remove the bungs and empty the sand back into the receptacle for re-use.

Note
Care must be exercised in handling the completed work as the annealed work remains soft and can only be hardened by work hardening, which is explained under that heading.

Loading method
It is advisable to have purpose-made hardwood plugs (bungs) with a fairly long taper to ensure maximum surface contact with the inside of the pipe. Figure 2.61 illustrates methods of sealing the pipe ends before the dry sharp sand is compacted by gently bumping the pipe on the floor and at the same time tapping the side of the pipe with a piece of wood. When the sand will not compress any more, remove sufficient sand to leave a void of approximately 30–40 mm. Drive in the wooden bung to seal the pipe, ensuring that there is no cavity under the bung. An alternative method is to use sand compressors as shown in Figure 2.62.

Vice holding method
When holding a pipe in a vice some form of protection is essential or the pipe will become either marked or even damaged. Purpose-made clamps of cast lead are made which accurately fit the various sizes of pipe (see Figure 2.63). An

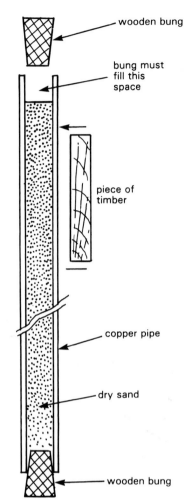

Figure 2.60 *Loose fill loading*

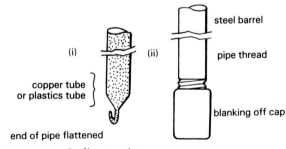

Figure 2.61 *Sealing on pipe*

alternative method is to make them from wood, but a much better method is to cut an old bending machine guide (slide) into pieces as shown in Figure 2.63.

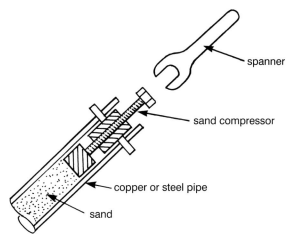

Figure 2.62 *Sand compressor*

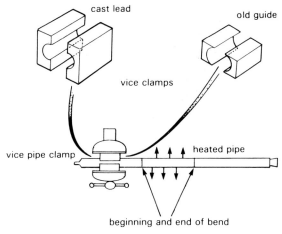

Figure 2.63 *Method of holding pipe while bending*

Forming bend to a template

As stated earlier a template could be made from a 4–6 mm diameter wire or it could be made from a piece of sheet metal. The type shown in Figure 2.64 is perhaps the easiest to make. It also gives the most accurate check on the finished bend. The required bend is set out on the sheet material which is then cut out accurately to fit the internal size of the bend. The pipe is then pulled to form the bend, the template being held in position as shown.

Plastics pipe bending

There are many different types of plastics on the market, some of which are not suitable for bending. Care must be exercised and consultation made with the manufacturers' literature if there is any doubt concerning the bending of a specific type of plastics. Where a particular plastics pipe is used which should not be bent the manufacturers have provided a comprehensive range of purpose-made fittings to enable all or most problems to be overcome.

There are several methods of heating and bending polythene and polyvinyl-chloride pipes. Perhaps the most commonly used method is that of dry sharp sand loading which is carried out in the manner described as for the loaded bending of steel and copper. The main difference is that plastics pipe does not change colour when heated and, because plastics material is a very poor conductor of heat, it is very easy to char or destroy the material completely.

Plastics memory

This is a term given to plastics material and can best be explained by using a straight length of plastics pipe as the subject. The plastics pipe when manufactured is in a straight length. It is then heated, softened and can be bent into the required position; it is held in this position until cold, whereupon it will remain in the new position so long as it remains below the temperature at which it was worked. Should the pipe be reheated the bend will tend to open and the pipe return to the original straight position. This is known as its memory, hence the term 'plastics memory'.

Accepted methods of heating plastics pipe prior to bending:

1 Immersing in boiling water (see Figure 2.65),
2 Heated air (see Figure 2.66),
3 Radiated heat,
4 Direct heat (see Figure 2.67).

Immersion in boiling water

Although satisfactory from the softening point of view, it has its limitations in practical use: the availability of boiling water and the difficulty of manoeuvring, filling and emptying a large container of water being the main problems. It is also very difficult to cater for long lengths of pipe so that this method tends to be restricted to bending short lengths.

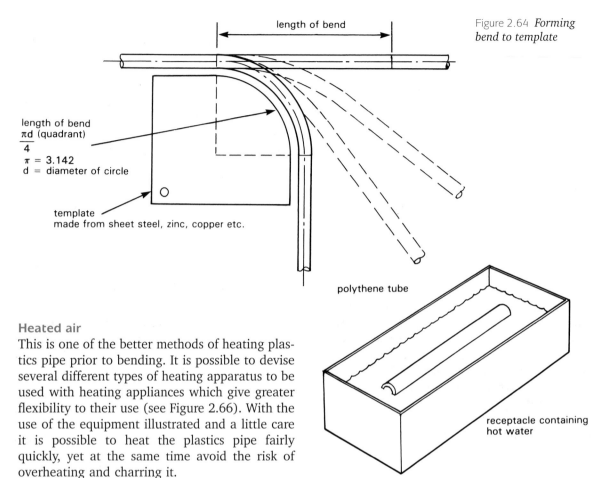

Figure 2.64 *Forming bend to template*

length of bend

length of bend
πd (quadrant)
────────
4
π = 3.142
d = diameter of circle

template
made from sheet steel, zinc, copper etc.

polythene tube

Figure 2.65 *Bending of pipes – methods of heating*

receptacle containing
hot water

Heated air

This is one of the better methods of heating plastics pipe prior to bending. It is possible to devise several different types of heating apparatus to be used with heating appliances which give greater flexibility to their use (see Figure 2.66). With the use of the equipment illustrated and a little care it is possible to heat the plastics pipe fairly quickly, yet at the same time avoid the risk of overheating and charring it.

Radiated heat

This form of heating is the safest method, but because of this it is very slow. It is also difficult to control the area being heated. It is therefore seldom used.

Direct heat

This is perhaps the most commonly used method. It is fast and effective but it must be stressed that it is very easy to destroy the nature of the plastics by overheating or charring it. As already stated, plastics are poor conductors of heat and this means that if a direct flame is played on to one point, even for a very short time, the result will be charring and irretrievable damage to the pipe. It cannot be over-emphasized how important it is to keep the flame continually on the move, at the same time revolving the pipe to ensure a slow heating of the pipe throughout its thickness and the length of bend.

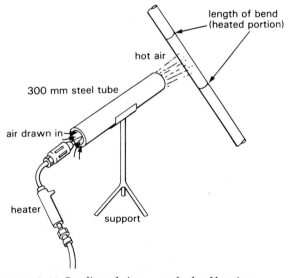

length of bend
(heated portion)

hot air

300 mm steel tube

air drawn in

heater

support

Figure 2.66 *Bending of pipes – methods of heating*

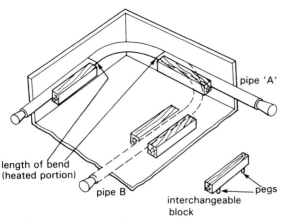

length of bend
(heated portion)

pipe B

interchangeable
block

pegs

pipe 'A'

Figure 2.67 *Purpose-made jigs*

Method of bending plastics pipes

The walls of the pipe must be fully supported both during the heating and bending process. This can be achieved by the use of a rubber insert which performs the function of a traditional steel-bending spring used for the bending of copper pipes. An alternative method is to support the walls of the pipe with a dry loose in-fill such as dry sharp sand. The application is fully described for the bending of copper tube.

Bending operation

1–4 As for copper tube

5   Heat the pipe very slowly, revolving it continuously, and constantly moving the torch flame backwards and forwards along the length of the bend.

     Note   This is a slow procedure and must not be rushed. Plastics have a very low conductivity and will very easily char. The pipe is ready for bending when it becomes pliable and floppy. A slight change in surface appearance may be detected: it takes on a more shiny look.

6   To form the bend place the heated pipe into a purpose-made jig (see Figure 2.67) or simply bend by hand. No force is required.

7   Cool the bend by applying cold water. This will set the bend in the required position.

8   Remove the bung and return the sand to its receptacle for re-use.

Polythene pipe

This is a plastics material used for conveying cold or at least cool liquids. It is a clean, lightweight, smooth bore pipe, unaffected by corrosion. It has a low conductivity which means it is a poor conductor of heat. Plastics have a high rate of thermal expansion which must always be taken into account when jointing and fixing plastics pipes. They also have considerable elasticity, particularly when warmed above a temperature of 60°C.

Bending polythene pipe

Polythene can be bent cold as long as the bend has an easy radius of approximately 12 × dia. of pipe. A bend with a large radius is not usually acceptable. It is therefore necessary to heat the tube so as to produce neat bends with smaller radii of approximately 4 × dia. of pipe.

Polyvinyl chloride tube

This type of tube is known simply as PVC. It is light in weight, smooth of bore, resistant to corrosion, all of which make it a very useful conduit for the conveying of cold or hot water.

Unplasticized polyvinyl chloride

This is PVC without additives. In this form it is more rigid and can be fractured if subjected to a severe blow. It is more rigid than plasticized polyvinyl chloride but when fixed, supported and used in the correct manner, should give trouble-free service. It is used for above- and below-ground drainage discharge systems. This type of plastic should not be bent, any change of direction being accommodated by purpose-made bends.

Plasticized polyvinyl chloride

This is the same plastic; manufactured with the addition of rubber and thus changing the rigid PVC into a more flexible material which makes it suitable for bending and more resistant to impact.

Machine bending

Bending machines are supplied in various forms suitable for bending all types of metal pipes both ferrous and non-ferrous, thin- and thick-walled varieties. They come under one of the following headings:

1 Compression bending,

2 Draw bending,

3 Push bending.

Compression bending

This method is used when bending thin-walled pipes as it gives the greatest support to the pipe at the point of bending. In this type of machine the centre former is fixed; the pipe is fitted into the groove of the former, and is surrounded and held in position by the guide. The former and guide together support the walls of the pipe and prevent it from collapsing, during the bending operation (see Figure 2.68).

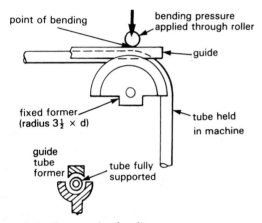

Figure 2.68 *Compression bending*

The pressure for bending is transmitted by a rotating lever arm and roller, positioned adjacent to the guide. The position of the roller is vital to the quality of the bend: too little pressure will result in a wrinkled bend; too much pressure will give a bend with excessive throating.

Draw bending

Although very effective, the type of machine used for draw bending is of a more specialist nature (see Figure 2.69). It can produce bends to a much smaller radius than those required in normal domestic work. This type of machine is not commonly used on site work. Due to the accuracy and perfection of the bending and the necessary tooling provided, these bending machines are also expensive to purchase.

Operation

In this type of machine the pipe is clamped to a centre-rotating former which when operated pulls the pipe forward, so forming the bend. The pipe is again fully supported by means of an internal

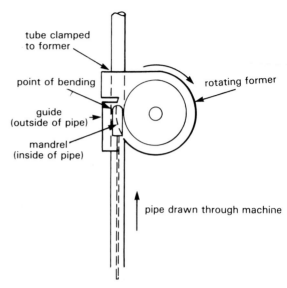

Figure 2.69 *Draw bending*

mandrel and an external guide, the adjustment and position of the mandrel being very critical.

Push bending

This type of bending is the simplest and requires the least skill or knowledge of pipe bending on the part of the operator. It is sometimes called 'centre point bending' because the bending pressure is applied at a single point in the centre of the bend.

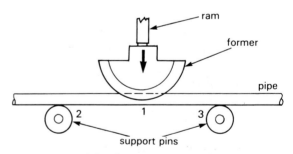

Figure 2.70 *Push bending*

This type of machine (see Figure 2.70) is also known as the three point bender because of its two support points in addition to the centre bending point. It is very satisfactory for bending heavy gauge (thick-walled) steel pipe, but unsuitable for light gauge (thin-walled) pipe, unless the walls of the pipe are again supported by the use of a suitable loose fill material.

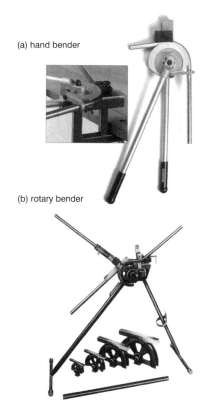

(a) hand bender

(b) rotary bender

Figure 2.71 *Machines in common use*

Figure 2.72 *Rothenberger bender (copper and steel tube)*

Types of bender

Light gauge copper tube can be easily bent by machines of which there are several different types to choose from.

Bending machines

Benders (a) and (b) in Figure 2.71 are by far the most commonly used, and are similar in their set-up and use. They can be used in either vertical or horizontal positions. Proficiency in their use is a must for all practising plumbers. Figure 2.73 shows an exploded view of a bending machine, detailing its component parts.

Note

When bending, the pipe is completely encircled by former and guide, thereby supporting the pipe wall and preventing deformation.

Setting the pressure indicator

Some of the smaller types of machines have fixed formers and guides with no pressure adjustment yet still give satisfactory performance, particularly when new. The better machines have incorporated in their design an adjustable pressure indicator and the correct positioning of this is very important if perfect bends are to be produced. Figure 2.74 indicates the correct position and also the reduced and increased pressure position.

The information given forms the basic guidelines about where to commence pipe bending, although with slightly worn machine parts and various grades of pipe it is always advisable to make a test bend first as slight adjustments may be required.

Figure 2.74(A) indicates the correct setting. This setting should give a perfect bend on light gauge copper pipe. Pressure indicator should be parallel with pipe.

Faults in bends

1 *Throat of bend rippled, heel flattens* This is caused by a reduced pressure setting as shown in Figure 2.74(B).
2 *Excessive throating* This is caused by increased pressure as shown in Figure 2.74(C).

Making a square bend on light gauge copper pipe

In this example we will be setting the bend to the outside of the former.

Method

1 Mark off pipe to the required distance.
2 Insert measured distance in the backside of the machine.

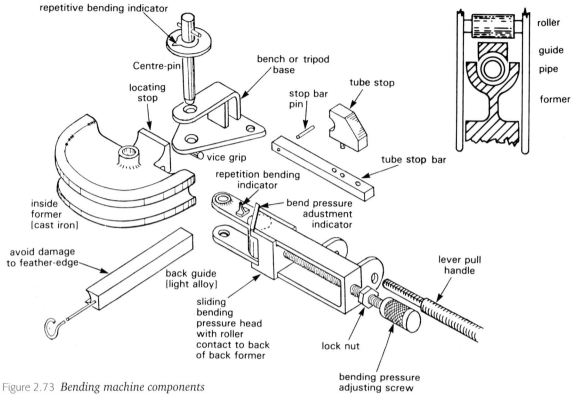

Figure 2.73 *Bending machine components*

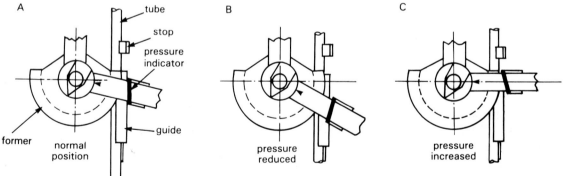

Figure 2.74 *Pressure adjustment*

3 Ensure that the pipe fits right into the former and on to the stop.

4 Place alloy guide around the pipe, tighten pressure slightly to hold pipe in position.

5 Place square against mark on pipe, adjust pipe until the square touches the outside of the former.

6 Adjust pressure to the correct bending position (if of the adjustable pattern).

7 The lever arm is then pulled around and the pipe bent to the required 90° angle.

8 The bend will need to be very slightly over-pulled to counteract the spring back in the bend.

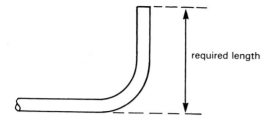

required length

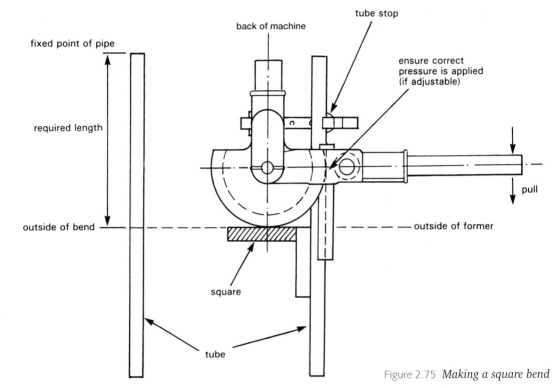

Figure 2.75 *Making a square bend*

In the example shown in Figure 2.76 the procedure is almost identical to that explained previously and shown in Figure 2.75, except that in this case we work to the inside of the bend and the inside of the former. Both methods are equally correct and will produce identical bends.

Method

1–4 As for previous bend.

5   Adjust pipe until the square touches the inside of the former.

6–7 As for previous bend.

Note
It must always be remembered that all bends *must* be placed at the *back* of the machine or errors in distances will occur.

Making an offset

An offset is also known as a double set, an ordinary single bend being known as a set. The method of setting up and operating the machine is as previously described.

Method

1 The first bend or set on the pipe is made at the required position. The angle of the bend is not critical but 45° is usually recommended as satisfactory.

2 As stated previously all the bends must be placed at the back of the machine. Adjust the pipe in the machine, holding a slight pressure on the lever handle to hold the pipe in place.

3 Place straight edge against the outside of the former and parallel with the pipe.

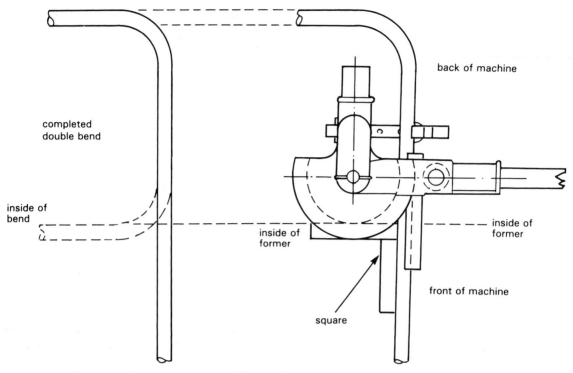

Figure 2.76 *Making a double bend (using the inside of the former)*

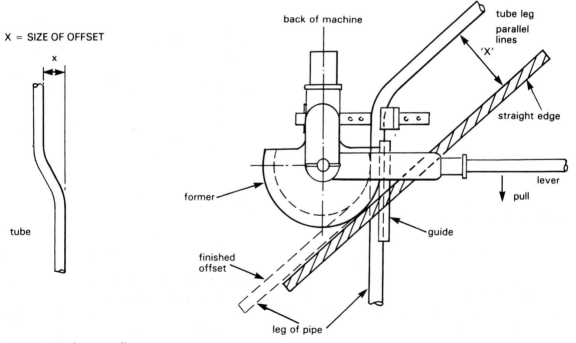

Figure 2.77 *Making an offset*

4 Adjust the pipe in the machine until the required measurement is obtained.

   Note To increase the size of the offset, push the pipe further through the machine. To decrease the size pull the pipe towards the front of the machine.

5 Apply pressure to lever arm and bend pipe until the legs are parallel.

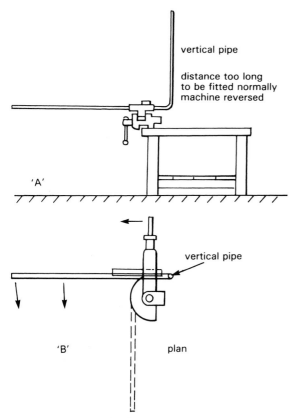

Figure 2.78 *Reverse bending*

Left-hand bending

It is sometimes necessary to form left-hand bends, such as when a number of bends are required on one length of pipe or it may be impossible to locate the pipe in the machine due to the length of the pipe fouling the bench or the floor. In these cases the machine stop bar and lever arm are changed to operate from the reverse side as shown in Figure 2.78. The long length of the pipe is now situated in a vertical position. In the case of a vertical positioned bender on tripods the problem is solved by

reversing the stop and placing the pipe in the bottom of the former and pulling upwards instead of downwards in the normal way.

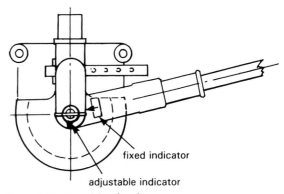

Figure 2.79 *Repetition bending*

Repetition bends

On some of the better types of machine there are attachments which enable bends to be accurately repeated such as:

1 Indicators

2 Stops

3 Graduated protractors.

Indicator method

This is as shown in Figure 2.79 where both fixed and adjustable indicators are an integral part of the bending machine. First, it is always necessary to make a trial bend, allowing for the spring back which will differ for different diameter pipes and machines.

1 Set and bend the pipe to the required angle.

2 Adjust the bend indicator to coincide with the fixed indicator on the handle mechanism.

3 Remove the pipe, ensuring the indicator is not altered. The machine is now set for repeat bends.

Stop method

An adjustable locking stop, which is situated in the bending quadrant of certain machines, is an alternative method of making repeat bends. The method of bending is as described in the indicator method. When the trial bend is correct the stop is then locked in that position; all repeat bends will now be bent to the same degree.

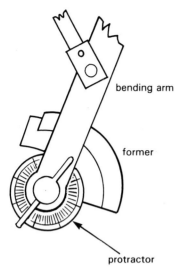

Figure 2.80 *Repetition bending*

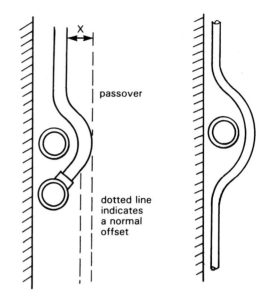

Figure 2.81 *Passover bend*   Figure 2.82 *Passover crank*

Protractor method

On some machines of the larger type the angles are shown on a protractor. On the top of the former in this instance the pipe is bent until the bending arm indicates the required angle on the protractor, as in Figure 2.80.

Passover bend

This is a double bend and another variation of the offset. This type of bend is used where a branch pipe is required to pass over another parallel pipe as shown in Figure 2.81. The method of making a 'passover' is the same as for an ordinary offset as shown by the dotted lines in the sketch. The size of the offset (C–C) is indicated by distance 'X'.

1 Set out passover bend to required dimensions.

2 Bend first bend to approximately 45°.

3 Place pipe in machine as previously shown to obtain second bend.

4 This second bend is over-pulled to give the required angled bend.

Crank

Another variation of the passover bend is shown in Figure 2.82 and is known as a 'crank'. This type of bend requires both skill and practice to ensure that it fits evenly to the wall and passes uniformly over the obstruction (pipe). The size of the former of the bending machine will govern the length of the 'crank'.

Method

1 Bend the pipe to form the first bend to an angle of approximately 90° (this angle is governed by the size of the obstacle) (see Figure 2.83).

2 Mark the two points for the second and third bends (make allowance for space between pipe and obstacle).

3 Line up mark with outside of former and pull the bend to the required angle (use template).

4 Reverse the pipe and repeat operation for the third bend; check alignment of pipe and clearance of obstacle by the use of a straight edge (see Figure 2.84).

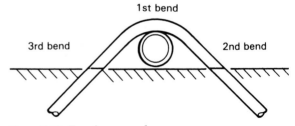

Figure 2.83 *Forming a crank*

Bending of steel pipe by machine

Steel pipe is manufactured in several grades and thicknesses depending on its use. Regardless of

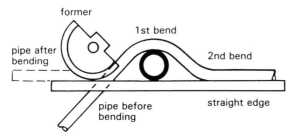

Figure 2.84 *Forming a crank*

thickness there are machines available to bend the pipes used in any domestic situation satisfactorily. In the case of thin-walled pipes the machine shown for bending copper pipe (see Figure 2.73) is equally suitable; for the thicker-walled pipes the machine commonly used is the centre point push bender (see Figure 2.85). Although the actual working of the hydraulic bender is very simple there are one or two points which the operator must observe.

1 Select the correct former for the pipe to be bent. (Different formers are used when bending copper pipe with the same bender.)

2 Locate the pins in the correct holes. (This is most important or serious damage may result.)

3 The vent or breather valve is in the open position for most machines during the operation.

4 Should the ram fail to operate, check the oil level. If it is low, replenish, using only the correct type of oil. Do not overfill.

5 If the pipe is over-pulled replace the former with a flattener (supplied with most benders, see Figure 2.86); reverse the pipe, relocate pins in the flattening position, then apply pressure in the normal manner.

6 The use of a 4 mm steel template rod is strongly recommended.

7 During the bending operation the pipe often becomes wedged in the former; do not hammer the former to remove it. Completely remove the pipe and the former from the machine and strike the end of the pipe sharply on a wooden block. Hold the former if possible to prevent damage when it is released from the pipe (see Figure 2.87).

Making a square bend

The method of making a square bend with a centre point bender will present no problem except when working from a fixed end of pipe. Even this is relatively simple once you appreciate that there will be an increase in length when the bend is pulled. This increase in length is equal in length to the diameter of the pipe being bent (see Figure 2.88).

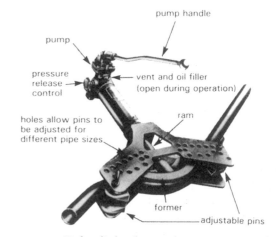

Figure 2.85 *Hydraulic bender*

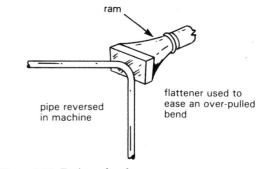

Figure 2.86 *Easing a bend*

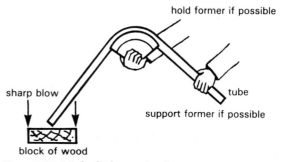

Figure 2.87 *Method of removing former*

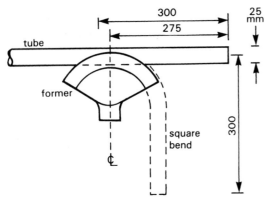

Figure 2.88 *Making a square bend*

As previously indicated the bend must be slightly over-pulled to allow for spring back when the pressure is released on the ram. Check the bend with the aid of a square or a steel template before removing the pipe from the machine.

Offsets, passovers and cranks

Making an offset

1 Set out full size the required bends (angles approximately 45°).

2 Bend the 4 mm steel template rod to fit the centre line accurately.

3 Mark the centre of the first bend on the pipe and place it centrally in the machine.

4 Bend the pipe to fit the template.

5 Mark the centre of the second bend on the pipe (either from the template or from the drawing).

6 Replace the pipe in the bender with the mark coinciding with the centre of the former; check with the template that the bend is being made in the correct direction and that the pipe is level.

7 Bend the pipe to fit the template.

8 Check accuracy of the bend with the template before removing the pipe from the machine; if satisfactory remove and give final check with the drawing, straight edge and rule.

Passover

It will easily be recognized that the passover bend (see Figure 2.90) is very similar to that of the offset shown in Figure 2.89, the difference being that the second bend is continued past the 45° angle until the required passover is obtained.

The method of setting out and bending is as already described for the offset.

Crank/passover

Differences in terminology have already been pointed out and here again we find two different names given to an object.

Figure 2.91 is a further progression from the passover shown in Figure 2.90. Once again the actual bending procedure is as described for Figures 2.89 and 2.90, the centres of the bends being marked, then lined up with the centre of the former.

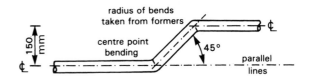

Figure 2.89 *Offset*

Figure 2.90 *Passover*

Figure 2.91 *Crank*

Emphasis on the use of templates cannot be over-stressed particularly when the bending becomes a little involved, i.e. in two planes or when the bending operation is to be performed some distance from the fixing location.

Materials

The installation of pipework systems for the supply of cold and hot water, heat and gas, together with the systems for the removal of surface, waste and foul water, form the major part of day-to-day work for plumbers. Included in the operations must be the manufacture and fitting of weathering components to roofs and the outsides of buildings. This means that it is essential for the plumber to have a comprehensive understanding of the characteristics, properties and performance of materials in current use.

British standards/European standards

The British Standards Institution was founded in 1901 to standardize industrial activities such as design, installation and manufacturing practice. The British Standards Specifications refer to standards of manufacture, for example low carbon steel tube is made to conform with the specification contained in BS/EN.

British Standard Codes of Practice are recommendations related to methods of good practice in installation work. By having standards which have been agreed by manufacturers and industrial experts, the process of obtaining the right materials and correct design and installation techniques or procedure is made easier, and leads to a better quality of completed job. Architects and others involved with design work need only to specify that material must conform to the relevant British Standard and that work is to be undertaken and installed in accordance with the relevant British Standard Code of Practice, to ensure that materials used and work undertaken is of a satisfactory and acceptable standard. Figure 2.92 shows the British Standard kitemark. Britain's membership of the European Community necessitates British workers having to familiarize themselves with European practices and customs identified by the marking **E N**.

Figure 2.92 *British Standard kitemark*

Properties of materials

Strength

One of the main properties of a material is its strength, i.e. its ability to withstand force or resist stress. Stress can be applied to a material in many ways (see Figure 2.93), for instance:

(a) tensile, or stretching;

(b) compressive, crushing, or squeezing;

(c) bending, which is both (a) and (b) on either side of an axis;

(d) shear, or cutting;

(e) torsion, or twisting.

$$Stress = \frac{load}{area}$$

and this means that the stress can be increased by either increasing the *load* or by decreasing the *area*. So that when a piece of material is cut out of a support it has the effect of increasing the *load* acting upon it. Therefore care must be taken when it becomes necessary to cut out structural support such as a joist or beam.

In addition to strength, materials have other properties which may affect their selection or use. These are:

Brittleness

This means that the material is easily fractured or broken. Brittleness in metals is usually associated with hardness, and it is often necessary to reduce the hardness of a material in order to give it greater strength.

Malleability

This is the property which means that the material is capable of being formed or shaped by the use of hand tools or machines (see Figure 2.94).

Ductility

This is the property of many metals and enables them to be drawn out into a slender wire or thread without breaking (see Figure 2.95). Ductile materials are easily bent.

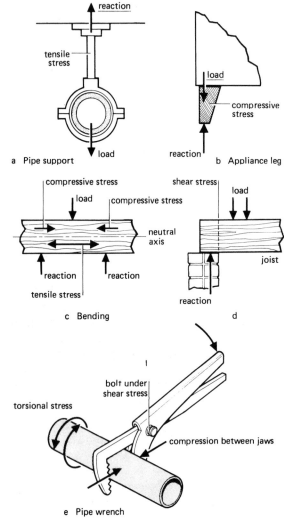

a Pipe support   b Appliance leg

c Bending   d

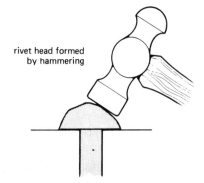

e Pipe wrench

Figure 2.93 *Types of stress: (a) tensile; (b) compressive; (c) bending; (d) shear; (e) torsion*

rivet head formed by hammering

Figure 2.94 *Malleability*

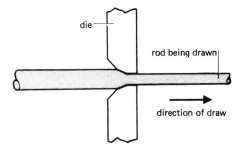

Figure 2.95 *Ductility*

Elasticity

The ability of a material to return to its original shape, or length after a stress (stretching) has been removed (see Figure 2.96). The 'elastic limit' is the greatest strain that a material can take without becoming permanently distorted.

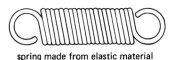

spring made from elastic material

Figure 2.96 *Elasticity*

Hardness

The property of a material to resist wear or penetration. This is an essential requirement for the cutting edge of a tool. The hardest naturally occurring substance is diamond.

Tensile strength

The tensile strength of a material is its ability to resist being torn apart. Samples are subjected to increasing loads until breakage occurs. The loading which causes breakage being the ultimate tensile strength.

Temper

The temper of a metal is the degree or level of hardness within the metal and can vary from dead soft to dead hard. Copper is a good example related to temper, being in a dead soft condition following annealing its temper changes as it is hammered or worked into a dead hard, cold-worked condition.

Work hardening

Describes the increase in hardness caused by hammering, bossing or other cold working techniques. These working techniques cause the

Table 2.1 *Properties of materials*

Material	Chemical symbol	Density (kg/rrr)	Coeff. linear exp. (°C)	Melting point (°C)
Aluminium	Al	2705	0.0000234	660
Copper	Cu	8900	0.0000160	1083
Iron (cast)	Fe	7200	0.0000117	1526
Iron (wrought)	Fe	7700	0.0000120	2200
Lead (milled)	Pb	11 300	0.0000293	327
Non-metallic sheet	–	1021	0.0000188	not applicable
Tin	Sn	7300	0.0000210	215
Zinc	Zn	7200	0.0000290	416

Note: Properties vary according to condition, that is, whether the material is cold-worked, hot-worked, hard or annealed.

grains or crystals which form the metal to become mis-shaped. This deformation of the metal structure reduces malleability and ductility which prevents further cold working.

Annealing

This is the description of the heat treatment applied to soften temper and to relieve the condition of work hardening. The metal is heated to a specified temperature, during which the deformed crystalline structure returns to its normal condition. This treatment softens the metal, thereby reducing internal stresses and allows further working processes to be carried out.

Corrosion/erosion

Both have the ability to destroy a metal by an eating away process. Iron is destroyed by corrosion, here you get a build up of rust. Lead is destroyed by erosion here you have an eating away process. These are chemical or electro-chemical action which causes the 'eating away' of some metals. Different metals react in varying ways to the different forms of corrosive attack, some having strong resistance, others need protection where attack may occur.

Categories of materials

Materials can be divided into different categories in a variety of ways. The easiest division is into: *metals* and *non-metals*.

Metals differ from non-metals in chemical properties as well as in the more obvious physical properties. There is not a rigid dividing line between the two categories. Some metals possess characteristics of both groups, while some non-metal materials behave like metal in some respects, for instance the conducting of electricity.

Table 2.2 gives examples of some of the differences between the two categories.

Metals can be subdivided into *element metals* and *alloys*.

Table 2.2

Metals	Non-metals
Characteristic appearance, 'metallic' sheen	No particular characteristic appearance
Good conductor of heat	Usually poor conductor of heat, may be insulator
Good conductor of electricity	Usually poor conductor of electricity, may be insulator (except carbon, silicon)
Electrical resistance usually increases as temperature rises	Electrical resistance usually decreases as temperature rises
Density usually high	Density usually low
Melting point and boiling point usually high	Melting point and boiling point usually low

The element or pure metals are those which do not have any other metal or materials mixed with them.

Alloys are produced when two or more metals, or a metal and a non-metal, are blended together, usually by melting (see Table 2.3). Many alloys have useful properties which their parent metals do not possess. For example, a much lower melting point, as in the case of solders.

Table 2.4 lists many metals which are used by the plumber.

Metals may be further classified as:

1 Ferrous metals, that is those which contain iron.

2 Non-ferrous metals, those that do not contain iron.

The basic difference between the two categories is that, with the exception of stainless steels, ferrous metals will rust when in contact with water and oxygen. Also, ferrous metals may be magnetized.

Non-ferrous metals do not rust, although they may corrode under certain circumstances, and they are non-magnetic.

Ferrous metals

Grey cast iron
This is a mixture of iron with 3.5 to 4.5 per cent carbon and very small amounts of silicon, manganese, phosphorous and sulphur. The carbon is in the form of 'flake graphite' and it is these flakes which make the material very brittle (see Figure 2.97). The iron fractures easily along the lines of the flakes.

Cast iron is hard and has a much higher resistance to rusting than steel. It is used for drainage and discharge system pipework, appliance components, manhole covers and frames. Its melting point is 1200°C.

Ductile cast iron
This has approximately the same carbon content as grey cast iron. The difference is in the shape, size and distribution of the carbon particles which are changed from flakes into ball-like nodules or 'spheroids' (see Figure 2.98). The change is due to alloying the iron and carbon with magnesium compounds before casting. Subsequent heat treatment at 750°C changes the brittle iron carbide into softer, ductile ferrite.

Used to make pipes for gas mains and drainage pipelines, ductile cast iron has the strength and ductility of lower grade steel and the high, corrosion resistance of cast iron.

Wrought iron
This is the purest commercial form of iron containing very little carbon. Made from heat-treated

Table 2.3 *Composition of common alloys*

Alloy	Main elements
Brass	Copper and zinc
Bronze	Copper and tin
Gunmetal	Copper, tin and zinc
Pewter	Tin and lead
Steel	Iron and carbon
Steel (stainless)	Iron, chromium and nickel
Soft solder	Tin and Silver

Table 2.4

Pure metal	Alloy
Aluminium	Brass
Chromium	Bronze
Copper	Chrome-vanadium steel
Gold	Duralumin
Iron	Gunmetal
Lead	Invar
Magnesium	Nickel silver
Mercury	Pewter
Nickel	Rose's alloy
Platinum	Solder
Silver	Stainless Steel
Sodium	White-metal (bearing metal)
Tin	
Tungsten	
Zinc	

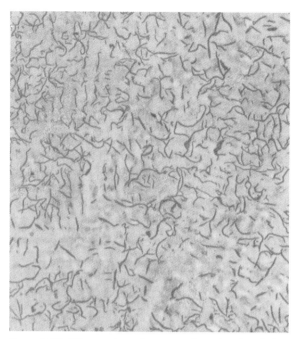

Figure 2.97 *Microsection of grey cast iron showing flake graphite*

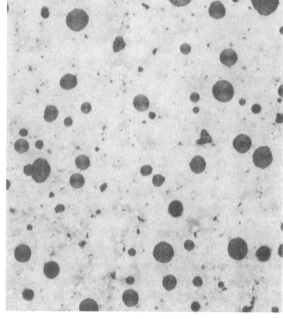

Figure 2.98 *Microsection of pearlite nodular cast iron*

cast iron it is rolled or hammered to give it the required grain structure. It is still used for chains, horse-shoes, gates and ornamental ironwork. Because of its high cost it is no longer used for pipes or fittings.

Malleable iron (malleable cast iron)

Malleable iron is white cast iron in which the carbon has combined with the iron to form iron carbide. This results in a very hard and brittle material which is annealed.

Most of the ferrous pipe fittings used in conjunction with low carbon steel pipe in internal installations are malleable iron.

Low carbon steel (LCS)

This is another mixture of iron and carbon. Its carbon content is between 0.15 and 0.25 per cent. The material has a wide use and is commonly known as mild steel, although its correct title is low carbon steel (LCS). In building, it is used for general structural work, pipes, sheets and bars. It can be pressed, drawn, forged and welded easily.

The LCS tube used by plumbers is manufactured in three grades of weight: light, medium and heavy.

The outside diameter of each grade of tube is similar, the difference being the tube wall thickness. The grades are identified by 50 mm wide colour band coding, brown, red and blue, to correspond with light, medium and heavy grades. Low carbon steel tubes used for domestic water supplies must be protected with a zinc coating; this is called 'galvanized'.

Tubes are generally available in 3 m and 6 m lengths, with bore size ranging from 6 mm to 150 mm.

Stainless steel

This material is mainly used in plumbing in tube form for domestic water and gas services or for the manufacture of sanitary appliances such as sink units, urinals, WC pans, etc.

The metal is an alloy and has a composition of chromium (18 per cent), nickel (10 per cent), manganese (1.25 per cent), silicon (0.6 per cent), a maximum carbon content of 0.08 per cent, the remainder being iron with small amounts of sulphur and phosphorous. The chromium and nickel provide the material with its shiny appearance and its resistance to corrosion, this being due to a microscopic film of chromium oxide, which forms on the surface and prevents further oxidation.

Stainless steel tube is available in nominal bore sizes ranging from 6 mm to 35 mm in one grade only (light gauge) with an average wall thickness of 0.7 mm. The outside diameter of the tube is comparable to copper tube so that most types of capillary and compression fittings can be used. The mechanical strength of stainless steel is much higher than that of copper giving it greater resistance to damage and a better ability to support itself and therefore will require less fixing positions during installation. Stainless steel tubes of 15 mm to 35 mm diameter are supplied in 3 m and 6 m lengths.

Non-ferrous metals

Aluminium

This is useful metal because it is so light. It is much lighter than steel, and some of its alloys are as strong as steel. It is also a good conductor of heat and electricity. Aluminium and its alloys are therefore widely used where lightness and strength are needed. The material does not rust as ferrous metals do when exposed to the atmosphere. Although the surface does oxidize, this oxide film acts as protection against corrosive attack. Aluminium is primarily alloyed with copper, magnesium, and silicon.

Aluminium is the most common metal in the earth's crust. It is found in common clay, although removal is a difficult and expensive process. The only suitable ore is bauxite, which contains alumina, or aluminium oxide. Producing aluminium from bauxite is very different from the usual methods of obtaining metals from their ores. First the bauxite is crushed and heated with caustic soda. The alumina dissolves and leaves behind the impurities. The pure alumina obtained is then dissolved in molten cryolite, and aluminium is produced when electricity is passed through this solution.

Aluminium in a soft condition is malleable and ductile, thus giving it good working properties, although it will work harden if cold worked. Annealing is easily completed, but because of its low melting point, care must be taken to avoid overheating.

Aluminium used for roofwork is obtainable in standard rolls of 8 m length, with widths of 150 mm, 300 mm, 600 mm, and 900 mm. Two thicknesses are available, these being 0.7 mm (standard) generally used for flashings, and 0.9 mm (heavy duty) which is suitable for bays and situations where foot traffic may occur.

Antimony

This is a hard, grey, crystalline metal which expands slightly when it solidifies. It is principally used in alloys of lead and tin (solder) to produce a degree of hardness and making the alloy more resistant to corrosion

Bronze

True bronzes are alloys containing copper and tin as the basic metals, but the name bronze has been so loosely applied that certain alloys containing no tin at all are sold as bronze.

The useful bronzes contain not less than 65 per cent of copper. As the tin content is increased the alloys become progressively harder, although bronzes containing above 25 per cent of tin become weak and brittle. On the other hand, if too much copper is present, the cast metal becomes porous. The properties of many bronzes may be greatly improved by the addition of a third element – lead. When zinc is present in this alloy the material is known as 'gunmetal'.

Phosphorous is added to some copper/tin alloys to deoxidize the metal prior to casting, these alloys are termed 'phosphor bronze' although the percentage of phosphorous added may be very low.

Chromium

This metal is plated on other metals and materials to protect them. It is hard and does not lose its shine through corrosion. Steel is made stainless by the addition of chromium and nickel. Nickel and chromium together form important alloys much used for plating and the manufacture of tools. Chromium is found as chromite, or chrome iron ore.

Copper

This is probably the most important of the non-ferrous metals and is used extensively in sheet, strip, rod, wire, tube and other fabricated forms. It is also employed in the making of a wide range of alloys, of which brass and bronze are best known.

Although copper is a comparatively inexpensive metal, physically and chemically it is closely allied to silver and gold, and there are many similarities in the properties of the three. The outstanding features of copper are its high malleability (particularly when in a soft or annealed state), its electrical and thermal conductivities and its good resistance to corrosion.

Copper exposed to air forms on its surface a natural protective skin or 'patina' which effectively prevents corrosive attack under most conditions. On copper exposed to the atmosphere the patina takes the form of a green covering of copper salts (mainly sulphate or carbonate). The formation of a somewhat similar protective film also takes place on the interior surface of copper water pipes, this film is insoluble in water, therefore making copper an ideal material for carrying water.

The first British Standard for copper tube was issued in 1936 to standardize the various sizes of tube on the basis of their outside diameter. Prior to this, copper tube had been used since the turn of the century, when jointing was achieved by screwed and socket fittings which required a tube wall thick enough to thread. As time progressed and methods of jointing which did not require threading of the tube developed, the wall thicknesses of the tube were reduced, and tubes were identified as 'light gauge'.

Copper tube

Copper tube for use in the construction industry is now currently manufactured to BS/EN 1057 this has brought about a number of changes notably in the sizes and temper of the tube.

Material tempers

Soft Copper tube shown in Table 2.5 as W and Y is now identified R 220.

Half Hard Copper tube shown in the table as Y and X is identified as R 250.

From the above it will be readily seen that there are two grades of tube identified as R 220 and two grades identified as R 250.

The table will clearly illustrate the changes including their diameters and wall thickness.

Under EN 1057, W and Y soft copper tube has the designated material temper reference R 220. X and Y half-hard tube becomes R 250.

To summarise: EN 1057 – R 250 is a general purpose half-hard copper tube for above ground services, including potable water, hot and cold services, central heating and sanitation.

EN 1057 – R 220 is a thick-walled tube available in soft condition coils or as half-hard condition in straight lengths for uses as above.

Decorative chrome plated copper tube is specified as EN 1057 – R 250.

The basic differences between these tubes is their temper and wall thicknesses, the wall thickness factor affecting the mass.

Tubes to BS/EN 1057, R 220 (annealed condition) can be obtained in standard coil lengths from approximately 10 m to 30 m (depending on diameter size). This tube in its soft condition can be bent by hand and is most commonly used for below ground service pipes, it can be obtained with a protective plastic covering fitted during manufacture for use in soils which have a corrosive nature.

Sheet copper is malleable and highly ductile so it can be easily worked particularly when in a soft or fully annealed state. Work hardening, however, occurs with cold working so as little manipulation as possible should be done.

As soon as the material becomes hard it should be annealed. This is carried out by heating to a dull red colour. The metal can then be allowed to cool naturally or may be doused with water to assist the cooling process.

Copper for weathering is available in the form of rectangular sheets or in strip form.

Strip copper is usually supplied in rolls or coils of varying length and width. The standard strip widths are 114 mm, 228 mm, 380 mm, 475 mm, 533 mm and 610 mm. The length is determined by the thickness and mass of the roll which is usually limited to 25 kg or 50 kg to assist safe handling.

Sheet copper is obtainable in a variety of sizes, although standard sizes of 1.83 m × 0.91 m and 1.83 m × 0.61 m are normally specified. All copper for weathering purposes is sold by mass.

Table 2.5a *EN 1057 – R220 Soft coiled tubes*

Outside diameter	Wall thickness (mm)					
	0.6	0.7	0.8	0.9	1.0	1.2
8mm	W		Y			
12mm			Y			
22mm						Y

Table 2.5b *EN 1057 – R250 Half hard straight length tubes*

Outside diameter	Wall thickness (mm)								
	0.6	0.7	0.8	0.9	1.0	1.2	1.5	2.0	2.5
8mm	X		Y						
12mm	X		Y						
22mm				X		Y			
35mm						X	Y		
54mm						X		Y	
76.1mm							X	Y	
133mm							X		

Key: W = BS 2871, Part 1, Table W   X = BS 2871, Part 1, Table X   Y = BS 2871, Part 1, Table Y

Lead

Lead is obtained from the ore galena and is mined and processed in many different parts of the world.

The name 'plumber' derives from the Latin name for lead which is plumbum. Lead is one of the six metals known to humanity from the early days of history, and the development of the use of lead in building work for sanitary, roofing and weathering purposes was also the development of the plumber's trade.

The ore galena is a compound of lead containing, in the pure form, 86.6 per cent lead and 13.4 per cent sulphur. Galena has a cubic crystalline structure, is very lustrous and of a dark grey colour. The lead produced from the smelting process usually contains small quantities of other metals, for example antimony, tin, copper, gold and silver, and these are removed as required by refining processes. The lead which is obtained possesses a very high degree of purity – a good commercial lead being over 99.9 per cent pure.

Lead is the softest of the common metals and has a very high ductility, malleability and corrosion resistance. It is capable of being shaped easily at normal temperatures without the need of periodic softening (lead does not appreciably work harden). Lead sheet and pipe are therefore easily worked with hand tools, and can readily be manipulated into the most complicated shapes. Lead is very seldom corroded by electrolysis attack when in contact with other metals.

Lead sheet is produced by either casting or by rolling. Cast lead is still made as a craft operation by the traditional method of running molten lead over a bed of prepared sand. A comparatively small amount is produced by specialist lead working companies, mainly for their own use, in particular for replacing old cast lead roofs and for ornamental leadwork. There is no British Standard for this material. The available size of sheet is determined by the casting table bed size.

Rolled sheet lead is manufactured in rolling mills. The process involves passing a slab of refined lead about 125 mm thick backwards and forwards through the mill until it is reduced to the required thickness.

Rolled sheet lead is manufactured to BS and is supplied by the manufacturer cut to dimensions as required or as large sheets 2.4 m wide and up to 12 m length. Lead strip is defined as material ready cut into widths from 75 mm up to 600 mm. Supplied in coils, this is a very convenient form of lead sheet for most flashing and weathering applications. The thickness of lead is designated by a BS specification code number, and by an identifying colour.

Zinc

This metal is obtained from several ores, the most common being zinc blende, sphalerite and calamine. Commercial sheet zinc has low ductility and is the least malleable of common roofing metals. This means that manipulation is difficult when compared with materials such as lead or aluminium, and jointing or shaping is achieved by folding, cutting or soldering. These disadvantages have led to the development of zinc alloys. Zinc alloys have good ductility, and a linear expansion rate of less than two-thirds that of commercial quality zinc sheet (overcoming the problems of creep often associated with zinc sheet). Two of these alloys are currently available, namely zinc/lead and zinc/titanium, these being produced under the respective trade names of Metiflash and Metizinc.

Commercial zinc is obtainable in sheets 2.438 m × 0.914 m with a recommended thickness of 0.6 mm for flashings and 0.8 mm for other roof areas. In cold weather zinc becomes brittle and should be warmed slightly before folding is attempted, otherwise cracking or fracture may occur. Such folds should not be too sharp – a rounded fold with a radius at least twice the thickness of the metal will be satisfactory.

An electrolytic action is set up when zinc is in contact with copper in the presence of moisture, and for this reason zinc and copper should never be allowed to touch each other in roofing or as water service piping.

Solders

A better understanding of the nature of the solders, and how to select one for a specific application, can be obtained by observing the melting characteristics of metals and alloys. Melting of pure metals is easy to describe as they transform from solid to liquid state at one temperature. The melting of alloys is more complicated as they may melt over a temperature range.

Solders are now manufactured as a lead free alloy, and are used for the joining of many metals. Impurities in tin/lead solders can result from carelessness in the refining and alloying operations, but can also be added inadvertently during normal usage. The soldering properties of tin/lead solders are affected by small traces of aluminium and zinc. As little as 0.005 per cent of either of these metals may cause lack of adhesion and grittiness. Above 0.02 per cent of iron in a tin/lead solder is harmful and will cause hardness and grittiness. The presence of above 0.5 per cent of copper will have the same harmful effects.

Antimony can play a dual role in tin/lead solders. Depending on the purpose for which the solder is to be used, it can be considered as either an impurity or as a substitute for some of the tin in the solder. When the amount of antimony is not more than 6 per cent of the tin content of the solder, it can be completely carried in solid solution by the tin. If the antimony content is more than the tin can carry in solid solution, tin/antimony components of high melting point crystallize out, making the solder gritty, brittle and sluggish.

Antimony content of up to 6 per cent of the tin content increases the mechanical properties of the solder but with slight impairment to the soldering characteristics. The use of lead/antimony/tin solder is not recommended on zinc or zinc-coated metals, such as galvanized iron. Solders containing antimony, when used on zinc or alloys of zinc, form an intermetallic compound, causing the solder to become brittle.

Tin/zinc solders are used for joining aluminium. Corrosion of soldered joints in aluminium (electrogalvanic) is minimized if the metals in the joint are close to each other in the electrochemical series. The addition of silver to lead results in alloys which will wet steel and copper. Flow characteristics, however, are very poor. The addition of 1 per cent tin to a lead/silver solder increases the wetting and flow properties and, in addition, reduces the possibility of humid atmospheric corrosion.

All solders are alloys, and as mentioned previously the melting point of an alloy varies with the different percentages of element metals comprising the alloy.

A solder with about 36 per cent lead and 64 per cent tin has the lowest melting point and changes from solid to liquid between 183°C and 185°C. This is known as the 'eutectic', which is the composition at which the alloy behaves as a pure metal. Adding either more lead or more tin raises the final melting point, although the alloy still begins to melt at about 185°C. In its intermediate stage it becomes 'plastic' and may be 'wiped' or formed into its required shape or position before it becomes completely molten, or solid.

Plastics

The word 'plastics' came into being as a general description for many materials of a similar nature, such as celluloid, casein and bakelite. The name merely signifies that the material is capable of being moulded into shape. In this sense the word is not a suitable description for the many materials that have since been developed.

The term 'plastics' describes a group of manmade organic chemical compounds which can broadly be divided into two types – *thermoplastics* and the *thermosetting plastics*.

Thermoplastics

Thermoplastics soften when heated and harden on cooling. They can be softened again afterwards provided that the heat applied is not sufficient to cause them to decompose. Among the thermoplastic materials used in plumbing work are polythene, polyvinyl chloride (PVC), polystyrene, polypropylene, acrylics (perspex) and nylon.

Thermosetting plastics

Thermosetting plastics are those which soften when first heated for moulding, and which then harden or set into a permanent shape which cannot afterwards be altered by the application of further heat. These are also called thermosetting resins, the most important of these are formaldehyde, phenol formaldehyde and polyester resins.

The advantages of plastics materials are their light weight, and the simple methods used in the jointing processes. Plastics materials used in plumbing have many common physical characteristics. Those used for pipes are invariably thermoplastics, all of which have a high resistance to corrosion and acid attack. They have a low specific heat, which implies that they do not absorb the same heat quantity as metals. They are also poor conductors of heat and electricity.

One of the disadvantages in the use of plastics is their high rate of linear expansion, although manufacturers make provision for this in the design of their components. Another disadvantage is that resistance to damage by fire attack is low. Plastics are not such stable materials as metals and are all, to some extent, affected by the ultra-violet rays of the sunlight. This causes long term degradation or embrittlement of the material.

Most plastics in their natural state are clear and colourless, and to reduce the effects of degradation, manufacturers add a darkening substance or agent to give the material colour and body.

It is important for a plumber to be able to identify the type of plastics being used, as the method of jointing suitable for one type may be unsuitable for another. Basic tests described will enable the plumber to correctly identify a particular material. Table 2.6 sets out typical properties of plastics used in building.

Polythene

Polythene pipes up to 150 mm nominal bore are manufactured to BS and are classified as 'low' or 'high' density. The low density pipe is comparatively flexible. High density polythene is more rigid and has a slightly higher melting temperature. Both types are resistant to chemical attack and are much used as a material for laboratory and chemical waste installations. Polythene is sufficiently elastic not to fracture if the water in the pipe should freeze, although normal frost precautions are recommended to prevent freezing and loss of supply.

Polypropylene

This material is in the same family group as polythene. It is tough, having both surface hardness and rigidity. It is able to withstand relatively high temperatures, and is in this respect superior to polythene, ABS or PVC.

Polypropylene can withstand boiling water temperature for short periods of time making it suitable material for the manufacture of traps. Methods of jointing are similar to those used for polythene, e.g. compression joints or 'O' ring couplings. Solvent welded joints are not suitable.

Polypropylene and polythene belong to the family of synthetic plastics known as polyofins. They have a waxy touch and appearance, and if ignited burn with a flame similar to a paraffin wax candle.

Polyethelene

The range of plastics used in the manufacture of pipes and fittings continue to grow, Polybutylene, polyethelene and Polysulphone are now extensively used. In recently developed building sites the water supply is brought into the building by means of a high density Polyethelene pipe known as MDPE. This pipe is manufactured blue for under-ground work, this pipe may be used above ground but must be protected from direct sunlight by being enclosed in ducts or have an outer covering, alternatively the polyethelene pipes are pigmented black.

Polyvinyl chloride

Generally abbreviated to PVC, this is possibly the most common plastics material used for drainage and discharge pipe systems. The material is a thermoplastic produced on the basic reaction of acetylene with hydrochloric acid in gas form in the presence of a catalyst. The material is rigid, smooth, light and resistant to corrosion.

PVC is often confused with polythene and polypropylene. One method of identifying the material is to drop a small piece of it into water, and if it sinks it is PVC which is heavier than water, whereas the other two materials are lighter and therefore float. PVC unlike many other plastics materials will not burn easily, another fact which can be used for its identification.

Unplasticized polyvinyl chloride (UPVC) is the basic material without softening additives. Plasticized polyvinyl chloride is produced by adding a small amount of rubber plasticizer to the basic material during the manufacturing process. The result is a slightly more flexible material which is more resistant to impact damage than UPVC. All UPVC pipes for cold water supply should comply with BS.

Polyvinyl chloride for sanitary pipework should conform to BS. This standard requires that the material should not soften below 70°C for fittings and 81°C for pipes and it should be capable of receiving discharge water at a higher temperature than these for short periods of time.

Jointing methods used include solvent welding, rubber 'O' ring joints and compression type couplings.

Acrylonitrile butadiene styrene

Known as ABS, this is a material used mainly for small diameter waste and discharge pipes or overflows. The material itself is a toughened polystyrene which can be extruded or moulded. It can withstand higher water temperatures for a longer

Table 2.6 *Typical properties of plastics used in building*

Material	Density kg/m	Linear expansion per °C	Coefficient mm/m	Max. temperature recommended for continuous operation °C	Behaviour in fire
Polythene* low density high density	910 945	20×10^{-5} 14×10^{-5}	0.2 0.14	80 104	Melts and burns like paraffin wax
Polypropylene	900	11×10^{-5}	0.11	120	Melts and burns like paraffin wax
Polymethyl methacrylate (acrylic)	1185	7×10^{-5}	0.07	80	Melts and burns readily
Rigid PVC (UPVC)	1395	5×10^{-5}	0.05	65	Melts but burns only with great difficulty
Post-chlorinated PVC (CPVC)	1300–1500	7×10^{-5}	0.07	100	Melts but burns only with great difficulty
Plasticized PVC	1280	7×10^{-5}	0.07	40–65	Melts, may burn, depending on plasticizer used
Acetal resin	1410	8×10^{-5}	0.08	80	Softens and burns fairly readily
ABS	1060	7×10^{-5}	0.07	90	Melts and burns readily
Nylon	1120	8×10^{-5}	0.08	80–120	Melts, burns with difficulty
Polycarbonate	1200	7×10^{-5}	0.07	110	Melts, burns with difficulty
Phenolic laminates	1410	3×10^{-5}	0.03	120	Highly resistant to ignition
GRP laminates	1600	2×10^{-5}	0.02	90–150	Usually inflammable. Relatively flame-retardant grades are available

Key: UPVC = unplasticized polyvinyl chloride   GRP = glass-reinforced polyester   PVC = polyvinyl chloride
ABS = acrylonitrile/butadiene/styrene copolymer

*High density and low density polythene differ in their basic physical properties, the former being harder and more rigid than the latter. No distinction is drawn between them in terms of chemical properties or durability. The values shown are for typical materials but may vary considerably, depending on composition and method of manufacture.

period of time than PVC and for this reason some manufacturers produce full ABS waste systems. It also retains its strength against impact at very low temperatures thus providing greater resistance to physical damage. ABS has a duller matt appearance than PVC, and if ignited burns with a bright white flame. The material is slightly more dense than water and will therefore not float.

Acrylic (perspex)

This thermoplastic is tough and durable with good resistance to abrasion. The material can be transparent or opaque, it is easily machined and parts can be joined together by cementing/adhesive. It is used mainly in plumbing for bathroom accessories.

Polystyrene

It is a white thermoplastic produced by the polymerization of styrene (vinyl benzine). This material is very light and brittle, and is mainly used for thermal insulation in granule, sheet or foam form.

Polytetrafluoro ethylene (PTFE)

This material can be used at temperatures up to 300°C, and because it is chemically inert it is used for lining pipes and components where chemical resistance is necessary. Used by the plumber in tape or paste form as a sealant or jointing material.

Nylon

This is a thermoplastic material which is produced from phenol or benzine. It is rot proof and strong. Nylon is widely used in the form of a solid plastic, often as valve seatings, taps, gears and bearings. Moving nylon parts need no oiling because they slide easily over each other. The word nylon was made up by its inventors Du Pont.

Synthetic rubber

There are many forms of synthetic rubber, some of which contain a proportion of natural rubber. The most common synthetic rubber used in plumbing is called Neoprene which is generally used for the manufacture of ring seals for various jointing techniques. Neoprene is a trade name. The substance resists attack from oil, grease and heat and is more stable against oxidation than natural rubber.

Ceramics

Sanitary equipment may be made from one of the three ceramic materials – fireclay, earthenware or vitreous china. These materials have different qualities and characteristics. Each is, therefore, suitable for different uses which may be divided broadly into public, industrial and private use.

Where rough or heavy usage is expected, strength is a most important factor, and this is the outstanding quality of both fireclay and vitreous china. For private purposes, where good appearance is important, vitreous china is used.

Fireclay

This ware is made from clay which can be fired at very high temperatures, resulting in an article both heavy and strong. Large articles can be made in fireclay with the minimum of distortion, which due to the shrinkage which occurs during firing is inevitable with all ceramic materials. Fireclay is used in the manufacture of large sanitary appliances such as sinks, urinal ranges, laboratory sinks and special hospital fittings. Fireclay is heavy in weight and has a buff coloured porous body protected by the hard glaze which covers it. It has the highest initial cost of all three ceramic materials.

Earthenware

Ball and china clays are the most important constituents of earthenware, which is considerably lighter in weight than fireclay. This material has a pleasing appearance due to the clean lines and sharp definition which characterize articles manufactured from this material. Earthenware has a white porous body protected by a hard impervious glaze. This material is cheaper than either fireclay or vitreous china.

Vitreous china

This material is made from the same clays as earthenware but feldspar is also included. This gives two additional qualities, great strength and a vitrified body impervious to water. It is lighter in weight and less costly than fireclay, and has the same clean lines as earthenware. Apart from industrial use it is particularly good for hotel and household use since it combines pleasing appearance and strength.

Since the body is vitrified, vitreous china does not rely upon its skin of glaze for its sanitary properties, and is glazed only to give it a smooth, glossy finish and to allow easy cleaning. Should the glaze become damaged, the appliance remains impervious to water, and, therefore, completely sanitary. For this reason, and because of its strength and moderate cost, vitreous china is the most economic pottery material, both from the point of view of initial cost and maintenance costs.

During firing, vitreous china tends to distort more readily than either fireclay or earthenware, and for this reason great care has to be taken to ensure that the finished articles are of good shape. For this reason some of the larger sanitary appliances such as sinks, urinals and hospital equipment are not manufactured in vitreous china.

Vitreous china lends itself to the manufacture of articles of curved and rounded design, which minimizes the production problem of good shape and results in articles that are practical in use and conform to contemporary ideas of good practice in design.

Weight for weight, vitreous china is the strongest of the three ceramic materials used in the manufacture of sanitary appliances.

Electricity supply

There is a growing need for mechanical engineering services operatives to have knowledge of electricity and electrical installation work. More and more plumbing systems and components depend on electricity to provide the energy for them to operate, and the NVQ plumbing scheme includes several objectives related to electrical systems and components.

It is necessary for operatives to acquire as quickly as possible an understanding of the relevant terminology which is described in the glossary at the back of this book. It must be mentioned that the glossary does not give a complete coverage of electrical engineering and electrical installation terminology but contains sufficient information to enable plumbing students to make reference to specific topics of interest to them and to the electrical technology covered in the following pages. Electricity is a form of energy which is produced by generating equipment at power stations. These generators may be powered by coal, oil, or gas turbines or a nuclear reactor. The power of flowing water may also be used to drive these generators (hydro-electricity). The Central Electricity Generating Board is responsible for the generation and primary distribution of electricity,

while area Supply Authorities or (SA) handle the regional and local distribution and supply of electricity to individual properties.

Electricity from the generators is supplied into the national grid (network of overhead cables on pylons) at very high voltages (electric pressure). This is transformed down in voltage and fed into the regional grids, which consist of both overhead and underground cables. Substations in the regional grid again reduce the voltage to suit the power requirements of the user.

Electricity from the grid is brought into a building to the meter position either by an overhead service cable from the Supply Authorities supply pole (common in rural areas) or an underground service cable connected to the Supply Authorities main, which is situated under the road. The service cable terminates at a sealed fuse unit which is connected via two short lengths of cable (live and neutral) to the meter fixed alongside. Electricity flows through the live wire and in order to complete the circuit returns along the neutral. In addition, an earthing connection is provided to an earth clamp, earth rod or an earth leakage circuit breaker. Some properties have a second meter which operates through a sealed time switch, to record the utilization of offpeak (cheaper) electricity.

The risks of electrocution and fire must be guarded against in all electrical installations. The precautions which are established practice are shown in principle in Figure 2.99.

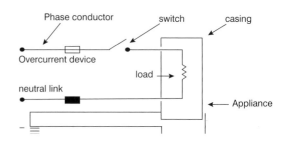

Figure 2.99 *Established precautions*

Supplies into consumers' premises

The *Electricity Supply Regulations* 1988 place a responsibility on all consumers of electricity to provide and maintain safe electrical installations. Such installations have to be designed to the requirements of the BS 7671 for electrical installations. It is usually the practice for a domestic consumer to be supplied with a single-phase, two-wire a.c. supply at a nominal voltage of 230 V and a frequency of 50 Hz. The basic arrangement is shown in Figure 2.100 where it will be seen that the single-phase supply is actually derived from a three-phase, four-wire system which operates at 400 V between its supply lines. Sometimes this voltage is used in domestic premises but it is more suitable for larger types of premises such as shops, offices and factories which demand more current and use a variety of three-phase equipment.

For single-phase supplies of 230 V and up to 100 A loading, the supply authority's low-voltage distribution cable will either be overhead or underground and will normally be installed along a public pathway or road, mostly at the front of consumers' property. The incoming supply to a particular consumer may be directly connected to the nearest point of the main distribution cable or it may be looped to an adjacent service point in nearby premises. This service termination will be accommodated either within the consumer's

premises or in an external meter cabinet. It will consist of a cutout *fuse* and an energy meter to record the number of units (kWh) of electricity.

The 'cut-out' contains a solidly bolted neutral conductor or a combined neutral-earth conductor known as a PEN conductor and a single-pole fuse-link which is normally a BS 1361 Type II high breaking capacity cartridge fuse of 100 A rating. This fuse provides protection against overcurrent between the output terminals of the 'cut-out' and the supply terminals. Figure 2.101 shows one method of wiring a consumer's premises.

The electricity meter or energy meter is an integrating meter and records the number of units (1 kWh is 1 unit) of electricity a consumer uses. There are basically two kinds of energy meter, the *dial meter* and *digital meter*. When reading a dial meter, you should note that the adjacent dials revolve in opposite directions. You should read the meter from left to right, ignoring the dial marked 1/10 which is used for testing purposes. Other points to note are. always write down the number the pointer has passed and if it is directly over a number, write this number down and underline it. If it is followed by a 9, reduce the underlined figure by 1 (e.g. 234911 becomes 233911). In the digital meter, the units used are

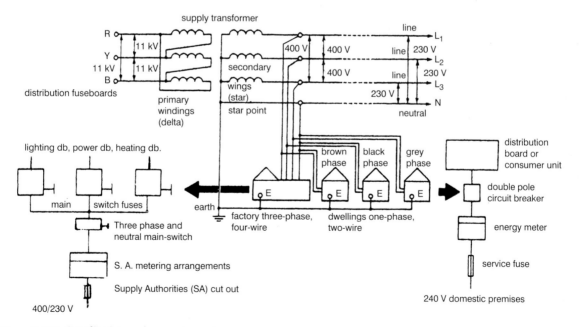

Figure 2.100 *Supplies into consumers' premises*

shown by a row of figures. If the premises are fitted with an Economy 7 meter, it will have two rows of figures, one for the lower-priced night-rate units and the other for the day-rate units.

Distribution arrangements

The connection between the supply authority's equipment and the next stage of protection is the responsibility of the consumer and it includes the meter tails, consumer unit and earthing and bonding arrangements. While the supply authority make every effort to provide an earthing terminal, where this is not possible the consumer has to install an earth electrode on the premises. This arrangement is known as a TT earthing system. Premises that have this system are supplied from overhead lines. Figure 2.102, shows this system, along with the supply authority's TN-S and TN-C-S systems. Both these latter systems allow the armouring of the service cable to be used as a means of earthing. The purpose of providing a good earth return from a TN-S and TN-C-S system is to encourage a large fault current to flow from the consumer's installation back to the supply transformer. If the earth loop impedance path is low enough then the returning fault current will automatically disconnect the circuit protective device, Protective devices for fixed circuits and socket outlet circuits have to meet two different disconnection times, namely 5 s and 0.4 s respectively. The faster operating time to disconnect socket outlets is because of the vulnerable nature presented by equipment connected through plugs and flexible leads, The BS 7671 provide a number of tables which give the maximum impedance values for different protective device.

To address the fundamental requirements for safety, the supply of electricity into a consumer's premises must satisfy three important conditions, namely:

- Control
- Circuit protection and
- Earth leakage protection.

The first condition is satisfied by isolation and switching so that the electrical installation can be cut off from its voltage source. The second condition is satisfied by circuit protective devices, designed to disconnect circuits automatically from excess current or overcurrent. The third condition will be satisfied using a residual device which can detect leakage currents and act as a means of protection against electric shock and fire by isolating the supply from the circuit.

Figure 2.103 shows the sequence of control and protection in two distinct types of premises. Where separate buildings exist it is essential they have their own means of control and protection.

There are numerous requirements in the BS 7671 for isolation and switching (see Regulation 130–06, Chapter 46, Sections 476 and 537). Where electrical equipment is fed from a final circuit, an isolator should be placed adjacent to that equipment but if the equipment is remotely placed, provision should be made to stop it being inadvertently closed during operation. This is often achieved with a padlock or removable handle. Besides functional switching for control purposes, some circuits might need switching off for mechanical maintenance or even emergency switching. The former is used to protect persons undertaking non-electrical tasks such as cleaning and maintenance, as the latter is used for the rapid cutting off of a circuit in order to remove a hazard. It should be noted that a plug and socket cannot be used for emergency switching; it is best achieved with a single switch or combination of devices initiated by a single action such as emergency push buttons of the latched type. Such devices should be clearly identifiable and preferably red in colour and must be manually operated and placed in a position of accessibility. Figure 2.104 is a typical emergency stop button control circuit.

Wiring systems

Chapter 52 of the *BS7671* concerns selection and erection of wiring systems. In practice, there may be several factors likely to influence the choice of wiring system for a particular electrical installation; for example, the type of building and its use. A *surface-type* wiring system may be acceptable in a factory or workplace but may be totally

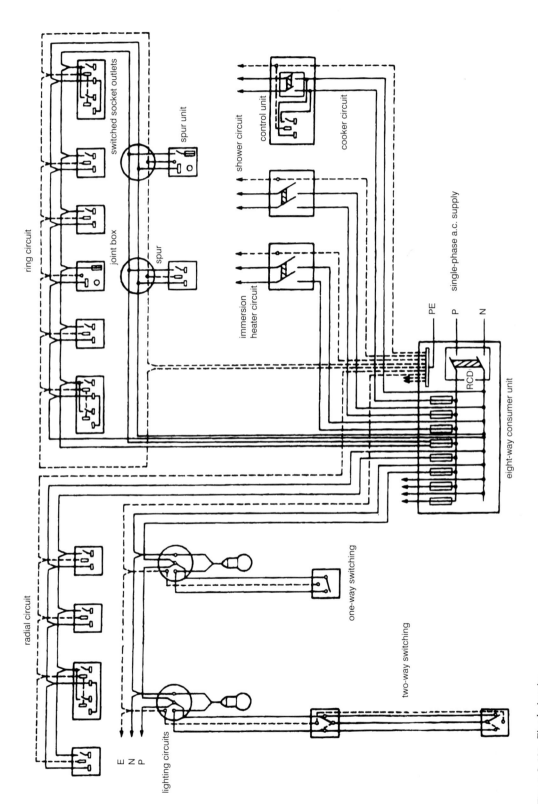

Figure 2.101 *Final circuits*

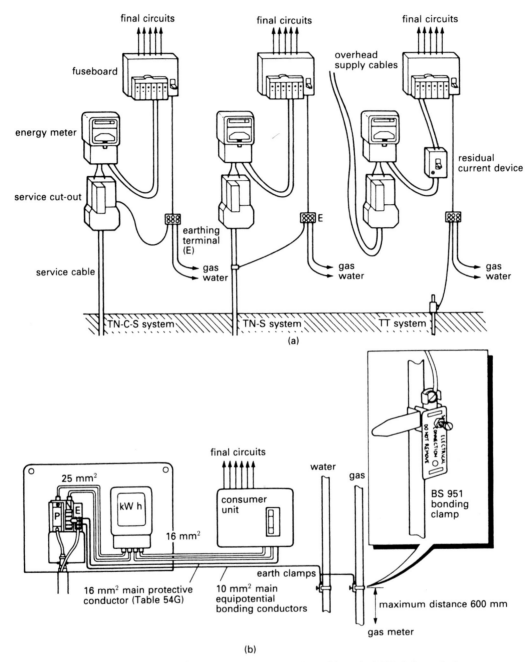

Figure 2.102 (a) Common methods of earthing a consumer's premises; (b) typical TN-C-S supply into a consumer's premises showing meter tails and earthing and bonding arrangements

unacceptable in an office block or hotel or even domestic dwelling, simply on the grounds of appearance and aesthetic taste. Here, it is usual to install a hidden or *flush-type* wiring system, such as PVC-insulated cables, mechanically protected under plaster. There is also consideration for the environment, such as excessive temperature, as well as any particular external influence from a corrosive atmosphere such as salt or dust. Some of these factors have already been

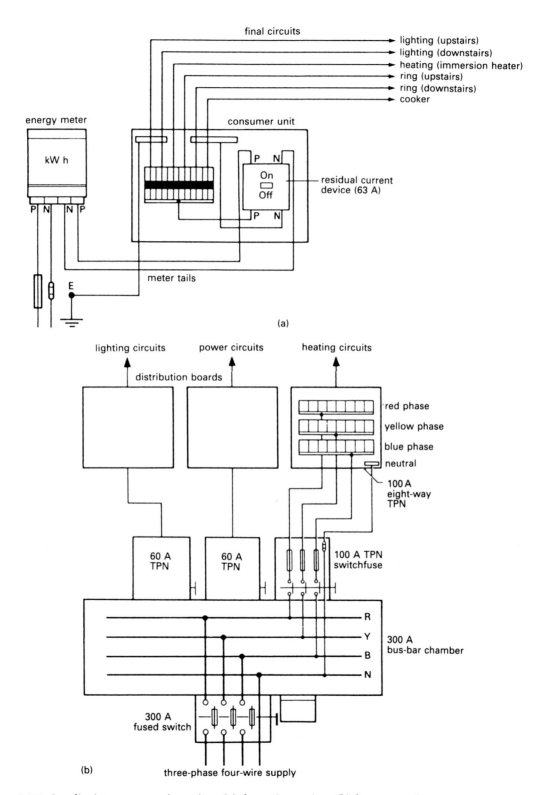

Figure 2.103 *Supplies into consumers' premises: (a) domestic premises; (b) factory premises*

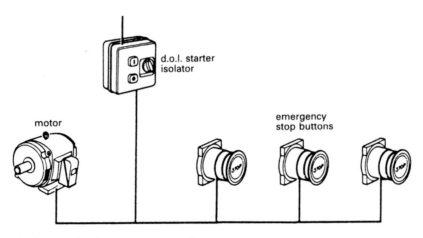

Figure 2.104 *Control of motor showing emergency stop buttons*

mentioned. Questions will need to be asked about the chosen system's durability, its mechanical protection and cost comparison with other favourable systems; not only might there be a material cost benefit but also an installation time benefit. Another important question is whether the chosen system needs to cater for any likely alterations and/or the installing of additional circuits. Fortunately, in many large premises, one is likely to come across numerous wiring systems and these are often integrated with each other so that alterations and modifications can occur.

Some common types of wiring system found today are:

- PVC-insulated PVC-sheathed cables
- PVC-insulated armoured PVC-sheathed cables
- Mineral-insulated metal-sheathed cables
- Metal and plastic conduit systems (incorporating cables)
- Metal and plastic trunking systems (incorporating cables)
- Bus-bar trunking systems.

It should be pointed out that there are various support arrangements for the above systems, such as cable tray, ducting and cable trench; see the methods in Table 9A of the *IEE Wiring Regulations*.

PVC-insulated PVC-sheathed cables

This wiring system, shown in Figure 2.105, has a general use and will be found listed in Table 4D1–2 of the BS 7671. The common PVC/PVC/ c.p.c. is widely used for surface wiring, where it needs to be clipped and supported to meet the requirements of the Regulations. There are requirements for the internal radii of bends and supports. The wiring system can be hidden from view, above a false ceiling, in joists or buried under plaster. In a false ceiling, the cables should be kept clear of sharp edges and be adequately supported. Where they are run through holes in wooden joists, the cables must be at least 50 mm from the top or bottom of the joist (Regulation 522–06–05), If cables are to be laid in existing notches, they should be provided with mechanical protection to prevent damage occurring from floor fixings. Figure 2.106 shows a typical arrangement through joists. It should be noted from Table 4D2 that these cables have a maximum conductor operating temperature of 70°C, and it will be observed that various correction factors apply for ambient temperatures in excess of 30°C. Chapter 42 of the BS 7671 deals with *protection against thermal effects*. It is generally recognized that a group correction factor does not apply to domestic final circuits but, where this is the case in other instances, Table 4B1 has to be used. At terminations, the outer PVC sheath should not be removed any further than is necessary and all conductors should be identified by their appropriate colour coding. Sections 526 and 527 of the BS 7671 cover *connections* and *selection and erection* to minimize the spread of fire, respectively. It is important to make sure that no mechanical stress is placed on the cables and that cable glands securely retain the outer sheath of the cable. On no account should conductor strands be cut in order to fit into a termination post.

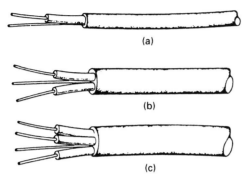

Figure 2.105 *PVC-insulated, PVC-sheathed cables 600/1000 V: (a) single-core; (b) two-core; (c) three-core*

PVC-insulated armoured PVC-sheathed cables

PVC armoured cables have a wide commercial and industrial application and there are several tables in the BS 7671 concerning both the copper and aluminum conductor types. Figure 2.107(a) shows a diagram of a typical cable; the armouring consists of galvanized steel wire secured between PVC bedding and a tough PVC outer sheath. The armouring is often used as a protective conductor but this should be checked by calculation using Regulation 543–01–03, An alternative method is to apply Table 2.7 which is derived from BS 6346 and shows the nearest smaller copper cable size equivalent to the armouring of the chosen cables.

The cables are often clipped on a surface using recommended wall cleats, or they can be installed on cable tray or run in a trench as outlined in Table 4A of the Regulations. Where they are buried in soil, one will find that the usual practice is to provide either cable tiles or yellow warning tape laid on top so as to denote their position. They must also be installed deep enough to avoid damage from any possible ground disturbance (see Regulation 522–06–03).

It is important to see that cable terminations are mechanically and electrically sound and that the cable gland and steel armouring make an effective earth connection. The use of an earth tag is recommended in order to give earth continuity between the gland and the steel enclosure to which the gland is fixed. All conductor cares of multicore cables should be identified properly using appropriate markers.

Mineral-insulated metal-sheathed cables

Mineral-insulated (MI) cables have a very wide commercial and industrial use. The common types are listed in Tables 4J1 to 4J2 of the Regulations, where they are divided between light-duty and

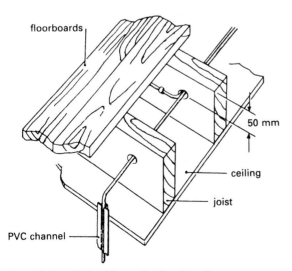

Figure 2.106 *PVC cables under floorboards*

Table 2.7

Nominal area of conductor (mm²)	Nominal area of copper conductor equivalent to armouring		
	Two-core	Three-core	Four-core
2.5	1.5	1.5	1.5
4.0	1.5	2.5	4.0
6.0	2.5	4.0	4.0
10.0	4.0	4.0	4.0
16.0	4.0	4.0	6.0
25.0	6.0	6.0	6.0
50.0	6.0	6.0	10.0

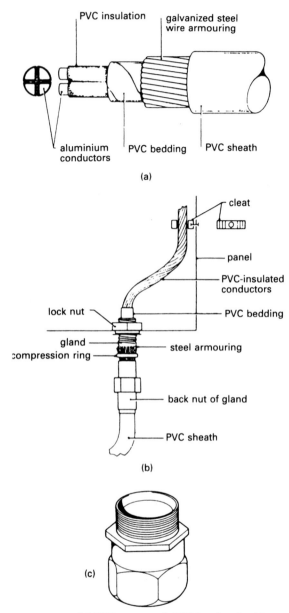

Figure 2.107 *(a) PVC-armoured, PVC-insulated cable;
(b) termination; (c) gland*

conductivity copper or aluminum and, as already indicated, both types may be provided with a PVC oversheath to give added protection against corrosion. The insulation medium between the conductor cores is a compressed mineral powder called magnesium oxide, and there have been improvements to reduce its hygroscopic nature (i.e. its ability to absorb moisture). Termination of the inner conductor cores involves several tools. The screw-on pot seal is widely used and this is attached to the sheath using a pot wrench. The correct temperature sealing compound is then inserted into the pot and the disc and insulating sleeving attached using a crimping tool.

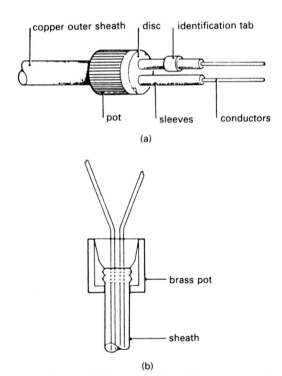

Figure 2.108 *(a) Cable termination; (b) screw-on seal*

heavy-duty use and the outer sheath either covered with PVC or not covered. These cables are ideal for 'hot' installations and their termination accessories such as sleeving, seals and compound can be designed for maximum sheath operating temperatures of 105°C. Figure 2.108 shows a typical MI cable termination. The cable conductors and sheath are generally constructed of high-

The cables are often referred to as mineral-insulated metal-sheathed (m.i.m.s.) cables and are fireproof, waterproof and oil-proof. They are also non-ageing and have higher current ratings compared with equivalent cable sizes. They are ideal for fire alarm circuits, boiler rooms and garage/petrol filling installations where dangerous atmospheres are likely to be present in Zone 1 hazardous areas. Here, they are fitted with flame-proof glands and the cables have an overall extruded covering of PVC. The metal sheath of these cables satisfies both shock protection and thermal protection requirements since it is approximately four times the cross-sectional area of the inside cores. To satisfy earthing connections, earth-tail pots are available having a cross-sectional area equal to the related size of the phase conductor(s). It is recommended that these should always be used for continuing the protective conductor function of the sheath through the earthing terminal at outlet points.

Metal and plastic conduit systems

Conduit systems are in wide use today either as surface or flush wiring arrangements. The common sizes are 16, 20, 25 and 32 mm. They provide the installed cables, such as the single-core PVC insulated or ethylene propylene rubber (e.p.r.) insulated cables, with additional mechanical protection. The advantage of conduit is the amount of flexibility it offers to final circuits since cables can be added or withdrawn. It is important to see that the conduits are erected first before the cables are drawn in (Regulation 522–08–01) and it is equally or more important that *metal conduit* systems are both electrically and mechanically sound, even if it is decided to install protective conductors for additional safety (Regulation 543–02–04). Further requirements are made as shown in Figure 2.109.

Conduits should be distinguished from other services by the colour orange (Regulation 514–03–01): in damp situations, galvanized conduit or plastic conduit should be used. With either method, it is essential to provide drainage facilities for the release of moisture (see Regulation 522–03–02). One of the most important requirements is Regulation 521–02–01 (*conductors of a.c. circuits installed in ferrous enclosures*): the reason behind this regulation is to stop *eddy currents* flowing in the metal conduit, which would result in them becoming hot and affecting the cables inside. It should also be pointed out that 3 m is the maximum length of span for heavy gauge steel conduit used for overhead wiring between buildings and it has to be a minimum size of 20 mm and be unjointed.

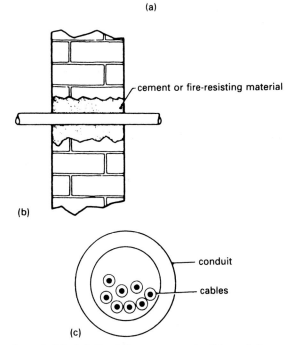

Figure 2.109 *(a) Conduit arrangements; (b) conduit passing through a wall; (c) cables in conduit*

With regard to *plastic conduit*, the system is of course non-rusting and is a protective measure in itself against indirect contact. Despite this fact, separate protective conductors must be enclosed in the system and whilst rigid PVC conduit is suitable for normal ambient temperatures it must be remembered that it has a greater coefficient of expansion than steel and therefore provision must be made to allow for movement. It is recommended that expansion couplers be fitted in the worst affected areas. Regulation 522–06–01 should be noted with regard to plastic boxes used for suspending luminaires.

Insulation

All conductors are covered with insulating material or supported on insulators within an earthed casing with a clear air gap round each conductor. Standards of insulation vary with voltage. If a number of wires carrying different voltages are enclosed in a trunking they must all be insulated to the standard of the highest voltage. To avoid this, systems at different voltages are usually run in separate trunkings (e.g. British Telecom wiring and electricity supply wiring). Figure 2.110 illustrates different types of conductor.

Fusing

Each section of wiring must be protected by having in the circuit a fuse wire which will melt if a current passes higher than that which is safe for the wiring (see Figure 2.111).

This prevents overheating of wiring with the possible risk of fire. Fuses may be of the traditional type where a fuse wire is stretched between terminals in a ceramic holder, or of the modern cartridge type where the wire is held in a small ceramic tube with metal ends.

The cartridge types of fuse are much easier and quicker to replace. It is also possible to use circuit-breakers instead of fuses (see Figure 2.112). These operate by thermal or magnetic means and switch off the circuit immediately an overload occurs. They can be reset immediately by a switch. They are more expensive than fuses but have the advantage that they can also be used as switches to control the circuits they serve. They are particularly valuable in industrial uses where circuits may become over-loaded in normal operation and the circuit-breaker will switch off but may be switched on again immediately the overload is removed. In domestic circumstances the blowing of a fuse or operation of a circuit-breaker usually indicates a fault.

Modern practice is to provide a fuse, or circuit-breaker, at the phase end of the circuit (called the line) and a simple link at the neutral end.

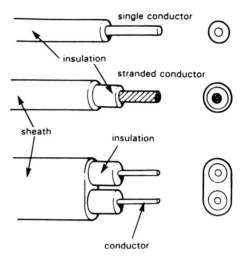

Single conductor, single core, insulated, non-sheathed.

Stranded conductor, single core insulated, sheathed.

Two-core flat, insulated and sheathed (also called flat twin).

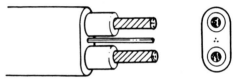

As above with an uninsulated earth continuity wire in the same sheath.

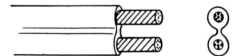

Parallel twin, cores easily separated without damage to insulation of either.

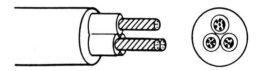

Three-core, insulated, sheathed, unarmoured.

Figure 2.110 *Electrical conductors*

Switch polarity

The position of the switch has the same effect upon safety as that of the fuse. If the switch is fitted on the neutral side of the apparatus this will always be live, even when the switch is turned off. Switches are therefore always fitted on the phase side of the apparatus they control.

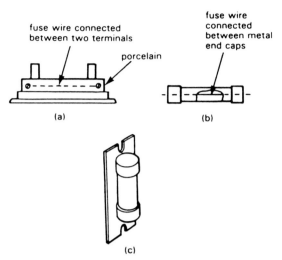

Figure 2.111 *Types of fuse: (a) rewirable, BS 3036; (b) cartridge, BS 1362; (c) high rupturing capacity, BS 88*

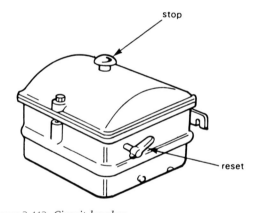

Figure 2.112 *Circuit-breaker*

Bathrooms

In bathrooms and similar situations where water is present and metal fittings or wet concrete floors provide a good passage to earth, special safety precautions are called for. No socket outlets or switches may be provided in the area at risk. They must be sited outside the area, or on the ceiling, operated by non-conducting pulls. Where any electrical apparatus is present, all metal, including not only the casing of electrical apparatus but also pipework, baths, etc, must be bonded together electrically and earthed.

Terminations and connections

The type of termination used depends upon the type and size of cable and the kind of connection to be made. Figure 2.113 illustrates termination to a plug.

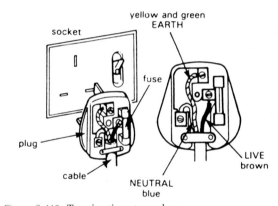

Figure 2.113 *Terminations to a plug*

Earthing

Any metalwork directly associated with electrical wiring could become live if insulation frayed or if wires became displaced. Anyone touching such a piece of apparatus would run the risk of serious electric shock. This is avoided by earthing the metalwork so that a heavy current flows to earth and the fuse is blown immediately the fault occurs. Although the neutral wire is earthed it will not serve for this purpose and a separate set of conductors for earthing are provided in almost all electrical installations. The earth connection itself is made locally in the building (see Figure 2.102).

Termination of cables and flexible cords

For flexible cords, the bare stranded conductors are usually twisted together and doubled back, if room permits, then screwed firmly in a pillar terminal, as shown in Figure 2.115.

For cables having single-stranded conductors which are to be terminated and connected under screwheads or nuts, it is convenient to shape the bare end into an eye to fit over the thread as shown in Figure 2.114.

Cables with more than one stranded conductor may be terminated in a similar manner to that shown for the flexible cord.

use a special claw washer to get a better connection. Lay the looped conductor in the pressing. Place a plain washer on top of the loop and squeeze the metal points flat

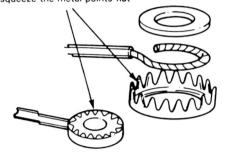

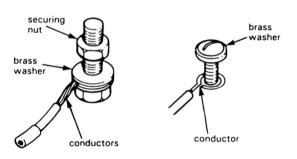

Figure 2.114 *Termination under screwheads or nuts*

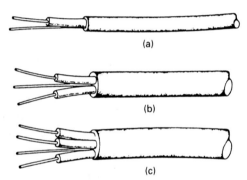

Figure 2.115 *Connection to terminal*

Sockets (soldered and crimped)

For conductors of larger cross-sectional area, a socket may be used. There are two methods of fixing a socket to a cable end. One method is by soldering a tinned copper socket on the cable end. The other method is to fasten a compression-type socket on the cable end using a crimping tool The two methods are shown in Figure 2.116.

The crimped socket is used to a large extent for terminating smaller cables. In smaller sizes they are even being used for flexible cords. Hand crimping tools are usually used for smaller size sockets which are capable of crimping a range of different sizes.

Termination of an aluminium conductor

Where an aluminium conductor is to be connected to a terminal it must be ensured that the conductor is not under excessive mechanical pressure. Further, an aluminium conductor must not be placed in contact with a brass terminal or

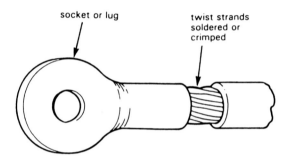

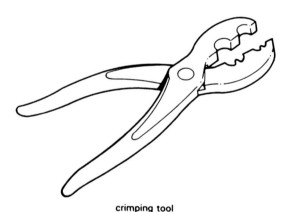

Figure 2.116 *Soldered and crimped sockets*

any metal which is an alloy of copper. This may only be done if the terminal has been plated, otherwise corrosion would take place.

Other types of cables

When cables having metallic braid, sheath or tape coverings are used, the sheath must be cut back from the end of the insulation. This is done to prevent leakage from live parts to the covering. However, this is unnecessary for mineral-insulated cables. The ends of mineral-insulated copper-sheathed cables must be protected from moisture by sealing, ensuring that the mineral insulation is perfectly dry before sealing.

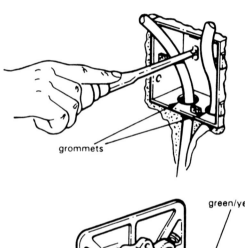

grommets

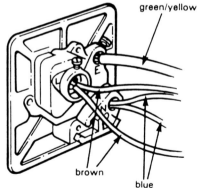

green/yellow

brown

blue

Figure 2.117 *Wiring a socket outlet*

Wiring to socket outlets, spur boxes and junction boxes

Wiring a socket outlet (see Figure 2.117)

1 Cut the cable loop to make two tails 75 mm long and thread the cables through grommets in the mounting box.

2 Screw the mounting box to the wall and strip back the cable insulation.

3 Twist together and connect:
   (a) Brown wires to terminal L.
   (b) Blue wires to terminal N.

4 Fit green/yellow insulated sleeving over both earth wires.

5 Twist earth wires together and connect to terminal E.

6 Tighten all securing screws.

Note Ensure that the wires are pushed into terminals up to the insulation.

Fused Connection Units (see Figure 2.118)
These are fitted in a similar manner to socket outlets but have two sets of terminals. The lead to the appliance must be clamped in the cord grip.

Junction boxes
Junction boxes (Figure 2.119) have four knockout sections for cable entry and exit. The central terminal is always used for the earth connection.

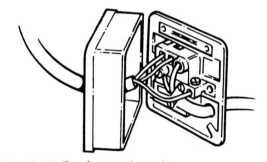

Figure 2.118 *Fused connection units*

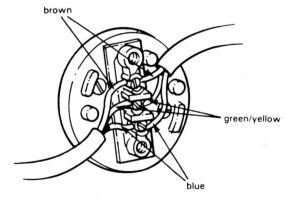

brown

green/yellow

blue

Figure 2.119 *Junction box*

It is good practice to run spurs from the back of a socket but junction boxes can be used for this purpose.

Fuses for domestic appliances

The watt is the unit of electrical power, and can be calculated by multiplying volts by amps. If we know the voltage of the supply and wattage of an appliance we can calculate how much current the appliance will need by dividing watts by volts.

Most electrical appliances have a label fixed to them giving the voltage with which they must be supplied and the wattage of the appliance. This is very useful in deciding what value of fuse should be used for a particular appliance.

Example

An electric heater is rated at 3 kW and is designed to be used with a supply voltage of 230 Volts

$$\frac{watts}{volts} = amps \qquad \frac{3000}{230} = 13.04 \text{ amps}$$

Therefore the current required by this appliance is 13.04 amps, and a correct value fuse must be fitted. If a 10 amp fuse is used the heater will attempt to draw more current than the fuse will pass, the fuse will blow (melt), breaking the circuit and cutting off the supply. On the other hand a 13 amp fuse will permit the heater to operate normally and still provide a safety factor if needed.

Self-assessment questions

Working processes

1 Galvanizing, the protective coating given to metals, is obtained by dipping the metal in:
   (a) zinc
   (b) tin
   (c) solder
   (d) aluminium.

2 When a metal is annealed it is made:
   (a) softer
   (b) harder
   (c) brittle
   (d) less malleable.

3 When removing a bending spring from a sharp set the correct procedure is to:
   (a) twist to tighten the coil of the spring
   (b) twist to loosen the coil of the spring
   (c) straighten the pipe as far as possible
   (d) pull it straight out.

4 Throat rippling on machine-made bends on copper pipe is most probably caused by:
   (a) an uneven pull on the machine handle
   (b) the machine not being level
   (c) pulling the bend too quickly
   (d) an incorrectly adjusted or badly fitted guide.

5 To obtain a 300 mm radius bend on 50 mm diameter low-carbon steel pipe, the most suitable method would be to use:
   (a) sand loading, cold, bending ring
   (b) rotary-type hydraulic bending machine
   (c) bending spring, hot
   (d) sand loading, hot, pin block.

6 Which of the following materials has the highest rate of linear expansion:
   (a) steel
   (b) copper
   (c) aluminium
   (d) UPVC.

7 An alloy is a:
- (a) solution of metals
- (b) mixture of metals
- (c) compound
- (d) an oxide of metals.

8 To ensure that a hydraulic bending machine gives maximum thrust it is essential that the:
- (a) correct former is used
- (b) oil level is maintained
- (c) machine is on a level base
- (d) oil pressure is adjusted.

9 Which tool is used for cutting thin sheet metal:
- (a) cold chisel
- (b) junior hacksaw
- (c) tin snips
- (d) wheel cutter.

10 Pressure should never be exerted on the side of a grindstone wheel because:
- (a) the stone may shatter
- (b) the motor will slow down
- (c) uneven wear will occur
- (d) redressing will be required more frequently.

11 The set on a hacksaw blade is provided in order to:
- (a) give the blade body clearance when cutting
- (b) allow the blade to cut on the backward stroke
- (c) make the blade more flexible
- (d) increase the tensile strength of the blade.

12 On building sites it is the employer's responsibility to provide a transformer for use with portable power tools having a voltage of:
- (a) 415 V
- (b) 240 V
- (c) 220 V
- (d) 110 V.

13 The tool allowing a small adjustment used for threading small diameter rods is known as a:
- (a) circular split die
- (b) solid die
- (c) split stock
- (d) solid stock.

14 The British Standard pipe thread is used for:
- (a) boiler tappings
- (b) thin walled brass tubes
- (c) gas meter unions
- (d) oxy-acetylene apparatus.

15 A metal which can readily be drawn into wire or rod is said to be:
- (a) ductile
- (b) malleable
- (c) tenacious
- (d) lustrous.

16 The thread form used on low carbon steel tubes is:
- (a) BSW
- (b) BSP
- (c) Unified
- (d) BSF.

17 To reduce the effect of heat on the blade and work when using a power driven hacksaw they should be:
- (a) lubricated during cutting
- (b) lubricated before cutting
- (c) quenched after cutting
- (d) allowed to cool slowly.

18 Screw threads are used for joining pipes, the external and internal threads are designed to:
- (a) fit uniformly together
- (b) slack fit to allow for jointing paste
- (c) reduce the effects of torsion
- (d) reduce the effects of compression.

19 Which saw would be best for cutting a floorboard:
- (a) junior hacksaw
- (b) hacksaw
- (c) tenon saw
- (d) shetack saw.

20 When drilling a hole in a small metal bracket the bracket should be held in:
- (a) a vice
- (b) a pipe clamp
- (c) between two pieces of wood
- (d) a pair of Stillsons.

3 Key plumbing principles, science and electrical supply

After reading this chapter you should be able to:

1 Define heat and temperature.
2 Understand the measurement of heat and temperature.
3 Define 'specific heat'.
4 Understand the term 'transmission of heat'.
5 Explain conduction, convection and radiation.
6 Understand the expansion characteristics of various substances.
7 Understand the nature of pressure in liquids and gases.
8 Calculate basic pressure calculations related to water.
9 Explain the terms 'density' and 'relative density'.
10 State the common causes of corrosion.
11 Demonstrate knowledge of how to prevent corrosion occurring.
12 Name and recognize fluxes used by the plumber.
13 Describe Dew Point, and how to prevent condensation.
14 State the properties of common insulating materials.
15 Describe the uses and application of insulating materials.
16 Define the terms 'anode' and 'cathode'.
17 Demonstrate knowledge of Archimedes' Principle.
18 Describe common examples of capillarity and surface tension related to building work.
19 Demonstrate knowledge of pressure in liquids and gases and relate this knowledge to plumbing situations.

Heat

Heat is a form of energy, and energy means a capacity for doing work.

All substances contain heat.

The molecules which make up substances are always vibrating to and fro. They need energy for this work, and heat provides it. The more energy the molecules possess, the more vigorously and further apart they will be able to vibrate. As the heat input into a substance increases the molecules step up their vibratory rate, and weaken their cohesive bonds, often bringing about a change of state, i.e. solid to liquid or liquid to gas.

For example, ice is water in a solid state. It only has a small heat content so its molecules hardly vibrate at all, and since they are close together the cohesion between them is strong and ice is therefore rigid. If heat is applied to ice, the molecules gain energy, vibrate, and weaken their cohesive bonds. The (solid) ice changes into (liquid) water. Further heating of the water increases the heat energy until the molecules vibrate so strongly that they actually jump out of the water to form a gas (steam).

The measurement of heat

As explained previously, heat is the name given to energy which is in the process of moving from one place to another as the result of a temperature difference between them. Since heat is a method of transferring energy, it is measured in *joules*, the same as any other kind of energy.

Heat capacity

The heat of a body is defined as the heat required to raise its temperature by 1°C. Therefore, the unit of heat capacity is the joule per degree C.

Specific heat capacity

If we take equal quantities of water and oil and warm them in separate containers, but by the same flame, we may find that the oil temperature may rise by 15°C in five minutes but the water may only rise by 8°C in the same period of time. Since the supply of heat is the same in both cases, it is clear that oil has a lower heat capacity than water (see Figure 3.1).

When comparing the heat capacities of different substances we talk of their *specific heat capacities*.

Definition

The specific heat capacity of a substance is defined as the heat required to raise unit mass of

Table 3.1 *Specific heat capacities in kJ/kg°C*

Water	4.186	Iron	0.460
Methylated spirit	2.400	Zinc	0.397
Ice	2.100	Copper	0.385
Air	1.046	Brass	0.380
Aluminium	0.887	Tin	0.234
Cast iron	0.544	Lead	0.125
Mild steel	0.502	Mercury	0.125

it through 1°C, and the unit of specific heat capacity is the joule per kilogramme degree C (J/kg°C).

Table 3.1 shows specific heat capacities for various substances. It will be seen that water has the unusually high specific heat capacity of 4.186J/kg°C. Very few substances have a higher value than this.

The specific heat capacities of all substances vary slightly as their temperature changes but Table 3.1 is sufficiently accurate for present needs.

From Table 3.1 it can be seen that only a small amount of heat is required to raise 1 kg of mercury 1°C. This means that it is very sensitive to temperature change, which is one reason why it is used in thermometers.

The knowledge of heat capacity and specific heat is necessary to complete calculations related to domestic hot water and central heating systems. The formula generally used is as follows:

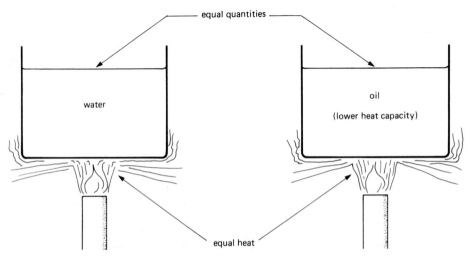

Figure 3.1 *Specific heat capacity of water and oil*

quantity of heat (kJ) =
mass (kg) × specific heat (kJ/kg) ×
temperature change (°C)

Example 1
Calculate the quantity of heat required to raise 80 litres of water from 10°C to 55°C.

quantity of heat = mass × specific heat × temperature change

= 80 × 4.186 × 45

= 15 069 kJ

Example 2
18 kg of cast iron is heated through 75°C. How many joules of heat energy are absorbed by the cast iron?

quantity of heat = mass × specific heat × temperature change

= 18 × 0.544 × 75

= 734 kJ

Example 3
25 litres of water cool from 62°C to 15°C. What quantity of heat is given off?

quantity of heat = mass × specific heat × temperature change

= 25 × 4.186 × 47

= 4918 kJ

In the above examples the answers are shown as whole. For the purpose of simplicity decimal portions have been omitted. No allowance was made for thermal efficiency of the water heating apparatus or for the loss of heat to the surrounding air while heating was taking place. These are, of course important points, the consideration of which affect the design and installation of hot water and central heating systems.

Temperature

The quantity of heat a substance contains and its temperature are two quite different things. Consider a bucketful of hot water and a red hot steel wire. The water contains more heat energy than the wire, but the wire is at a higher temperature (see Figure 3.2).

The temperature of a substance is its degree of hotness, and this is measured by means of a thermometer.

Thermometers
There are several different types of thermometer available. The most common depend on the expansion of a liquid when heated, others on the expansion of a compound strip of two metals.

Most students will be familiar with mercury or alcohol thermometers. These usually have spherical bulbs and are mounted on metal, boxwood or plastic scales. Thermometers used in laboratories have cylindrical bulbs for easy insertion through holes in corks and have their scales engraved directly on the stem.

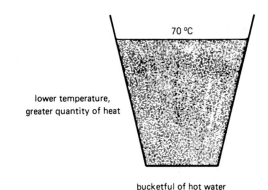

lower temperature, greater quantity of heat

70 °C

bucketful of hot water

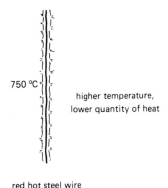

750 °C

higher temperature, lower quantity of heat

red hot steel wire

Figure 3.2 *Comparative temperatures of water and wire*

The principle of the graduation on all types of thermometers is to choose two easily obtainable fixed temperatures and use these upper and lower fixed points and to divide the interval between them into a number of equal parts or degrees. The upper fixed point is the temperature of steam from water boiling under standard atmospheric pressure. The temperature of the boiling water itself is not used for two reasons. First, any impurities which may be present will raise the boiling point. Second, local overheating may occur accompanied by bumping as the water boils. The temperature of the steam just above the water will be constant.

The lower fixed point is the temperature of melting ice. The ice must be pure, since the impurities will lower the melting point.

Temperature scales

The difference between the two fixed points is divided into 100 equal parts, each called a degree. The ice point is 0°C, and the steam point 100°C. This method of subdividing was suggested by a Swedish astronomer named Celsius, and is now called the Celsius scale (see Figure 3.3).

Another method of dividing the difference between the two fixed points is to use the absolute or Kelvin scale (Figure 3.4). Each single degree on these scales equals the same temperature interval as each single degree on the Celsius scale but freezing point is 273.15 K with boiling point 100 degrees higher at 373.15 K.

Absolute zero is theoretically the lowest possible temperature that can ever be reached. The conversion of temperature from °C to K is completed by adding 273.15. Conversion from K to °C is the reverse, i.e. subtract 273.15.

No temperature on the Kelvin scale is negative but temperatures on the Celsius scale become negative once they drop below 0°C.

Thermometers enable us to determine the temperature of substances with very great accuracy. The liquid most commonly used in a thermometer is mercury, because:

1 A small increase in its temperature causes a comparatively large expansion of the mercury.

2 Equal increases in its temperature result in equal amounts of expansion.

3 It remains liquid over a wide range of temperature.

4 The mercury is easily visible in the thermometer.

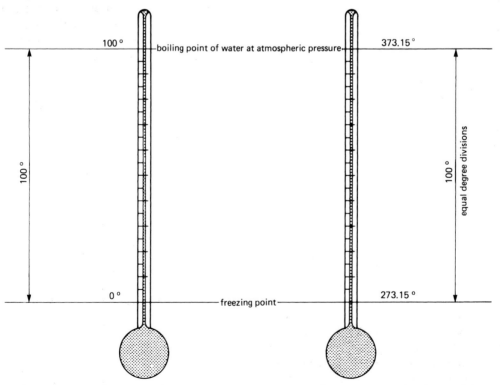

Figure 3.3 *Thermometer graduations: Celsius*

Figure 3.4 *Thermometer graduations: Kelvin*

Measurement of temperatures in a workshop

Many of the processes carried out by a plumber require materials or components to be heated to certain temperatures. The methods used to measure these temperatures may have to be precise or of an approximate value dependent upon the circumstances.

Some metals, such as steel, have noticeable colour changes as their temperature alters.

In the case of copper the only noticeable effect is the change in the degree of redness, i.e. from dull red to a light orange. Other metals such as aluminium have no noticeable colour change and this method cannot be applied to them.

Transmission of heat

If a steel rod is placed into a flame and left there for a time the section of the rod not in the flame becomes warm. Heat travels through the metal by a process called *Conduction*. This process is complex and differs between metals and non-metals, and only a brief explanation is given here.

When a metal is heated the free electrons which it contains begin to move faster: the hot electrons move towards the cooler parts of the metal and at the same time there is a slower movement of cooler electrons in the reverse direction.

To a much smaller extent, heat is transmitted through a metal by vibrations of the atoms themselves, passing on energy from one to the other in the form of waves. These waves are in tiny packets and are called phonons. In non-metals which have no free electrons, heat energy is conducted entirely by phonons.

Temperature

Since the introduction of SI units, the use of the British Fahrenheit scale for reading temperature has been superseded by the Celsius (centigrade) scale. Owing to the length of time required to complete the change-over and because there are still a great many components that are calibrated in degrees Fahrenheit it is desirable, if not even essential, to explain each of the scales, including the conversions from one scale to the other.

Table 3.2 *Useful temperatures*

Celsius	Location	Fahrenheit
0	Freezing point (water)	32
4	Maximum density (water)	39.2
20	Average room temperature	68
36.8	Blood temperature	98.4
43.3	Bath water	110
60	Washing-up water	140
65	Primary return	149
85	Primary flow	185
100	Boiling point (water)	212

Conversion by mathematics

The fixed points on a thermometer are the freezing and boiling points of water. Other important and useful temperatures are shown in Table 3.2.

As stated previously it may be necessary to convert degrees Celsius to degrees Fahrenheit or vice versa. By examination of the thermometers in Figure 3.5 it will be seen that on one thermometer there are 180 divisions while on the other there are 100 divisions,

i.e.   180 divisions Fahrenheit
        100 divisions Celsius

$$\frac{\cancel{180}^{9}}{\cancel{100}_{5}} = \frac{9}{5}$$

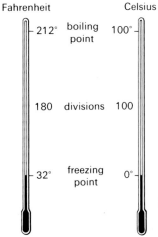

Figure 3.5 *Fahrenheit and Celsius thermometers*

This means that 9 divisions (degrees) Fahrenheit are equal to 5 divisions (degrees) Celsius, or that one degree Celsius is 9/5 times greater than one degree Fahrenheit. It must also be understood that the Fahrenheit scale starts at 32° whereas the Celsius scale starts at 0° (freezing point).

1 The rule when converting a Fahrenheit temperature to Celsius is:

degrees Celsius
$$= \text{degrees Fahrenheit} - 32 \times \frac{5}{9}$$

2 The rule when converting degrees Celsius to Fahrenheit is:

degrees Fahrenheit
$$= \text{degrees Celsius} \times \frac{9}{5} + 32$$

These statements may be better understood once the following examples have been studied.

Example 1
Convert 212 Fahrenheit to degrees Celsius.

Rule $\text{degrees Fahrenheit} - 32 \times \dfrac{5}{9}$

$$212 - 32 \times \frac{5}{9}$$

$$= \overset{20}{\cancel{180}} \times \frac{5}{\cancel{9}_1}$$

$$= 100°C$$

Answer $212°F = 100°C$

Example 2
Convert 20 Celsius to degrees Fahrenheit.

Rule   Degrees Celsius $\times \dfrac{9}{5} + 32$

$$\overset{4}{\cancel{20}} \times \frac{9}{5} + 32$$

$$= 4 \times 9 + 32$$

$$= 36 + 32$$

$$= 68°C$$

Answer $20°C = 68°F$

Solve the following problems:

1 Convert 85° Celsius to Fahrenheit.
(Answer = 185°F)

2 Convert 140° Fahrenheit to Celsius.
(Answer 60°C)

Conversion by graph

An alternative method to that of solving the problem by the use of mathematics is to use a simple straight line graph. The accuracy of the graph will depend on the size of the graph, i.e. the scale of the drawing, and also on the accuracy of the drawing itself.

Figure 3.6 shows a possible graph with 100 divisions Celsius on the horizontal axis and 180 divisions Fahrenheit on the vertical axis.

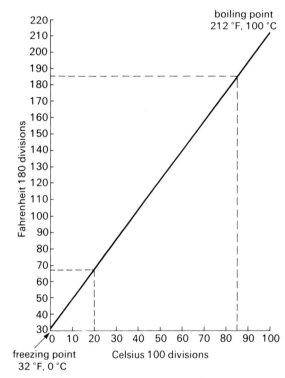

Figure 3.6 *Temperature conversion graph*

Remember that the fixed points on the thermometer are the freezing point, 0° Celsius and 32° Fahrenheit, and the boiling point, 100° Celsius and 212° Fahrenheit. A straight line is drawn through the intersection of these two points. It is now an easy matter to read off any other temperature, as shown by the dotted lines, i.e. 20° Celsius is equivalent to 68° Fahrenheit and 185° Fahrenheit is equivalent to 85° Celsius.

Most metals are good conductors of heat. Copper is exceptionally good. Other substances such as wood, cork and air are bad conductors.

Good and bad conductors have their uses. The bit of a plumber's soldering iron is made of copper, so that as it is used and its tip cools through contact with the work, heat is rapidly conducted along the body of the bit to restore the temperature of the tip and maintain it above the melting point of the solder being used.

Bad conductors have a very wide application. In plumbing work these materials with low thermal conductivity are generally used to prevent heat loss from pipes or to insulate systems to prevent freezing during very cold weather.

Conduction of heat through liquids and gases

All common liquids, with the exception of mercury, are poor conductors. Nevertheless, heat can be moved very quickly through liquids such as water by a different process called *convection*. When a vessel containing water is heated at the bottom (Figure 3.7) a current of hot liquid moves upwards and is replaced by a cold current moving downwards. Unlike conduction, where heat is passed on from one section of the substance to another, the heat here is actually carried from one place to another in the liquid by the movement of the liquid itself.

The same process occurs when gas is heated, although gases are far poorer conductors of heat than liquids (see Figure 3.8).

Convection

When a quantity of water near the bottom of a vessel is heated it expands. Since its mass remains the same, it becomes less dense, and therefore it rises. Thus a warm convection current moves upwards. On the other hand, if some water in a vessel is heated at the top, the liquid there expands and stays floating on the denser water beneath. Convection currents are not set up, and the only way heat can travel downwards under these conditions is by conduction.

This explains why heated water circulates in hot water supply and gravity heating systems. The movement of warmed air in a room follows

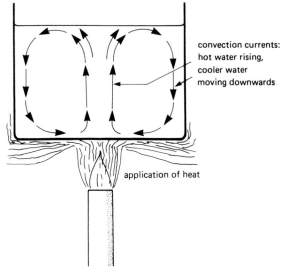

convection currents: hot water rising, cooler water moving downwards

application of heat

Figure 3.7 *Convection currents*

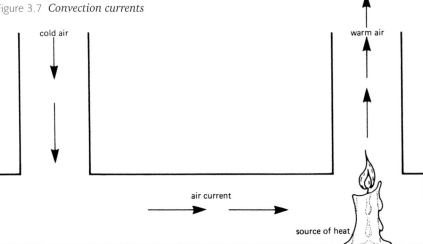

cold air

warm air

air current

source of heat

Figure 3.8

the same principles, and will do so as long as there is a difference of temperature between the rising and failing gas streams.

When the temperatures become equal the process ceases. It therefore follows, that the greater the temperature variation, the quicker will be the circulation (see Figure 3.9) within the liquid or gas. Both conduction and convection are ways of conveying heat from one place to another and require the presence of a material substance either liquid, solid or gas.

There is a third method of heat transmission which does not require a material medium.

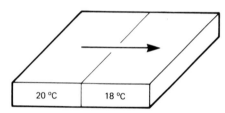

small temperature difference = low rate of heat movement

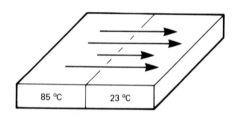

large temperature difference = high rate of heat movement

Figure 3.9 *Heat movement*

Radiation

Radiation heat consists of invisible waves or lines of heat energy which are able to pass through a vacuum. They will also pass through air without appreciably warming it. These waves are partly reflected and partly absorbed by materials and objects upon which they fall. The part which is absorbed becomes converted into heat, this process is called *radiation*.

The rate at which a body radiates heat depends on its temperature and the nature and area of its surface. A body absorbs most heat when its surface is dull black and least when its surface is highly polished (see Figure 3.10). The polished surface is therefore a good *reflector* of heat.

The radiators of a hot water central heating system, despite their name, in fact emit their heat mainly by conduction and a smaller amount by radiation and convection.

The effect of heat on solids, liquids and gases

With a few exceptions, substances expand when they are heated, and very large forces may be set up if there is an obstruction to the free movement of the expanding or contracting material.

Heat may also bring about a change of state; for example, solder, which is normally a solid metal, becomes a liquid when sufficiently heated, and changes its physical state.

Heat can also accelerate or bring about a chemical change as is produced when hydrogen and oxygen are burned in the correct proportions to produce water.

Expansion of various substances

When rods of different substances but of the same length are heated evenly, experiment shows that their expansion is not equal (see Figure 3.11). Aluminium expands about twice as much as steel. Brass expands about one and a half times as much as steel. An alloy of nickel and steel known as invar has a very small expansion when its temperature rises and for this reason is used in thermostats and watches.

Plastics materials such as polyvinyl chloride and polythene are now used extensively in plumbing systems, and it is worth noting the comparatively high rates of expansion of these and similar materials. Small as the changes in material size appear, the fact remains that they happen and can exert a considerable pull or push on other substances that try to restrain this movement. Unless suitable allowances (as shown in Figure 3.12) are made and precautions taken to accommodate thermal movement, damage and inconvenience will occur.

Applications of thermal expansion

Although expansion can be troublesome in plumbers' work it often proves very useful. For example, the bimetallic strip has many useful applications including flame failure devices, electric and gas thermostats.

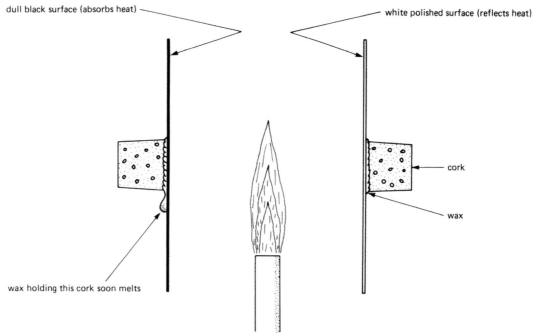

Figure 3.10 *Heat reflection*

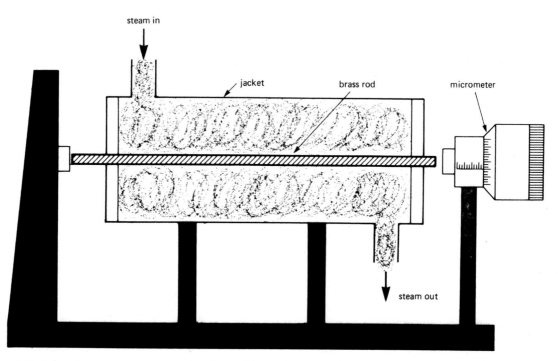

Figure 3.11 *Laboratory apparatus to measure thermal expansion*

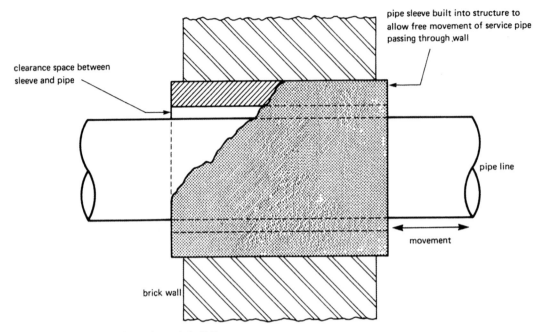

Figure 3.12 *Sleeving of pipelines through building structure*

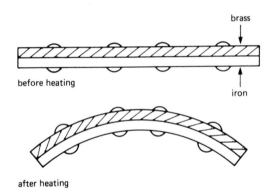

Figure 3.13 *Bimetallic strip*

Figure 3.13 shows how the different rates of expansion of iron and brass can be made use of to form a bimetal strip. Equal lengths are riveted together. On heating, the brass expands more than the iron and the strip forms a curve with the brass on the outside.

Figure 3.14 shows the principle of a thermostat which makes use of the different rates of thermal expansion in metals. This type of thermostat could be used for controlling the temperature of a room, or the water in a hot water storage vessel.

Coefficient of thermal expansion

The amount a substance expands when it is heated depends upon the properties of the substance: whether it is solid, liquid or gaseous, and upon the amount of heat it absorbs.

The amount that solids expand for each °C rise of temperature is fairly constant and therefore easily measured. Liquids and gases do not respond

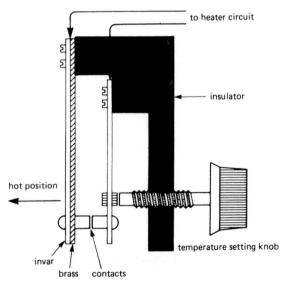

Figure 3.14 *Thermostat*

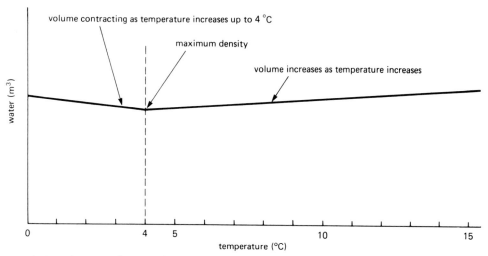

Figure 3.15 *Variation of water volume with temperature*

quite so conveniently, and water behaves in a manner which is termed anomolous (Figure 3.15).

Thermal expansion affects all the dimensions of a material. Length, width and thickness all increase as the temperature of the material rises. But since most of the materials we deal with are much longer than they are wide or thick, at this stage only length or linear expansion will be dealt with.

Definition

The coefficient of linear expansion of a solid substance is the fraction of its original length by which the substance expands per degree rise in temperature, or will contract when its temperature decreases by 1°C.

The length can be measured by using any convenient unit, although generally metres are preferred. Whatever unit length is used, the fraction that the material expands for 1°C change of temperature must also be measured as a fraction of that unit. Table 3.3 shows the coefficients of linear expansion of various materials.

To calculate the amount of expansion or contraction which occurs in materials as their temperature changes, the following formula may be used:

change in length = length of material × temperature change in material × coefficient of linear expansion

Table 3.3 *Linear coefficients of thermal expansion*

Material	Variation per unit of length for one degree Celsius temperature change
Lead	0.000 029
Zinc	0.000 029
Aluminium	0.000 026
Tin	0.000 021
Polythene	0.000 018
Copper	0.000 016
Iron	0.000 011
Cast iron	0.000 011
Mild steel	0.000 011
Mercury	0.000 005

The following examples will show the application in relation to plumbers' work.

Example 4

A copper hot water pipe 15 m long is filled with water at 10°C. By how much will the length of this pipe increase when it carries hot water at 70°C?

change in length = 15 000 mm × 60°C × 0.000 016
$$= 900\,000 \times 0.000\,016$$
$$= 14.4\,\text{mm increase in pipe length}$$

Example 5
A polythene waste pipe has received a discharge of hot water at 55°C, and is allowed to cool to 15°C. When the pipe was at its hottest it measured 4 m in length. How much will it shorten or contract?

change in length = 4000 mm × 40°C × 0.000 18
$$= 160\,000 \times 0.000\,18$$
$$= 28.8\,\text{mm decrease in length}$$

Expansion of liquids
Different liquids have different thermal expansions (see Figure 3.16) and unlike solids have no fixed length or surface area, but always take up the shape of the containing vessel. Therefore in the case of liquids we are only concerned with volume changes when they are heated.

Definition
The coefficient of expansion of a liquid is the fraction of its volume by which it expands per degree rise in temperature.

Any attempt at very accurate measurement of the expansion of a liquid is complicated by the fact that the vessel which contains the liquid also expands.

However, since all liquids must always be kept in some kind of vessel or container it is just as useful to know the apparent expansion of a liquid. This is the difference between its real expansion and the expansion of the vessel, and is accurate enough for plumbers' work.

The unusual expansion of water
Not all substances expand when they are heated. Over certain temperature ranges they, contract. Water is an outstanding example. If we take a quantity of water at 0°C range and begin to apply heat the water contracts over the temperature range 0°C – 4°C (Figure 3.15).

At about 4°C the water reaches its smallest volume which means it is at maximum density. If we continue to apply heat to raise the temperature the water expands.

The peculiar expansion of water has an important bearing on aquatic life during very cold weather (see Figure 3.17). As the temperature of a pond falls, the water contracts, becomes denser and sinks. A circulation is set up until all the

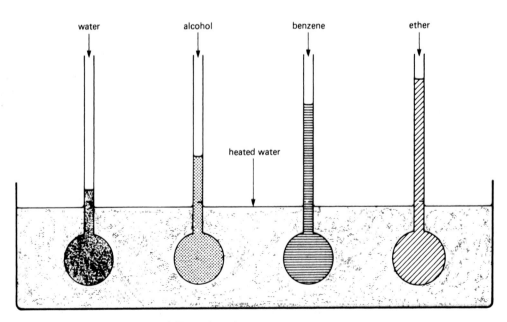

Figure 3.16 *Comparison of expansion for different liquids*

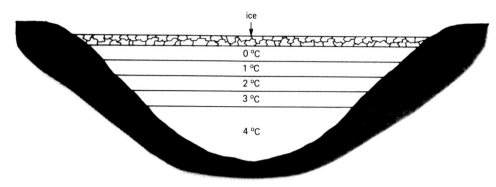

Figure 3.17 *Temperatures in an ice covered pond*

water reaches its maximum density at 4°C. If the temperature continues to drop, any water below 4°C will stay at the top due to its lower density. In due course ice forms on the top of the water, and after this the water beneath can only lose heat by conduction. This explains why only very shallow water is likely to freeze solid.

As mentioned previously a rise in temperature causes objects to expand and a fall of temperature causes contraction. This rule applies to gases, most liquids and solids, but the effect is much more marked in the case of gases than in the case of the other two.

Expansion of gases

In the case of gases, however, the situation is different. The expansion which occurs in gases is very much larger than that of solids, and consequently the value of the coefficient obtained will depend on the starting temperature. Therefore, in order to make an accurate comparison between different gases, the coefficient is always calculated in terms of an original volume at 0°C.

The volume coefficient of expansion of a gas is defined as follows.

Definition

The coefficient of expansion of a gas at constant pressure is the fraction of its volume at 0°C by which the volume of a fixed mass of gas expands per °C rise in temperature.

The original experimental work on this subject was carried out towards the end of the eighteenth century by the French scientist Jacques Charles. The results obtained are generally known as Charles's Law:

The volume of a fixed mass of gas at constant pressure expands by 1/273 of its volume at 0°C per °C rise in temperature.

Pressure in gases and liquids

The nature of gas pressure

Any gas consists of a collection of molecules of a particular kind which are in a state of rapid motion. The fact that the molecules are in motion is evident from the fact that if a small quantity of an odorous gas, such as acetylene or propane, is liberated at any point in a workroom or laboratory the smell of the gas soon pervades the whole room.

If the gas is confined in a closed vessel, some of the moving molecules strike the sides of the vessel and each impact exerts a small force upon the side. The number of molecules of gas inside the vessel will normally be very large and, on the average, the same number of molecules will strike a given area on the sides of the vessel each second, so producing a steady pressure.

If the gas contained is compressed by some mechanical means, its molecules will be pressed closer and closer together and the cohesive force between them becomes stronger. Eventually, if sufficient pressure is applied so that the intermolecular cohesive force is strong enough to hold the molecules close together, the gas will change to a liquid. If the compression force is released, the liquid will return to its gaseous state.

This fact is used to separate the gases which make up the atmosphere. If air is sufficiently compressed under controlled conditions of pressure and temperature, it will change from a gaseous to a liquid state. When the compression forces are released, the liquid reverts to a gas. As the elements which make up air become gases at different temperature, it is possible to collect them separately, and store them in cylinders for use in welding, and other industrial processes. In this way oxygen, argon and nitrogen gases are obtained.

Pressure can also affect the process of a physical change. For example, water boils and changes from a liquid to a gas at 100°C at standard atmospheric pressure ($100 \, kN/m^2$). If a pressure greater than this is acting on the water, then it will not boil until it has reached a much higher temperature and will not boil until it has gained sufficient heat energy to do so. Likewise reduced pressures (below standard atmospheric pressure) reduce its boiling point temperature.

Definition

Pressure is defined as the force acting normally per unit area. (Here the word 'normally' means vertically.)

The SI unit of pressure is *newton per metre*^{2} (N/m^2).

Relation between pressure and volume

The pressure of a gas decreases in the same proportion as its volume increases. From this statement it is clear that if we multiply the varying volumes of a given mass of gas by the corresponding pressures, any decrease in the value of one of them will be exactly counterbalanced by the increase in the value of the other, and the result will always be the same.

Expressed mathematically, this may be stated as:

$P_1V_1 = P_2V_2$ (temperature constant)

where P_1 and P_2 are two pressures and V_1 and V_2 the corresponding volumes of a given mass of gas.

This result is known by the name of its discoverer as Boyle's Law.

Boyle's Law

The volume of a given mass of gas is inversely proportional to its pressure, if the temperature remains constant.

'Inversely proportional' is the mathematical expression of the fact that as the pressure *increases* the volume *decreases* in the same proportion.

Water pressure

Water pressure is caused naturally by the weight of water which, under the influence of the earth's gravitational force, exerts pressure on all surfaces on which it bears (atmospheric pressure).

The molecules of which water is composed are held together by cohesion. This cohesive force is stronger in liquids than in gases but not as strong as in solids. This fact means that water molecules can move with relative freedom, and the force of gravity tends to pull them down in horizontal layers so that the surface of a liquid subject to the pressure of the atmosphere is horizontal. Therefore water flows to find its own level in irregular shaped vessels, pipes and cisterns (see Figure 3.18).

The fact that the liquid stands at the same vertical height in all the tubes whatever their shape confirms that, for a given liquid, the pressure at a point within it varies only with the vertical depth of the point below the surface of the liquid.

Liquids exert a pressure in the same way that air does.

The pressure in a liquid increases with depth. This may be shown by means of a tall vessel full of water with side tubes fitted at different heights (Figure 3.19). The speed with which water spurts out is greatest for the lowest tube, showing that pressure increases with depth.

Intensity of pressure

This is defined as that force created by the weight of a given mass of water acting on 1 unit of area (m^2). Or:

$$intensity \; of \; pressure = \frac{force}{area}$$

and, since force is measured by newtons:

$$intensity \; of \; pressure = \frac{newtons}{area}$$

The newton is of very small value in numerical terms, and it is generally more practical to use

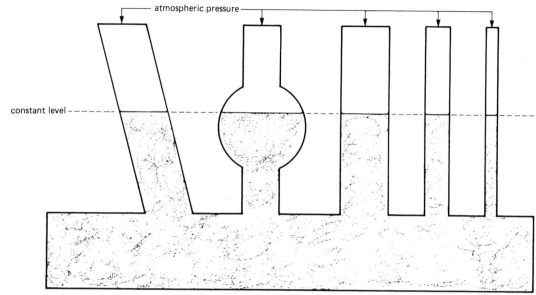

Figure 3.18 *Apparatus used to show that water flows to find its own level*

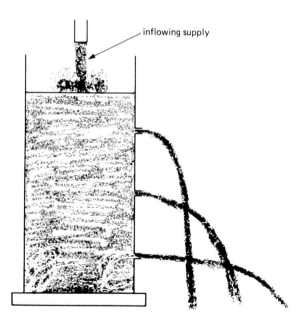

Figure 3.19 *Spout can, indicating that pressure in a liquid increases with the depth of the liquid*

the kilonewton (kN = 1000 newtons) when dealing with pressures.

One litre of water weighs 1 kg and the force produced by this mass of water equals 9.8 newtons. If we consider a cube of water measuring 1 m × 1 m × 1 m (1 m³), this volume of water con-

tains 1000 litres and weighs 1000 kg. Therefore, the intensity of pressure on the base of this cube which has an area of 1 m² will be 1000 × 9.8 newtons or 98 kN/m².

So, if we visualize a column of water 1 m in height or depth and call this distance head we can derive the following basic formula for calculating pressure:

intensity of pressure = head (m) × 9.8 kN/m²

Example 6
Calculate the intensity of pressure on the base of a hot water tank subjected to a head of 4 m.

$$\begin{aligned} \text{intensity of pressure} &= \text{head (m)} \times 9.8 \text{ kN/m}^2 \\ &= 4 \times 9.8 \\ &= 39.2 \text{ kN/m}^2 \end{aligned}$$

The example shows that a 4 m head of water will exert an intensity of pressure of 39.2 kN on a base area of 1 m². What must be considered now are areas larger or smaller than 1 m² which leads us to the total pressure acting upon an area.

If the pressure shown in the example had been acting upon a base of 2 m² then the total pressure acting upon the whole area would be 2 × 39.2 kN or 78.4 kN. At this point it should be noted that the area symbol (m^2) is left out of the answer because with total pressure calculations we are not relating the pressure to $1\,m^2$ but to an area larger. The same rule applies to areas smaller than $1\,m^2$, which leads to the formula for calculating total pressure.

total pressure = intensity of pressure ×
area acted upon

Example 7
Calculate the total pressure on the base of the hot tank in Example 6, if the base has an area of $0.3\,m^2$

$$\text{total pressure} = \text{intensity of pressure} \times \text{area acted}$$
$$\text{upon}$$
$$= 39.2 \times 0.3$$
$$= 11.76\,kN$$

Density and relative density

Definition

The density of a substance is defined as its mass per unit volume.

Equal volumes of different substances vary considerably in mass. For example, aluminium alloys are as strong as steel, but volume for volume weigh less than half as much. This lightness or heaviness of different materials is known as *density*.

One method of finding the density of a substance is to take a sample and measure its mass and volume. The density is then calculated by dividing the mass by the volume.

The densities of all common substances, solids, liquids and gases, and all chemical elements have been determined and are to be found in books of chemical and physical data, although some of particular interest to the plumber are listed in Table 3.4.

Water has a density of 1 g/cm^3 or 1000 kg/m^3 at 4°C.

It is important for construction workers to appreciate density, particularly when related to various building materials and claddings. The known volume of any part of the structure can be multiplied with the density of the material to give the mass, and hence the weight.

Table 3.4 *Densities in kg/m^3*

Aluminium	2.7×10^3	Mercury	13.6×10^3
Brass (varies)	8.5×10^3	Steel (varies)	7.8×10^3
Copper	8.9×10^3	Water (at 4°C)	1.0×10^3
Ice (at 0°C)	0.92×10^3	White spirit	0.85×10^3
Lead	11.3×10^3	Zinc	7.1×10^3

Relative density

Definition
The relative density of a substance is the ratio of the mass of any volume of it to the mass of an equal volume of water. Or:

$$\text{Relative density} = \frac{\text{mass of any volume of the substance}}{\text{mass of an equal volume of water}}$$

In normal weighing operations the mass of a body is proportional to its weight, so it is also true to say:

$$\text{Relative density} = \frac{\text{weight of any volume of the substance}}{\text{weight of an equal volume of water}}$$

This will explain why relative density has also been called *specific gravity*: the word gravity implying weight. At the present time, the term relative density is recommended rather than specific gravity. Note that relative density has no units: it is simply a number or ratio. On the other hand, density is expressed in kg/m^3 or g/cm^3.

Fluxes

A soldering flux is a liquid or solid material which, when heated, is capable of promoting or accelerating the wetting of metals with solder. Table 3.5 lists some of the fluxes commonly used by plumbers. The purpose of a soldering flux is to remove and exclude oxides and other impurities from the joint being soldered. Anything interfering with the attainment of uniform contact between the surface of the base metal and the molten solder will prevent the formation of a sound joint. An efficient flux removes tarnish films and oxides from the metal and solder and prevents re-oxidation of the surfaces when they are heated. It is designed to lower the surface tension of the molten solder so that the solder will flow readily and adhere to the metal. The flux should be readily displaced from the metal by the molten solder.

Surfaces to be soldered are often covered with films of oil, grease, paint, heavy oxides or atmospheric grime which must be removed. When clean metal surfaces are exposed to the air, chemical reactions occur, depositing fresh surface films. These reactions are generally accelerated as the temperature is raised, and although nitrides, sulphides, and carbides are formed in some instances, the prevalent reaction is oxidation. The rate of oxide formation, its structure and resistance to removal with a flux, varies with each base metal. Aluminium, stainless and high alloy steels, aluminium and silicon bronzes, when exposed to air, form hard adherent oxide films – highly active and corrosive fluxes are used to remove and prevent the reforming of these films during soldering. Lead, copper and silver on the other hand form less tenacious films and at a slower rate, so that mild fluxes remove them easily and prevent them from reforming.

Flux action

In most soldering operations the flux removes the oxide film from the base metal and solder by dissolving or loosening the film and floating it off into the main body of the flux. Because of the refractory nature of many oxide films it has been suggested that the flux wets, coagulates and suspends the oxide which has been loosened by a penetrating and reducing action. The molten flux then forms a protective blanket over the bare metal which prevents the film from reforming. Liquid solder displaces the flux and reacts with the base metal to form an intermolecular bond. The solder layer builds up in thickness and when the heat is removed it solidifies.

Types of flux

A functional method for classifying fluxes is based upon the nature of their residues. They are classified into three main groups:

Table 3.5 *Fluxes used by the plumber*

Name	Uses	State	Notes
Zinc chloride (killed spirits)	All forms of copper bit soldering	Liquid	Produced by dissolving zinc in hydrochloric acid. The flux is actively corrosive and must be removed by washing the soldered joint following the jointing process.
Zinc-ammonium chloride	All forms of copper bit soldering	Liquid	This flux is active at lower temperatures than zinc chloride. This is helpful when soldering metals with a relatively low melting point. The flux is corrosive and must be removed by washing the soldered joint to remove all traces of the flux.
Tallow	Lead soldering	Solid	Tallow is obtained from the fat of animals. These organic fats contain glycerin, which makes them mildly acidic.
Resin	Tinning (brass and copper)	Powder or paste	Resin is obtained from pine tree bark. It is a gum-like substance. This flux is usually applied in powder form, and is sprinkled on the surfaces to be tinned. It is mildly corrosive at soldering temperature and non-corrosive when cold.

Note: Patent fluxes are available for most situations.

1 high corrosive

2 intermediate

3 non-corrosive fluxes.

Good soldering practice requires the selection, from the three main groups, of the mildest flux that will perform satisfactorily in a specific application.

The *highly corrosive* fluxes consist of inorganic acids and salts. These fluxes are used to best advantage where conditions require a rapid and highly activated fluxing. They can be applied as solutions, pastes, or as powders. Corrosive fluxes are almost always required when using the higher melting temperature solders.

The corrosive fluxes have one distinct disadvantage in that the residue remains chemically active after soldering. This residue, if not removed, may cause severe corrosion at the joint.

Zinc chloride is the main ingredient in the majority of corrosive fluxes. It can be prepared by adding an excess of zinc to concentrated hydrochloric acid, or it can be purchased as fused zinc chloride which is readily available and more convenient to use.

A water solution of ammonium chloride may be used as a flux. It is less effective than zinc chloride.

A combination of one part of ammonium chloride to three parts of zinc chloride forms a eutectic flux mixture which is considerably more effective than either constituent when used alone.

The *intermediate fluxes*, as a class, are weaker than the inorganic powder types. They consist mainly of mild organic bases and certain of their derivatives. These fluxes are active at soldering temperatures but the period of activity is short because of their susceptibility to thermal decomposition. They are useful, however, in quick spot soldering operations and, when properly used, their residue is relatively inert and easily removed with water. Intermediate fluxes are useful where sufficient heat can be applied to decompose or volatilize fully the corrosive constituents contained within the flux.

Non-corrosive fluxes

White resin dissolved in a suitable organic solvent is the closest approach to a non-corrosive flux.

The active constituent, abietic acid, becomes mildly active at soldering temperatures.

This form of flux is widely used in plastic form in cored solder wire.

Because of the slow fluxing action of resin on anything but clean or precoated metal surfaces, a group of stabilized and activated resin fluxes have been developed, aimed at increasing the fluxing action without altering the non-corrosive nature of the residue.

Paste fluxes

In the plumbing trade it is sometimes convenient to have the flux in the form of a paste. Paste fluxes can be more easily localized at the joint and have the advantage of not draining off the surface or spreading to other parts of the work where flux may be harmful. The paste-forming ingredients may be water, petroleum jelly, tallow or lanolin, with glycerine or other moisture-retaining substances. If the pastes contain inorganic salts, such as zinc or ammonium chloride they are classified as corrosive fluxes. Paste fluxes have been developed for universal application containing resins dissolved in butyl cellosolve and with plasticizers added to provide the flux activity. Resin pastes have also been developed which are non-corrosive and meet the requirements of the electrical industry.

Lead-free solders

To comply with the new Water Regulations lead-free solder must be used when soldering joints on copper pipes, where these pipes are to be used for conveying water for drinking and the preparation of food. A range of lead-free solders is available from stockists as sticks, bars and solid wire. The standard paste fluxes, can be used with these lead-free solders.

Condensation

Principles involved in condensation

The amount of water vapour that air can contain is limited and when this limit is reached the air is said to be saturated. The saturation point varies with temperature – the higher the temperature of the air, the greater the weight of water vapour it can contain. Water vapour is a gas and when the air is cooled it cannot hold the same amount of water, it is then deposited on the surface of the pipe, wall, etc. This is known as 'The Dew Point'. This can be overcome by air movement, i.e. ventilation.

Atmospheric condition (dew point)

When the warm moist incoming air comes in contact with a cold wall or cold water pipe surface which is below its dew-point, water will condense upon them (see Figure 3.20) but as the walls warm up and eventually reach the dew-point condensation ceases and the condensed moisture evaporates.

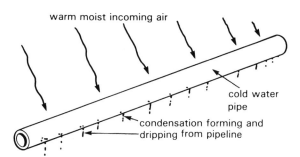

warm moist incoming air

cold water pipe

condensation forming and dripping from pipeline

Figure 3.20 *Action of condensation*

Corrosion

In plumbing, examples of corrosion of metals are not difficult to find. When we consider that pipes carrying water (which may itself be corrosive) are laid in soil with strongly corrosive properties, i.e. clays, clinker or ash, it is not surprising that corrosion takes place. What is sometimes difficult to determine is the cause of corrosion. The chief causes of corrosion are:

1 The effects of air and water
2 The direct effect of acids
3 Electrolytic action.

When metals are exposed to the atmosphere they form a layer of oxide on their surface. The speed at which the oxide forms varies with each metal. For atmospheric corrosion to take place it is not necessary for the metal to be constantly wet or even exposed to rain for long periods.

The gases present in the atmosphere that have the greatest effect on metals are oxygen, carbon dioxide, sulphur dioxide and sulphur trioxide, together with the water vapour in the atmosphere.

Oxygen produces a film of oxide on metals. Carbon dioxide may mix with rainwater to form a weak solution of carbonic acid and in contact with metals it tends to promote the formation of carbonate films, as with sheet copper when it produces a basic carbonate film (patina). Sulphur dioxide is probably the greatest accelerator of atmospheric corrosion. It is a gas, ejected with flue gases, which when mixed with rainwater forms a weak sulphurous acid.

Sulphur trioxide also combines with water to form sulphuric acid. These gases are found in considerable concentrations in industrial areas and iron rusts three to four times as fast and zinc six times as fast in industrial areas as in rural areas. Seaside towns also suffer from atmospheric corrosion because of the sodium chloride (salt) particles present in the air and derived from the sea.

Copper

The bright lustre of copper sheet changes after a time to a greenish shade (patina). This discoloration is due to the combined effect of carbon dioxide, and sulphur dioxide which form carbonates and sulphates. Copper tubes that have been exposed to ashes containing sulphur exhibit the same hard green coating. The green covering on sheet copper is a hard and tough protective layer which prevents any further attack by the atmosphere.

Lead

Lead also forms a tough non-scaling film (lead oxide) on its surface when exposed to the atmosphere. If the atmosphere is an industrial one, the surface of the lead appears to be ingrained with sticky tar-like deposits.

Lead is also affected by the acids present in some timbers. Oak boards in particular contain acids that attack lead. The action of alkali in cement upon lead flashings or weatherings can also be detrimental in moist or permanently damp situations.

Zinc

This is a very stable metal in dry air, and in moist air forms a film of oxide and carbonate. The carbonate is formed from the carbon dioxide present in the air. It is liable to attack from sulphurous gases in the atmosphere – this results in the zinc being converted to zinc sulphate, a whitish compound. In addition, any solutions containing ammonia tend to attack the metal quite rapidly.

Aluminium

This metal also forms a hard and durable oxide film that protects the underlying metal from further attack. If exposed to alkali attack the material should be protected with paint.

Any weaknesses in the oxide film on aluminium are exploited by corrosive elements in the atmosphere. This results in 'pitting', which is a localized form of attack that is not serious: it does not remain as a steady rate of corrosion but falls off as the oxide film builds up. Alloys of aluminium are liable to attack by salt solution. Seaside atmosphere is inclined to be aggressive because of its high salt content.

Electrolytic action

Almost any hot water system offers many examples of electrolytic action. The cold water cistern may be made of steel covered with zinc (galvanized), the hot water cylinder of copper, the pipework of copper or steel, often with brass or gunmetal connections or wiped soldered joints. It has been found that certain types of water are capable of dissolving small quantities of copper in hot water systems. This copper-bearing water comes in contact with zinc coatings and some of the zinc is dissolved by electrolytic action between the two metals. Once the zinc coating has been perforated then the attack of the steel underneath proceeds at a rapid pace. The ability of the water to dissolve copper is important and it is known that any increase in the carbon dioxide content is liable to increase the copper-solvency of the water.

In hard water districts the formation of scale on pipes, boiler and cylinder, provided it is an unbroken film, often prevents attacks due to electrolytic action.

In general it is best therefore to construct the hot-water system of one metal only – unless previous experience in the district shows that corrosion problems do not arise from mixing metals in an installation. Before the introduction of the plastics flushing cistern, with nylon siphon and float valve, it was common to find a flushing cistern made of iron, with a brass float valve; a copper float with a soldered seam and perhaps a lead-alloy siphon. It is not surprising that in this confined space the life of the soldered seam on the copper float was short, especially when the local water tended to be slightly acid.

The use of other metals in addition to lead as a roof covering have made it necessary for the plumber to be most careful when fixing roof coverings. An aluminium-covered roof provides a weathertight finish, but its efficiency would soon be affected if a copper or iron rainwater pipe discharged water on to it from an old lead- or zinc-covered roof. The aluminium roof would soon be pitted and perforated as a result of the electrolytic action between the dissimilar metals.

Corrosion in central heating systems

Steel is a man-made alloy and it is skill and knowledge that make it possible to convert the oxidic ore into iron and steel.

Common red rust is probably the best known of all the corrosion products of iron. Others are white, green and black. The black oxide, also known as ferrous oxide or magnetite, is most commonly found within central heating systems in the form of a black sludge (see Figure 3.21). Red rust requires moisture and generous supplies of free oxygen for its formation, while the black oxide of iron has a lower oxygen content in its molecule. and it will form when there is very little free and dissolved oxygen available in the water.

Hydrogen gas is a by-product of this corrosion process, and the frequent necessity to bleed the gas from radiators (see Figure 3.22) clearly indicates that corrosion is taking place.

A simple way to determine if the water in a central heating system is corrosive is to carry out the corrosion test as shown in Figure 3.23.

Corrosion is the 'eating away' of a substance by an attacking influence, which is usually external. The term corrosion is often misapplied

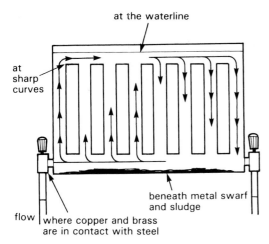

Figure 3.21 *Corrosion attack in a radiator*

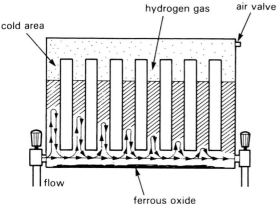

Figure 3.22 *Build-up of hydrogen gas and reduced hot water circulation due to corrosion*

(1) fill a small jar with water drawn from a radiator vent.

(2) add a few clean steel nails (not galvanised nails) to simulate the steel of the radiators. Close the jar and leave for three days.

Figure 3.23 *A simple test for corrosion*

to cases of encrustation or deposition where no actual corrosion has occurred.

The more common corrosive agents with which we in the plumbing trade are concerned are:

1 Air containing moisture, carbon dioxide, sulphur dioxide, sulphuric acid, or combinations of these

2 Water containing dissolved air, mineral or vegetable acids, alkalis and certain salts.

All acids and the strong alkalis are corrosive, but as plumbers we are mainly concerned with those agents likely to attack plumbing pipework, roof work, sanitary fittings, etc.

Many soils are slightly acid or alkaline, and with the inevitable moisture have detrimental effects on many metals.

Atmospheric corrosion

Pure air or pure water acting independently have practically no corrosive action. Moist air and water with dissolved air attack iron and steel very quickly, producing the familiar oxide known as 'rust'. If this corrosive action is unchecked the metal will be completely destroyed.

If sulphur dioxide or carbon dioxide are present in the air, copper is attacked, covering the metal with a film of basic sulphate and/or carbonate.

This film protects the underlying copper; it is easily identified as the green coating seen on copper roofs. Zinc, while withstanding air and moisture, is subject to quick deterioration in the acid air of industrial towns, so also is brass – especially brass with a high zinc content.

Lead and tin withstand atmospheric corrosion well. Aluminium is corroded by the atmosphere to the extent of surface dullness, but is seriously damaged by alkaline solutions. This also applies to tin and lead solders.

Corrosion by water

The corrosive effects of impure water are very important. The case of iron and steel have already been referred to. So called 'soft' waters have a pronounced action on lead. The strength of a lead pipe is not greatly affected by this minute corrosion, but since a very small quantity of lead in domestic water supplies (plumbo-solvency) is highly dangerous to health, the matter assumes great importance.

Very few waters attack copper, but with highly acidic waters, green staining of sanitary fittings may occur (cupro-solvency). Generally, in the case of neutral or hard waters, the tubes become coated internally with a thin protective film.

Corrosion resistance

Gold and the rare metals of the platinum group can be regarded as incorrosible for all practical purposes. To a lesser degree, nickel and chromium, much used for ornamental finishes, are resistant to corrosion.

The alloys of the 'stainless' steel group (steel and nickel) are reasonably immune to corrosion when blended together to form 'stainless steel'.

Electrolytic corrosion

This is caused when two very dissimilar metals, e.g. a galvanized tube and a copper fitting, are in direct metallic contact in certain types of water. This combination is in effect a primary electric cell and the currents induced, although small, cause one or other of the metals (in this case, the zinc) to be corroded and dissolved with considerable rapidity.

Specks of iron rust resting in a brass tube may cause perforation of the tube in certain types of water. This form of corrosion can take place in water systems or in damp soil. Electrolysis or 'galvanic corrosion' requires four things in order for it to take place. These are:

1 An anode – the corroding area,

2 An electrolyte – the means of carrying the electric current (water or soil),

3 A cathode – the protected area,

4 A return path – for the corrosion currents.

An electric current is generated at the anode and flows through the electrolyte to the cathode. The current then flows through the return path back to the anode again (see Figure 3.24).

The principal causes of electrolytic corrosion are:

1 Different metals joined together and both in contact with the electrolyte, for example, mild steel radiators and copper pipe in a wet central heating system. Or a steel service pipe connected to a cast iron main. In both examples the steel is the anode or corroding area. The metals do not have to be completely different to set up electrolysis. It is sufficient to

have clean, pure metal at one point and scale, impurities or scarring at another. Corrosion can take place between iron and particles of graphite or carbon in the same metal.

2 Differences in the chemical environment of the metal. For example, in buried pipes, a lack of oxygen or a concentration of soil chemicals or bacteria at the anode point (see Figure 3.25).

3 Stray electric currents. This may occur where bonding is ineffective and a gas service pipe acts as an electrical earth return. It happens on gas mains when in contact with other authorities' plant or electrified railway systems (see Figure 3.26).

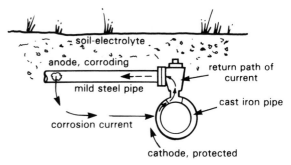

Figure 3.24 *Electrolytic corrosion: dissimilar metals*

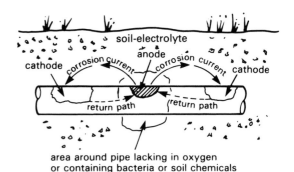

Figure 3.25 *Electrolytic corrosion: difference in environment*

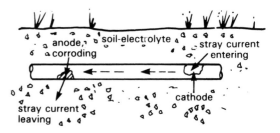

Figure 3.26 *Electrolytic corrosion: stray currents*

The electro-chemical series

The farther apart two metals appear in Table 3.6 the more active the corrosion will be when they are placed in contact in a slightly aqueous solution.

Cathodic protection

Cathodic protection is a form of corrosion control designed and arranged to combat the chemical effect of electric current flows induced by electrolytic action. It is the protection of a cathodic metal (i.e. copper) by a sacrificial metal (i.e. zinc). The sacrificial metal is known as the anode and is destroyed over a period of time by the chemical effect of the electrical current.

Briefly summarized, the electro-chemical decomposition of metals is as follows:

1 Two dissimilar metals are involved.

2 The two metals or poles must have contact so that current flow can take place.

3 The 'poles' must be immersed in an electrolyte, that is, a liquid or moist substance capable of conducting electricity.

The simplest example of electrolysis, or corrosion due to the chemical effect of electric current flow, is the voltaic cell. This comprises a jar to contain the electrolyte and the two 'poles' or dissimilar metals, say steel and copper. If the 'poles' are connected, by a wire or similar connector, outside the electrolyte then an electric current will flow around the circuit.

This simple arrangement (see Figure 3.27) produces an electric cell or 'battery' capable of producing electrical energy and this can be measured on suitable instruments. The current generated is very small, as also is the voltage, but if the current flow is allowed to continue, there will be evidence of the steel 'pole' seemingly being dissolved away. This electro-chemical decomposition is the form of corrosion which cathodic protection aims to inhibit or stop.

Zinc is said to be anodic to copper, or copper is cathodic to zinc, and a feature of electrolytic decomposition, or corrosion, is that the anodic metal is the one which is corroded by the chemical effect of the electric current passage as just outlined.

Figure 3.6 *Electro-chemical series*

Metal			Chemical symbol
Cathodic	1	Gold	Au
	2	Platinum	Pt
	3	Silver	Ag
	4	Mercury	Hg
	5	Copper	Cu
	6	Lead	Ph
	7	Tin	Sn
	8	Nickel	Ni
	9	Cadmium	Cd
	10	Iron	Fe
	11	Chromium	Cr
	12	Zinc	Zn
	13	Aluminium	Al
Anodic	14	Magnesium	Mg

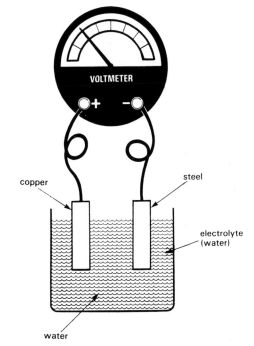

Figure 3.27 *Example of electrolysis. Copper and steel, immersed in water and connected to a voltmeter confirm that electricity is being generated. Steel is the 'sacrificial' element in this instance. How water, or the addition of impurities to the water, will increase electrolytic corrosion and thus the voltage*

Table 3.7 *Electro-chemical reaction of metals*

	Metal	*Potential difference*	*Volts*
Cathodic	Copper	0.35	positive
	Lead	0.13	negative
	Nickel	0.25	negative
	Chromium	0.71	negative
	Zinc	0.75	negative
	Aluminium	1.70	negative
Anodic	Magnesium	2.38	negative

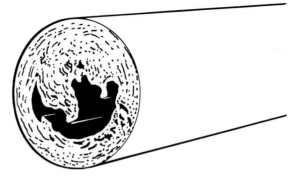

Figure 3.28 *Galvanized steel water pipe blocked by corrosion deposits*

Different metals have varying capacities of current flow. These are referred to as their potential differences and enable a table to be drawn up to indicate the likely electro-chemical reaction one might expect when any two dissimilar metals are being considered.

It will be seen from Table 3.7 that each metal is cathodic to all those metals listed beneath it.

Corrosion resistance

Water pipes

1 Suitable for hard water
Copper and galvanized low carbon steel.

2 Suitable for soft water
Copper, stainless steel, galvanized low carbon steel.

Lead

Water regulations
Prohibit the use of lead as a water supply pipe material.

Galvanized steel

Galvanized low carbon steel pipes are given an additional protection against corrosion by the formation of a scale, calcium carbonates, from hard waters. However, some hard water with a high free carbon dioxide content will form a loose deposit (see Figure 3.28) which gives no protection.

Copper

Most types of water will dissolve minute particles of copper from new pipes. This may be sufficient to deposit green copper salts in fitments. Where this occurs (neutral and hard waters) it will usually cure itself as it forms a protective film over the internal surface.

These small copper particles may also cause corrosion to galvanized steel cylinders, etc. and pitting in aluminium kettles.

Corrosive soldering flux residues can often cause pitting. No more flux than is necessary should be used in making capillary soldered joints, and all surplus flux should be removed on completion of jointing process.

Corrosion by building materials

Some types of wood have a corrosive action on lead, and latex cements and foamed concrete will effect copper. Some wood preservatives contain copper compounds and can cause corrosion of aluminium. Copper is not affected by cement or lime mortar, but should be protected from contact with magnesium oxychloride flooring or quick setting materials such as Prompt cement.

Lead is not affected by lime mortar, but must be protected from fresh cement mortar.

Galvanized coatings are not usually attacked by lime or cement mortars once they have set. Aluminium is usually resistant to dry concrete and plaster after setting, but it is liable to attack when they are damp.

Corrosion of water cisterns and tanks and the exterior of underground pipes may often be prevented by cathodic protection. This involves a natural small electric current passing through the water or soil between the metal to be protected and a suitable anode. If the anode is of magnesium or zinc alloy it is connected to the metal to be protected. The two metals, the pipe and the anode, act as an electric cell and a current passes between them, all the corrosion taking place at the anode (sacrificial metal), which has therefore to be replaced eventually. Permanent non-corrodible anodes are sometimes used, but the electric current has then to be provided from an external source, through a transformer and rectifier.

Fittings

Galvanized fittings should always be used with galvanized pipe. Brass fittings may be cast or hot pressed. Cast fittings are usually of alpha brass. Hot pressed fittings are of 'duplex brass' which some waters will affect with dezincification. Dezincification may cause:

1 Blockage by build up of corrosion products,

2 Mechanical failure due to conversion of brass to porous copper,

3 Slow seepage of water through the material.

Cisterns

Galvanized steel cisterns will be affected by soft water with a high carbon dioxide content or, where copper is present in the water, some waters are cupro-solvent. Where this is likely to occur the painting of the inside of the cistern with a suitable paint is recommended. They may also be protected by means of magnesium anodes (Figure 3.29) or other forms of cathodic protection.

Copper vessels should not be soft soldered as this may give rise to corrosion of the tin in the solder.

Hot water tanks

Galvanized steel

This is a suitable material in hard water districts but may be affected by soft water which has a high carbon dioxide content. Failure of this material is usually due to:

1 High copper content in the water,

2 Excessively high water temperature,

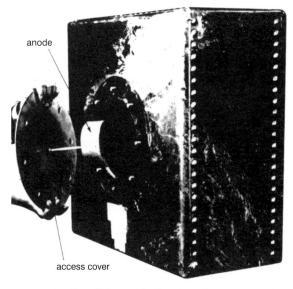

Figure 3.29 *Sacrificial anode fitted to a hot water tank*

3 Debris and metal filings, etc. on the bottom of the tank,

4 Damage to the protective galvanized coating.

Copper circulating pipes or cold feed services should not be used with galvanized steel tanks, although magnesium anodes may be used to provide protection in areas where failure might occur.

Copper cylinders

These cylinders should not be brazed with any copper-zinc alloy which is susceptible to dezincification.

Pitting may occur in the dome of a cylinder if the top connection protrudes too far into the vessel or if the cylinder top has been dented, thus forming an air pocket in which carbon dioxide liberated from the water during the heating process can accumulate. Fittings for cylinders should be manufactured from brass, or where dezincification may occur, of copper or gun metal.

Boilers

Domestic boilers are usually of cast iron, which resists attack from the water inside and combustion products outside. In soft-water areas, cast iron may be susceptible to attack. In this case Bower Barffed boilers should be used.

Copper boilers will resist corrosion in waterways, but are more readily affected by combustion products. In soft-water areas, copper-steel boilers are sometimes used. These have an internal surface of copper and an external one of steel.

Aluminium bronze boilers are claimed to have a high resistance to both corrosive waters and combustion products.

Radiators

Corrosion to light-gauge steel radiators depends upon the presence of dissolved oxygen in the water. In closed systems this is soon reduced to a very low quantity and therefore these radiators may be safely used. They should not, however, be used in open circuits where fresh water containing oxygen is replenished.

Towel rails are usually of chromium-plated brass or copper.

Impingement attack

The rapid flow of turbulent water (see Figure 3.30) has a characteristic form of damage, small deep pitting, often of horseshoe shape. It occurs more often in heating systems than cold-water systems due to rapid pumping or local turbulence set up by partially opened valves or abrupt changes of pipe size, sharp elbows, etc.

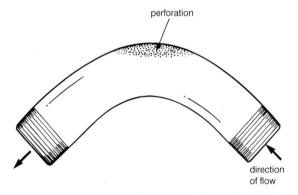

perforation

direction of flow

Figure 3.30 *Impingement damage to a low carbon steel bend*

Copper can suffer attack at speeds of above 2 m/sec. For marine work other materials have been developed to overcome this attack. Aluminium and brass will stand speeds of up to 3 m/s and cupro-nickels will stand still higher speeds.

Underground corrosion

Certain types of soil will affect pipes laid underground. Heavy clay may cause trouble, especially if waterlogged, as it may then contain sulphate-reducing bacteria which will corrode steel, lead and copper. Made-up ground containing cinders is very corrosive, and pipes laid in this or any other corrosive soil should be wrapped in one of the proprietary tapes sold for this purpose. Pipes should not be wrapped in hessian or similar material as this may lead to microbiological attack producing corrosive acids. Copper may be protected by the use of plastics covered tube.

Fittings

Dezincification of duplex brass fittings may be caused by some soils. In this case, gun metal or copper fittings should be used (these fittings are identified as CR corrosion resistant or DZR dezincification resistant). Cast iron may corrode in some soils by graphitization. Connected to non-ferrous metals it will generally suffer initial corrosion, but after its surface has suffered graphitization the corrosion potential between the two is reversed and the non-ferrous material becomes attacked.

Electric leakage

An electrical leakage from faulty apparatus earthed to the main water supply can cause severe corrosion to pipes.

Roofing

Lead will corrode if laid directly on oak or unseasoned timber. The corrosion will be on the underside in the form of a white powder and is due to the liberation of acetic acid from the wood. Lead should always be laid on an approved underlay.

Copper

Once green patina, a form of corrosion on copper roofs, has formed, it affords protection against further attack.

Another form of corrosion may occur from water dripping from rusting steel, slates that contain pyrites, or lichen-covered slates or tiles as this water is highly corrosive.

Zinc

Zinc may suffer 'white rusting' on the underside if condensation can occur there. It is also readily attacked by a polluted industrial atmosphere.

Aluminium

Aluminium is corroded by the soluble corrosion products of other metals. It is important that no drippings be received from any copper or copper alloy structure, e.g. copper gutters, lightning conductors, etc.

Soot

Soot deposits from chimneys will cause corrosion on all types of metal roofing (a) because a corrosion cell is set up between soot particles and the metal; and (b) because the soot will contain sulphur acids formed by the combustion of the fuel.

Rainwater goods

Cast iron is generally satisfactory. Gutters should be cleaned and painted regularly both inside and outside.

Galvanized gutters and pipes have good resistance to corrosion, but should not be used in conjunction with copper roofing materials.

Aluminium gutters and pipes have good resistance except where allowed to receive drippings from copper.

Magnetism

A magnet is a piece of metal which can attract to itself, and hold, pieces of iron. The invisible force that enables magnets to attract other objects is called magnetism.

Magnetic materials

A material is said to be magnetic if it is attracted to a magnet. The main magnetic materials are the metals, iron, nickel and cobalt.

Alnico – an alloy containing aluminium, nickel, iron, cobalt and copper – is used to make permanent magnets.

Plumbers use magnets to clean swarf and steel filings out of steel cisterns and tanks, following the cutting of holes for pipe connections.

Magnets are also used in some types of automatic control and are a useful aid for holding objects in place or position during working processes.

Forces between magnetic poles are shown in Figure 3.31 which illustrates repulsion at (a) and attraction at (b). The lines of magnetic force create a 'magnetic flux' or total force between the poles. These lines always form complete loops and never cross each other. Figure 3.32 shows the lines of magnetic flux which form the magnetic 'field' around a bar magnet.

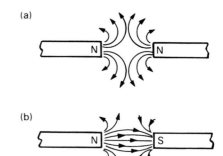

Figure 3.31 *Forces between magnetic poles: (a) like poles; (b) unlike poles*

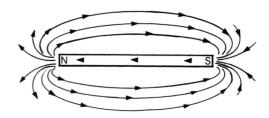

Figure 3.32 *Magnetic field around a bar magnet*

Capillarity and surface tension in building

One example where capillarity is put to good use is in the capillary type of joint used for copper tubing. This type of coupling usually has a channel on the inner face of each socket (see Figure 3.33) which contains a solder ring. After the cleaning of the joint surfaces and fluxing, the joint is slid together and heated, the solder ring melts and is drawn by capillarity into the space between pipe and socket. On cooling, a sealed watertight joint is created. An alternative to this form of joint is the 'end feed' pattern which requires the solder to be fed in from the end of the joint socket, as shown in Figure 3.34.

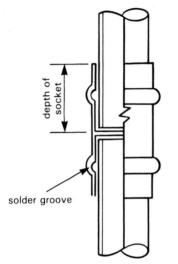

depth of socket

solder groove

Figure 3.33 *Capillary fitting with integral solder ring*

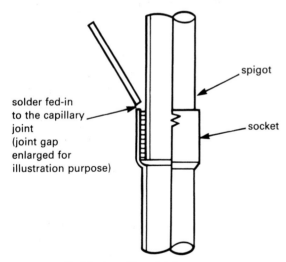

solder fed-in to the capillary joint (joint gap enlarged for illustration purpose)

spigot

socket

Figure 3.34 *End feed capillary joint*

Another way that surface tension affects liquid is in the formation of drops. When a small drop of mercury is placed on a flat surface it forms a ball. This is the result first, of surface tension, and second, of the fact that it does not adhere to the surface, if a small quantity of water is dropped

on a sheet of clean glass, adhesion causes it to spread out. If, however adhesion is prevented, say by the surface being oiled, round droplets are formed which will easily roll off. From this it can be seen that rainwater will be shed more easily as round droplets from surfaces that do not offer good adhesion. To achieve this, the surface may be treated by oiling (hardwoods, metals), waxing (masonry), painting and the application of various sealers. The practice of oiling the surfaces of moulds to receive concrete is particularly important since it prevents the concrete from adhering while hardening.

In some circumstances, however, adhesion is an advantage; in painting, there must be sufficient adhesion for the wet paint to spread evenly, in a continuous film. If there is no adhesion, the paint will form into globules on the surface and the work will be unsatisfactory. If a wall which is to be plastered or rendered does not give a good mechanical key, then it must offer a degree of 'suction' (capillarity) in order to get adequate adhesion between the surfaces. This suction must not be too powerful, since this would rapidly withdraw the mixing water so that the surface would harden before it could be finished.

For the jointing of metal surfaces by soldering the surfaces must be clean or the molten solder will not adhere and will simply roll off the surface in droplets.

Capillarity provides the exception to the rule 'that liquids will find their own level' as shown in Figure 3.35.

On slated roofs a 'tilting fillet' is used to permit a reasonable gap between the slates and chimney back-gutter to break any passage of water drawn

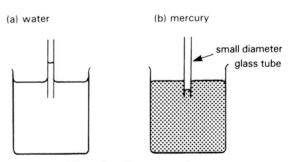

(a) water   (b) mercury

small diameter glass tube

Figure 3.35 *Effect of capillary attraction*

up from the lower edge. (Plain lap tiles for roofing are manufactured with a camber to prevent capillarity.)

Damp-proof courses (horizontal and vertical) are used in buildings to form a continuous impervious layer to prevent moisture from penetrating the structure from either high or low level.

Capillary means 'like a hair', so a capillary tube is one with a very small bore.

If a very small diameter glass tube, open at both ends, is placed vertically in a glass of water, the water will be seen to rise up the tube (see Figure 3.35(a)). This is due to 'capillarity' or capillary attraction.

The water rises because its molecules have a greater attraction to the glass than they have to each other. With mercury (see Figure 3.35(b)) the reverse is true and so the level in the capillary tube is below the surface of the liquid, not above it.

Any substances with pores of sufficient size are able to suck up water, for example, tissues, sponges, blotting paper. Moisture rises in the roots and stems of plants by the same means.

The height to which the water will rise depends on the diameter of the tube and becomes greater as the bore is diminished. This can be seen if two glass plates are held vertically and a small distance apart in the liquid (see Figure 3.36).

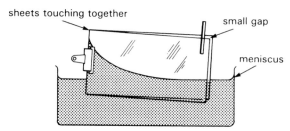

Figure 3.36 *Experiment to show the effect of size of the gap on the capillary attraction*

Thermal insulation

Properties of insulating materials

1 The material must have a *low thermal conductivity*. This means that the material must resist the passage of heat through it. A poor conductor used as an insulator will reduce heat losses by conduction from the warm surfaces of pipes, cylinders, etc.

2 *Porosity* is another important property. Many modern insulating materials incorporate, and use, still air. Most of these materials are sponge-like or fibrous in structure – containing millions of tiny air cells which trap air within the material and hold it still.

3 Insulating materials should be *incombustible*, or at least they should not catch fire easily. The reasons for this are obvious, especially for insulation in roof spaces and other inaccessible places where fire could cause havoc if encouraged to spread.

4 The material must be *resistant to fungal attack*. Fungi thrive in damp situations, and some insulating materials are made from organic materials which could nourish fungi and encourage them to grow. Where conditions may be damp, care must be taken to avoid using organic materials such as hairfelt. A better choice would be an inorganic material such as glass fibre, mineral wool, or foamed plastics material such as expanded polystyrene.

5 Although *weight* is not tremendously important in pipework insulation, it must be considered in cases where an additional load would be undesirable, such as if the material were being laid above ceilings.

6 The material should always be *resistant to vermin*, for the obvious reasons that vermin can give rise to unsanitary conditions. Anti-vermin preparations can be applied to most materials likely to suffer from an infestation.

7 It is important that insulating materials should *not absorb moisture*, especially where they have to be used out of doors or in damp situations. If an open cellular material becomes waterlogged the air will be forced out of the cells and the insulating value of the material will be seriously affected. Many open-celled materials can be rendered moisture-proof by using a special water repellant covering or jacket.

8 A *good surface finish* is desirable in an insulating material. A surface finish which can

Table 3.8 *Thermal insulating materials in plumbers' work*

Type		Application
A	*Loose fill* (In granular or fibrous form)	Infill to pre-formed cavities, for example between hot and cold store vessels and their prepared casings.
B	*Flexible* (In strip or blanket form)	Wrap around covering for curved surfaces such as pipes and cylinders.
C	*Rigid* (In pre-formed, ready to-fit sections, moulded polystyrene, etc.)	Made specially for pipework. Also in flat sections for hot water tanks and cold storage cisterns.
D	*Mouldable* (Workable until set hard compound lagging)	For covering awkward or irregular shapes such as uncased boilers, large cylinders and valves.

be painted or decorated easily will help the insulation to blend with the general decorations and so be less conspicuous

9 *Ease of application* is very important. The fixing of thermal insulations generally demands very careful attention, since coverage must be complete and uniform if it is to be fully effective. Materials that are fitted easily are more likely to be fixed with interest and care than those which are difficult to apply.

Modes of heat transfer

Heat flows from a substance at a higher temperature to one at a lower temperature, and this effect is known as 'heat loss'. Heat loss continues until both substances are at the same temperature – unless some form of thermal insulation is used to isolate one substance from the other and thus prevent heat from being transferred.

Heat can travel as a result of one of the following 'modes of heat transfer': conduction, convection, and radiation, or by a combination of two or more of them.

Conduction occurs when heat flows through or along a material (see Figure 3.37) or from one material to another in contact with it. Conduction increases with the difference in temperature between the hot and cooler materials in contact. Some materials, such as metals, are good conductors of heat; others which resist the passage of heat by conduction are called poor conductors

(thermal insulators). A good thermal insulating material must, among other things, be a poor conductor of heat.

Convection occurs in liquids and gases. When the liquid or gas is heated, the warmed particles expand and become lighter, the cooler heavier particles fall by gravity and push the lighter, warmed ones upward (see Figures 3.38 and 3.39). In this way convection currents will continue so long as heat is applied or so long as there is some difference in temperature in the liquid or gas.

Radiation heat is given off from a hot body to a colder one in the form of heat energy rays.

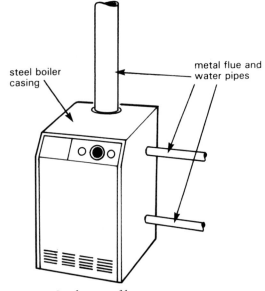

Figure 3.37 *Conductors of heat*

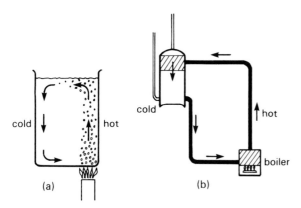

Figure 3.38 *Convection currents in liquid: (a) in an open vessel; (b) in a closed water heating circuit*

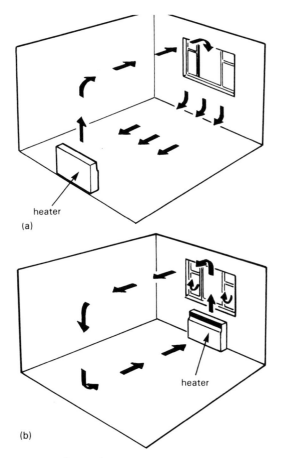

Figure 3.39 *Convection currents in a room: (a) heater at opposite end to a window – air cooled by the window becomes a cold draught across the floor; (b) heater under window – cold air is warmed and carried up by the convection currents*

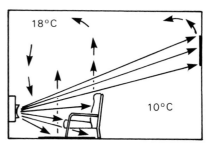

Figure 3.40 *Room heated by radiation from a gas fire*

Radiant heat rays do not appreciably warm the air through which they pass, but they do warm any cooler solid body with which they come into contact (see Figure 3.40). Radiant heat falling on a surface is partly absorbed and partly reflected. A dull black surface is a much better absorber of radiation than a polished surface. The latter is therefore a good reflector of heat.

Insulating values for building materials

It is important for the plumber to understand the thermal insulating values of building materials, and certain purpose-made insulating materials, if he or she is to be able to advise the customer on the best and most economic ways of using those materials to protect a plumbing system and keep the building warmer.

Transmittance coefficient (symbol U) is a term used to denote the quantity of heat which will flow from air on one side of the wall, roof, etc., to air on the other side, per unit area and for unit air temperature in unit time. The units employed are:

Area	square metres (m^2)
Temperature	degrees Celsius (°C)
Time	seconds (s)
Quantity of heat	watts (W)

The definitions are simply applied as shown in Figure 3.41, where $1\,m^2$ of 230 mm thick brickwork is shown – it can be seen that, for a 1°C difference in temperature inside and outside, heat will pass from air on the warmer side, through the wall thickness and be lost to the cooler air on the other side at the rate of *2.7 watts/m^2/°C in 1 second*. This, then, would be referred to as the

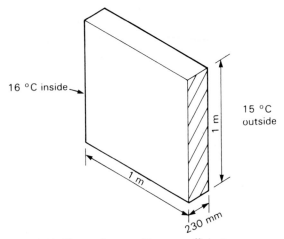

Figure 3.41 *Thermal transmittance coefficient*

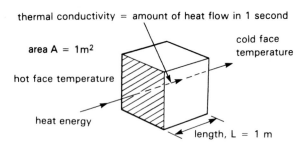

Figure 3.42 *Thermal conductivity*

Table 3.9 *Thermal conductivities*

Material	W/m°C
Metals (at 18°C)	
Copper	384.2
Brass	104.6
Aluminium	209.2
Steel	48.1
Cast Iron	45.6
Lead	34.7
Building materials	
Brick	1.15
Concrete	1.44
Plaster	0.58
Glass	1.05
Deal boards	0.12
Fluids (at 0°C)	
Methane	0.029
Hydrogen	0.16
Carbon dioxide	0.014
Steam	0.015
Air	0.022
Water	0.054
Oil	0.18
Mercury	8.37
Insulating materials	
Slag wool	0.042
Aluminium foil	0.042
Granulated cork/bitumen slab	0.15
Glass silk mats	0.040
Mineral wool slab	0.034
Fibre board	0.059
Vermiculite	0.067
Firebrick	0.61

thermal transmittance coefficient or 'U-value' for unplastered brickwork 230 mm thick.

The thermal transmittance or U-value of a wall, roof or floor of a building is a measure of its ability to conduct heat out of the building: the greater the U-value, the greater the heat loss through the structure. The total heat loss through the building fabric is found by multiplying U-values and areas of the externally exposed parts of the building, and multiplying the result by the temperature difference between inside and outside.

The capacity of a material to conduct heat is called its 'thermal conductivity'. This is a measure of the amount of heat energy which can be conducted in 1 second through an area of $1\,m^2$ across a length of 1 m for 1°C difference in temperature between the two ends, or 'faces' (see Figure 3.42).

Thermal conductivity

$$= \frac{\text{Heat flow} \times \text{thickness}}{\text{area} \times \text{temperature difference}}$$

$$= \frac{\text{Watts} \times \text{metres}}{\text{metres}^2 \times {}^\circ\text{C}} = \frac{\text{Wm}}{\text{m}^2{}^\circ\text{C}}$$

This can be simplified to W/m°C.

Table 3.9 shows the thermal conductivities of common materials. The figures shown in Table 3.9 are the results of various experiments and are only approximate. Nevertheless, they serve to show the considerable difference in conductivity between the insulators and the gases on the one hand and the solid materials and particularly the metals on the other. Copper is obviously the best and CO_2 is the worst of those shown.

Insulators generally have cellular, granular or matted thread construction. These forms of structure break up a solid path for heat flow and trap small pockets of still air which offer considerable resistance to conduction.

Latent heat

There are three forms of state in which a substance can exist:

1 solid

2 liquid

3 gas.

Water can exist in three states, solid (ice), liquid (water), gas (steam), but because of the temperature normally prevailing we generally see the substance in only one of its states.

The change of state from solid to liquid or vice versa is termed the *lower change of state* and the change from liquid to gas is termed the *upper change of state*.

A solid material, brass for instance, can be changed to a liquid by the application of heat. Excessive heat will cause the zinc in the brass to be driven off in the form of a white vapour.

Gases can also be turned into liquid and solid states. Liquid oxygen is used as a propellent, while liquid butane gas is used in plumbers' blowlamps. Carbon dioxide is used to freeze the water in a pipe when carrying out an alteration or a repair.

Often when a change of state occurs, heat is being put into or given out by a substance without there being any change in the temperature of the substance.

Heat that brings a change in temperature is usually called sensible heat. Heat that does not bring about a change in temperature is usually called hidden or *latent heat*. This latent heat is used to 'unbind' from each other the particles that go to make up a substance. In a solid these are tightly packed giving the solid its characteristic shape. The extra energy (heat) moves the particles

further apart and allows them to move about more freely, as in a liquid. Further energy (heat) causes them to become more 'unbound' and they can move even more freely as in a gas. There are therefore two changes of state and two values of latent heat for any substance. These are:

1 Latent heat of fusion

2 Latent heat of vaporization.

Experiment to determine sensible and latent heat

If a piece of ice was placed in a pan and heated its temperature would gradually rise until it reached 0°C. Then the ice would begin to melt. If the heating was continued until all the ice was melted, the pan would then contain water at 0°C. There would be no increase in temperature while the ice was melting.

If heating was still continued, the temperature of the water would rise until it reached 100°C. Then the water would begin to boil. The temperature would stay at 100°C until all the water was turned into steam. If the heating process was continued with the steam, the temperature would begin to rise again.

Figure 3.43 shows, on a graph, what is happening.

From point A to point B the temperature is rising steadily. The heat being absorbed during that period is called 'sensible heat' because it can be 'sensed' by the thermometer.

From B to C it is apparent that heat is still being absorbed but it is not visible on the thermometer, so it is called 'latent heat' because it cannot be seen.

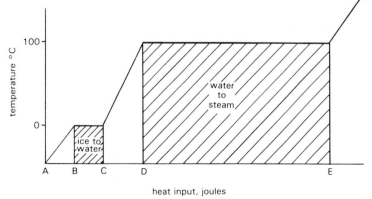

Figure 3.43 *Graph of sensible and latent heat of water (not to scale)*

From C to D the thermometer shows the second increase in sensible heat and from D to E shows latent heat again.

The same sort of thing happens with most substances. When a solid melts into a liquid or a liquid cools to a solid the heat absorbed or given off is called the 'latent heat of fusion'.

When a liquid evaporates into a vapour or a vapour condenses into a liquid the heat involved is called the 'latent heat of vaporization'.

Different substances require different amounts of heat to bring about these changes in state. For example:

Latent heat of fusion of ice = 334 kJ/kg
Latent heat of vaporization of water = 2250 kJ/kg

Archimedes' Principle

When anything is placed in a liquid it is subjected to an upward force or *upthrust*.

A simple but striking experiment to illustrate the upthrust exerted by a liquid can be shown by tying a length of cotton to an object. Any attempt to lift the object by the cotton fails through breakage of the cotton, but if the object is immersed in water it may be lifted quite easily. The water exerts an upthrust on the object and so it appears to weigh less in water than in air.

Experiments to measure the upthrust of a liquid were first carried out by the Greek scientist Archimedes. The result of his work was a most important discovery which is now called *Archimedes' Principle*. In its most general form, this states:

> When a body is wholly or partially immersed in a fluid it experiences an upthrust equal to the weight of the fluid displaced (see Figure 3.44).

To verify Archimedes' Principle for a body in liquid

A eureka (or displacement) can is placed on the bench with a beaker under its spout (see Figure 3.45). Water is poured in until it runs from the spout. When the water has ceased dripping the beaker is removed and replaced by another beaker which has been previously dried and weighed.

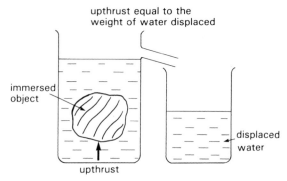

Figure 3.44 *Archimedes' Principle*

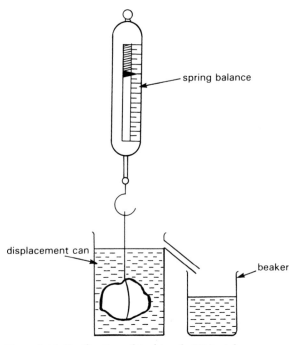

Figure 3.45 *Verification of Archimedes' Principle*

Any suitable solid body, e.g. a piece of metal, is suspended by thin thread from the hook of a spring-balance and the weight of the body in air is measured. The body, still attached to the balance, is then carefully lowered into the displacement can. When it is completely immersed its new weight is noted. The displaced water is caught in the weighed beaker. When no more water drips from the spout the beaker and water are weighed.

The results should be set down as follows:

Weight of body in air   =   g
Weight of body in water   =   g

Weight of empty beaker   =   g
Weight of beaker + displaced water   =   g
Apparent loss of weight of body   =   g
Weight of water displaced   =   g

The apparent loss of weight of the body, or the upthrust on it, should be equal to the weight of the water displaced, thus verifying Archimedes' Principle in the case of water. Similar results are obtained if any other liquid is used.

To measure the relative density of a solid by using Archimedes' Principle

Earlier we explained the meaning of the term relative density (or specific gravity), and its importance in the accurate measurement of density.

Relative density of a substance

$$= \frac{\text{mass of any volume of substance}}{\text{mass of an equal volume of water}}$$

or, since weight is proportional to mass

$$\text{RD} = \frac{\text{weight of any volume of substance}}{\text{weight of an equal volume of water}}$$

Archimedes' Principle gives us a simple and accurate method for finding the relative density of a solid. If we take a sample of the solid and weigh it first in air and then in water, the apparent loss in weight, obtained by subtraction, is equal to the weight of a volume of water equal to that of the sample. Therefore:

Relative density of a substance

$$= \frac{\text{weight of a sample of the substance}}{\text{apparent loss in weight of the sample in water}}$$

Density

By density we mean the weight of that substance whether it be a solid or a liquid, and is expressed as its mass per unit volume e.g. Kg/m^3.

One often hears the expression, 'as light as a feather' and 'as heavy as lead'. Equal volumes of different substances vary considerably in mass. Aircraft are made chiefly from aluminium alloys, which are as strong as steel but, volume for vol-

Table 3.10 *Table of densities of common metals*

Materials	Chemical symbol	Density	Relative density
Lead	Pb.	11300 kg/m^3	11.3
Copper	Cu.	8900 kg/m^3	8.9
Iron	Fe.	7860 kg/m^3	7.8
Tin	Sn.	7310 kg/m^3	7.3
Zinc	Zn	7130 kg/m^3	7.1
Aluminium	Al.	2705 kg/m^3	2.7

ume, weigh less than half as much. In plumbing we refer to the lightness or heaviness of different materials by the use of the word *density*.

The density of a substance is defined as its mass per unit volume

One way of finding the density of a substance is to take a sample and measure its mass and volume. The density may then be calculated by dividing the mass by the volume. The symbol used for density is the Greek letter ρ (rho). Thus:

$$\text{Density} = \frac{\text{mass}}{\text{volume}} \quad \text{kg/m}^3$$

or in symbols $\rho = \dfrac{m}{v}$

The densities of all common substances, solids, liquids and gases, and all chemical elements have been determined and are to be found listed in books of physical and chemical constants.

Water has a density of about 1 g/cm^3, or 1000 kg/m^3 owing to the fact that the kilogram was originally intended to have the same mass as 1000 cm^3 of water at 4°C. Mercury is a metal which is a liquid at ordinary temperatures and it has the very high density of 13.6 g/cm^3. It is a very useful substance in scientific laboratories and plays a part in many experiments.

Importance of density measurements

Architects and engineers refer to tables giving the densities of various building materials when engaged in the design of building works.

From the plans drawn up, they can calculate the volume of any part of the structure, which, multiplied by the density of the material, gives the mass and hence the weight. Such information is essential for calculating the strength required in foundations and supporting pillars.

Simple measurements of density

Liquids
A convenient volume of the liquid is run off into a clean, dry previously weighed beaker, using either a pipette or burette. The beaker and liquid are then weighed and the mass of the liquid found by subtraction.

Solids
The volume of a substance of regular shapes, e.g. a rectangular bar, cylinder or sphere, may be calculated from measurements made by vernier callipers or a micrometer screw gauge.

The volume of an irregular solid, e.g. a piece of lead, may be found by measuring its displacement volume in water. For solids soluble in water, e.g. certain crystals, some liquid such as white spirit would be used in the measuring cylinder. The mass of the solid is found by weighing.

In each case the density is calculated from:

$$\text{Density} = \frac{\text{mass}}{\text{volume}}$$

Relative density (formerly called specific gravity)
By *relative density* we mean its comparative weight of water is taken as the standard and is known as '1' all other substances are related to it. For example, the density of lead is known as 11.3 this means that it is 11.3 times heavier than water volume for volume.

In the last experiments described we had to make two measurements to find the density, namely, a mass and a volume. Now we can always measure mass more accurately than volume, and so, in the accurate determination of density, scientists have overcome the necessity to measure volume (hence eliminating one source of error) by using the idea of *relative density*.

The relative density of a substance is the ratio of the mass of any volume of it to the mass of an equal volume of water, or

$$\frac{\text{relative}}{\text{density}} = \frac{\text{mass of any volume of the substance}}{\text{mass of an equal volume of water}}$$

In normal weighing operations the mass of a body is proportional to its weight, so it is also true to say

$$\frac{\text{relative}}{\text{density}} = \frac{\text{weight of any volume of the substance}}{\text{weight of an equal volume of water}}$$

This will explain why relative density has also been called specific gravity: the word gravity implying weight. At the present time, by international agreement the term relative density is recommended rather than specific gravity.

Note that relative density has no units: it is simply a number or ratio. On the other hand, density is expressed in kg/m^3.

Example
1 cubic foot of water weighs 1000 g
1 cubic foot of lead weighs 1000 × 11.3 g
= 11 300 kg/m^3

Pressure

Definition of pressure
'Pressure' is the same as stress, but whereas stress applies to solid objects, pressure is more concerned with fluids, that is, liquids or gases.

Fluids have the capacity to press themselves against the surface of the vessel that contains them (see Figure 3.46b) and so exert a force. The intensity of this force can be measured in relation to the area of the surface. Pressure is the relationship between force and area.

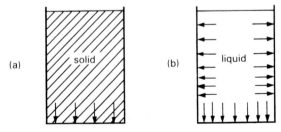

Figure 3.46 *Difference between pressures exerted by solids and liquids*

$$\text{Pressure} = \frac{\text{force}}{\text{area}} = \frac{\text{newtons}}{\text{square metres}}$$

$$\text{(in units)} = \frac{N}{m^2} \text{ or } N/m^2$$

The unit N/m^2 has the special name 'Pascal' after the French mathematician. It is in more common use on the European continent and has the symbol 'Pa'.

Pressure in fluids

Figure 3.46 illustrates the difference between solids and liquids. At (a) there is a solid block which just fits into a container. The only pressure is that exerted on the base.

At (b), the container is filled with liquid. This now exerts a pressure on the sides as well as the base. In fact, the pressure will increase as the depth increases.

If the weight of the block is 15.4 kg, then in the different positions the pressures exerted by the base would be different also.

It follows that, for the same weight of object, the smaller the base area, the higher the pressure that it will exert. For example, a boiler of shape (a) would exert more pressure per unit of area than a boiler of shape (b) or (c), as shown in Figure 3.47.

Pressure exerted by a solid

A solid object has the ability to exert a pressure on the floor on which it stands, but this is only a downward direction. Fluid pressure acts in all directions at the same time.

The pressure exerted by a solid object is the weight of the object divided by the area of its base. Take as an example the object shown in Figure 3.47.

Units of pressure

Pressure can be calculated either by measuring the force exerted on a unit of area or by measuring the height of a column of liquid supported by the force. So its units are either those of force per unit area, e.g. Newtons per square metre, or they are metres or millimetres height or 'head' of liquid.

There are alternative units. In some areas of activity kilograms force per square centimetre may be used and in the gas industry, for example, pressure will be measured normally in 'bars' and 'millibars'. The symbols are 'bar' and 'mbar'.

Where liquids are used in pressure gauges, water is the most common for low pressures and mercury, which is 13.6 times as dense, for higher pressures. The gauges have scales graduated in millibars.

It may sometimes he necessary to convert a pressure reading from force/area units to height units or vice versa and Table 3.11 shows the comparison between them.

Atmospheric pressure

The earth is surrounded by an envelope of air, held to the earth's surface by gravity. The weight of air creates a pressure on the earth's surface of about 1 bar or $101.325 \ N/m^2$ at sea level.

Although our bodies are subjected to this pressure they have an internal pressure which normally exactly balances atmospheric pressure so that we are unaware of it.

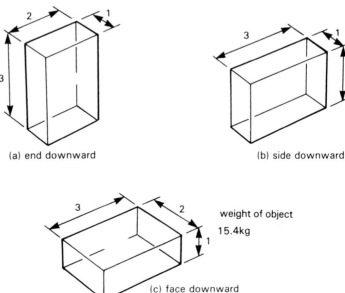

Figure 3.47 *Pressure exerted by a rectangular block: (a) end downward; (b) side downward; (c) face downward*

Table 3.11 *Comparison of units of pressure*

Height	Bars	Force/area
1 m head of water	98 mbar	9800 N/m²
10.2 mm of water	1 mbar	100 N/m²
1 atmosphere or 760 mm of mercury or 30 in of mercury	1013.25 mbar (1 bar)	101.3 kN/m² or 101.325 N/m²

Atmospheric pressure must, however, be taken into consideration in a number of calculations involving gas pressure.

One effect of atmospheric pressure can be shown by a simple experiment (see Figure 3.48). Take a metal can which is open to the air so that the pressure inside is the same as that on the outside. Put a small quantity of water in the can, place it over a gas flame and boil the water. The steam produced will push the air out of the can.

When the can is full of steam only, seal the neck with a bung. Then place the can in a stream of cold water.

The cold water will cause the steam to condense into water in the can. The sudden reduction in volume will create a partial vacuum or 'negative' pressure and the atmospheric pressure on the outside will crush the can. This is called 'implosion' (which is, of course, the opposite of explosion).

Precautions have to be taken to prevent any negative pressures occurring in gas supplies otherwise atmospheric pressure can damage gas meters or similar components.

Atmospheric pressure varies with the weather. When pressure is high the day will be fine and dry. As pressure falls the weather becomes changeable, rainy and finally stormy.

Atmospheric pressure also varies with the height above sea level at which the reading is taken. On the top of a mountain the pressure is less than the pressure on the beach.

Barometers

Atmospheric pressure is measured by a 'barometer'. The simplest form of barometer is the mercury barometer devised by Torricelli (Figure 3.49).

This consists of a glass tube, sealed at one end, which is first filled with mercury and then inverted into a trough also containing mercury.

The mercury will begin to pour out of the tube, leaving a vacuum at the top, until the height of the column of mercury exactly balances the atmospheric pressure which is being exerted on the surface of the mercury in the trough.

The reason for Torricelli using mercury becomes obvious when you calculate the length of tube which would be required if water was to be used.

$$h = \frac{101.3}{9.81} = 10.3 \text{ metres}$$

The mercury barometer offers a very accurate means of measuring atmospheric pressure and

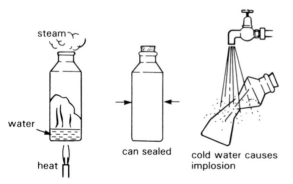

Figure 3.48 *Experiment to show the effect of atmospheric pressure*

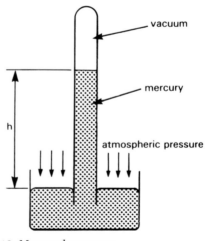

Figure 3.49 *Mercury barometer*

there are a number of ways in which provision can be made for the scale to be set to zero and compensation made for altitude. Most domestic barometers are, however, aneroid barometers.

Aneroid barometer

'Aneroid' means 'not liquid' and this type of barometer consists of a cylindrical metal box or bellows almost exhausted of air. The box has a flexible, corrugated top and bottom and is very sensitive to changes in atmospheric pressure (Figure 3.50). Such changes cause inward or outward movements of the flexible top and bottom sections and this movement is made to rotate a pointer on a scale by means of a suitable mechanism attached to the bellows.

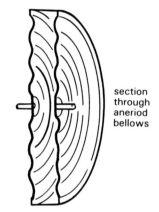

section
through
aneriod
bellows

Figure 3.50 *Aneroid bellows*

Absolute pressure

'Absolute' pressure is the pressure from zero to that shown on a gauge. Figure 3.51 shows a container under three different conditions. In all three cases the container is subjected to atmospheric pressure on all sides.

At (a) the air has been pumped out and the valve closed. The pressure inside is zero.

At (b) the valve is opened and air rushes in to fill the vacuum. The pressure inside is atmospheric pressure 1013 mbar.

At (c) the container has been connected to a gas supply at a pressure of, say, 20 mbar. Some gas will be forced in and the pressure will now be

atmospheric pressure plus gas pressure
= 1013 + 20
= 1033 mbar, absolute pressure

So,

atmospheric pressure + gas pressure
= absolute pressure

Any pressure gauge will have atmospheric pressure inside it before the gas is turned on, like the container at (b). So it will only indicate the additional pressure due to the gas. For an absolute pressure reading, atmospheric pressure must be added on:

gauge pressure + atmospheric pressure
= absolute pressure

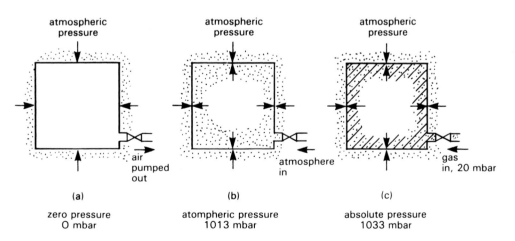

atmospheric pressure

atmospheric pressure

atmospheric pressure

air pumped out

atmosphere in

gas in, 20 mbar

(a)
zero pressure
0 mbar

(b)
atomphceric pressure
1013 mbar

(c)
absolute pressure
1033 mbar

atmospheric pressure + gas pressure = absolute pressure

Figure 3.51 *Absolute pressure*

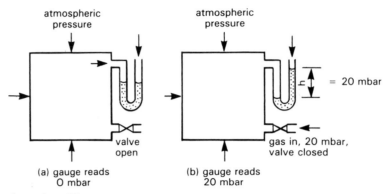

Figure 3.52 *Pressure shown by a 'U' gauge*

Figure 3.52 shows a water gauge attached to the container. When gas pressure is introduced the height of water indicates the gas pressure only.

Measurement of pressure

Pressure gauges, like barometers, can indicate height or force/area so they can be liquid or dry types. The liquid types commonly use water or mercury. There are two main differences between these liquids, so far as their use in gauges is concerned.

1 Mercury has a specific gravity of 13.6, so it will measure pressures 13.6 times higher than the same height of water.

2 Water adheres to the side of the gauge tube and mercury does not. Figure 3.53 shows what effect this has on the 'meniscus' or liquid level which in both cases is crescent-shaped. When reading a gauge the true value comes from the bottom of a water meniscus and the top of a mercury meniscus (see Figure 3.54).

The reason for the difference in the meniscus lies in the difference in the force of attraction between the molecules. There are two forces on the molecules.

1 The attraction of one molecule of the substance to another of the same substance, 'cohesion',

2 The attraction of a molecule of the substance to a molecule of another substance, 'adhesion'.

In water, the adhesive quality is stronger than the cohesive quality. So water adheres readily to other substances. In mercury the reverse is the case. An example of the effect of adhesion and cohesion can be seen in capillary attraction.

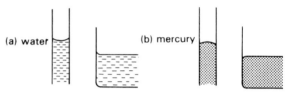

Figure 3.53 *Effect of adhesion and cohesion*

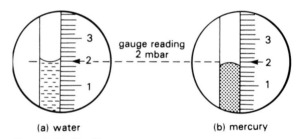

Figure 3.54 *Reading a gauge*

Pressure in water heating systems

Hydrostatic pressure

'Hydrostatic pressure' is simply the pressure exerted in the system by the weight of the water.

Figure 3.55 shows a domestic water heating system consisting of a cistern A supplying cold water to a cylinder B, heated by a boiler C and with bath, basin and sink draw off taps at D, E and F.

Typical pressures in an average house are shown in Table 3.12. From the table the following points can be seen:

1 The pressure at the sink tap, F, is more than twice the pressure at the basin tap, E. So you would expect to get a much faster flow of

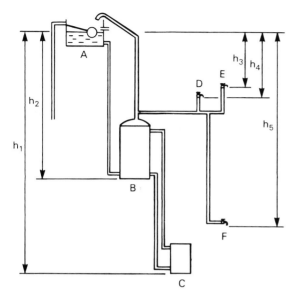

Figure 3.55 *Pressures in a domestic hot water system*

Table 3.12 *Pressures of different heads of water*

Head	Head metres	Force/ area kN/m^2	Equivalent in millibars
h_1	5	49(50)*	490
h_2	2.5	24.5(25)	245
h_3	1.5	14.7(15)	147
h_4	2	19.6(20)	196
h_5	4	39.2(40)	392

*Figures in brackets are approximate.

water at the sink (depending on the pipes and the size of taps).

2 If the area of the base of cylinder was 0.8 square metres, then the force on the base would be:

25 kN/m^2 (pressure) × 0.8 m^2 (area)
= 20 kN or 20 000 Newtons

3 If the area of the base of the boiler was 0.5 square metres, then the force on the base would be:

50 kN/m^2 × 0.5 m^2 = 25 kN

This shows why makers of boilers and water heaters always state a maximum head of water up to which their appliances may be fitted. If subjected to higher pressures the appliances could be damaged.

Circulating pressure

Water 'circulates' or goes round and round in water heating or wet central heating systems. The pressure which causes this circulation is naturally called 'circulating pressure'.

The circulating pressure in a system depends on two factors:

1 The difference in temperature between the hot water flowing from the boiler and the cold water returning to it,

2 The height of the columns of hot and cold water above the level of the boiler.

The example in Figure 3.56 shows part of a simple water circulation from a boiler to a radiator and back again.

Generally, central heating systems now rely on pumps to provide the main circulating pressure so that smaller bore pipes may be used. But gravity can be used and is still the motive force in many water heating systems. (Figure 3.56 shows the boiler connected to the cylinder by flow and return pipes, and the radiator in the figure could be replaced by a cylinder. The principle is still the same although temperatures are lower.)

All fluids expand when heated and water is no exception. When a fluid expands, the same weight of fluid occupies a bigger volume. So its density (weight ÷ volume) is reduced.

Put simply, a given volume of hot water weighs less than the same volume of cold water.

Tables are available giving the density of water at various temperatures and, if the flow and return temperatures and the circulation height are known, it is possible to calculate the circulating pressure in a system operating on gravity circulation.

Circulating pressure is proportional to the temperature difference between flow and return water and the circulating height.

Circulating pressure will increase:

1 If the temperature difference increases.

2 If the circulation height is increased.

The circulating pressure in Figure 3.56 is approximately 5 millibars. It can be obtained from the formula:

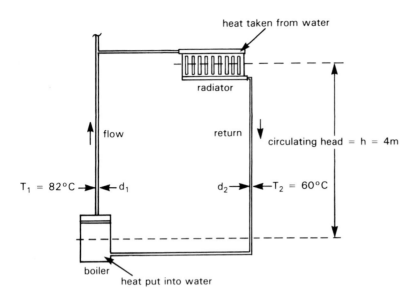

heat taken from water

radiator

flow   return

circulating head = h = 4m

$T_1 = 82°C$ →|← d_1 d_2 →|← $T_2 = 60°C$

boiler

heat put into water

Figure 3.56
Circulating pressure. On a water heating system the cylinder would be in the position occupied by the radiator. T1 would be 60°C and T2 would be about 10°C when heating began

$$p = h \frac{(d_2 - d_1)}{d_2 + d_1} \times 196.1$$

where p = circulating pressure in millibars
       h = circulation height in metres
      d_1 = density of water in flow in kg/m^3
      d_2 = density of water in return in kg/m^3 and 196.1 is a constant

From tables, d_1 at 82°C = 970.40 kg/m^3

             d_2 at 60°C = 983.21 kg/m^3

Therefore p = 4 × 983.21 – 970.40/983.21 + 970.40 × 196.1

       = 5.14 mbar (about 5 mbar)

Other tables are available which give a reading of circulating pressure per metre height directly from a temperature difference.

Plumbing mechanics (levers)

A plumber needs to know a great deal about the properties of different materials and the basic rules which govern their behaviour. All materials are inert, that is, the material cannot move of its own accord, and it is important to remember that water and air in particular will not move unless some force (or forces) makes them move.

Force is not a property of a material; it is something *which moves* or *tries to move* an object and is usually defined as 'that which changes, or tends to change, a body's state of rest or uniform motion in a straight line' (Figure 3.57). The *force* in levers is also known as the *effort*.

Load is the term given to the work to be overcome by the effort, i.e. moving an object, closing off the water by the valve in a tap, etc.

A *lever* is a form of simple machine, and may be defined as 'a rigid bar free to rotate or move about a fixed point known as the fulcrum'. The fulcrum is also known as the pivot.

A lever is used to magnify a force applied to move an object (Figure 3.58). Its length is measured from the point where the force is applied to the centre of the fulcrum. The simplest (machine) form of lever is a straight rigid bar; the handle of a pipe wrench (Stillson) is such a lever. When the wrench is held by the end of the handle

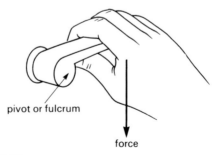

pivot or fulcrum

force

Figure 3.57

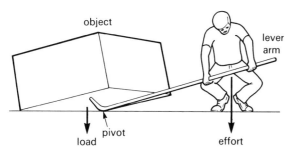

object

lever
arm

pivot

load

effort

Figure 3.58

(see Figure 3.59) the amount of pressure that can be exerted on the pipe or fitting gripped by the wrench will be much greater than if the wrench was held in the middle of the handle (lever).

The amount by which the lever increases a force depends partly on where the pressure is applied in relation to the fulcrum, and partly on the position of the object to be moved, in relation to the fulcrum. The lever magnifies the effort applied to move an object, and is said to give a *mechanical advantage*. Mechanical advantage is defined as the ratio of the load (L) to the effort (E). Therefore

$$\text{Mechanical advantage} = \frac{\text{Load}}{\text{Effort}}$$

Mechanical advantage is a ratio of like quantities and therefore has no units.

Example 1
A simple machine requires an effort of 20 N to raise a load of 160 N. What is the mechanical advantage?

$$\text{Mechanical advantage} = \frac{\text{Load}}{\text{Effort}}$$

$$\text{Mechanical advantage} = \frac{160}{20}$$

$$= 8$$

Answer = 8 N

Example 2 (see Figure 3.60)
A crowbar requires an effort of 12 N to move a boiler which weighs 132 N. What mechanical advantage does the crowbar give?

$$\text{Mechanical advantage} = \frac{\text{Load}}{\text{Effort}}$$

$$= \frac{132}{12}$$

$$= 11$$

Answer = 11 N

As the above examples show, a lever is an instrument used to overcome resistances, or to lift a load to the advantage of the person operating the lever.

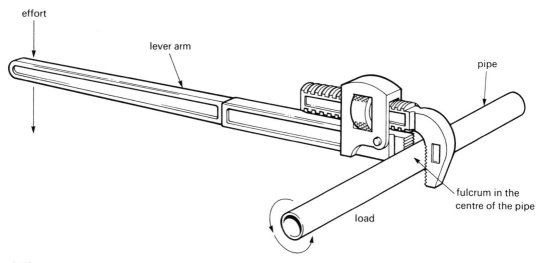

effort

lever arm

pipe

fulcrum in the
centre of the pipe

load

Figure 3.59

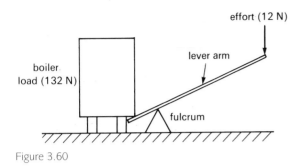

Figure 3.60

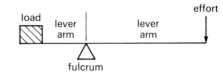

Figure 3.61

Calculations involving levers

Load × Length of lever arm
= Effort × Length of lever arm

Levers are classified according to where the fulcrum, load and effort are situated. There are three kinds, known as first-order, second-order and third-order levers. Figure 3.62 illustrates examples of each kind found in everyday use. After careful examination of the diagrams you should be able to identify and appreciate how each lever works.

As previously stated, the purpose of a lever is to make work easier to do and this will be demonstrated by the following examples.

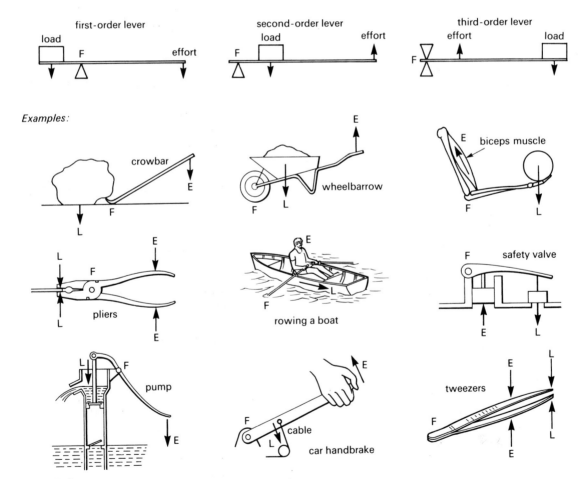

Figure 3.62 *Levers*

Example 3
Using the first order of levers as shown in Figure 3.63, calculate the effort required to raise a boiler weighing 132 N.

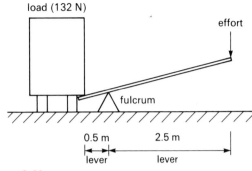

Figure 3.63

$$
\begin{aligned}
\text{Effort} \times \text{Lever arm} &= \text{Load} \times \text{Lever arm} \\
E \times 2.5\,\text{m} &= 132\,\text{N} \times 0.5\,\text{m} \\
E &= \frac{132 \times 0.5}{2.5} \\
E &= 26.4\,\text{N}
\end{aligned}
$$

Answer Effort = 26.4 N

Self-assessment questions

1 The poisonous gas liberated because of incomplete combustion of natural gas is:
   (a) methane
   (b) carbon dioxide
   (c) hydrogen
   (d) carbon monoxide.

2 The main reason why thermal insulation is incorporated into a building is to prevent:
   (a) vapour penetration
   (b) sound transference
   (c) condensation
   (d) heat flow.

3 The abbreviation 'PVC' refers to:
   (a) protective varnish
   (b) polythene
   (c) polypropylene
   (d) polyvinyl chloride.

4 Steel cuttings left in a galvanized storage cistern are likely to cause:
   (a) electrolytic action
   (b) plumbo-solvency
   (c) dezincification
   (d) electrostatic action.

5 Heat that does not bring a change in temperature is usually called:
   (a) measured heat
   (b) latent heat
   (c) sensible heat
   (d) heat loss.

6 Which one of the following combinations is likely to produce the most electrolytic action?
   (a) iron and zinc
   (b) copper and lead
   (c) lead and brass
   (d) copper and zinc.

7 Atmospheric pressure is usually measured on a:
   (a) barometer
   (b) thermometer
   (c) spirit level
   (d) hydrometer.

8 The attraction of one molecule to another of the same substance is called:
   (a) adhesion
   (b) bonding
   (c) condensation
   (d) cohesion.

9 A by-product of corrosion in a pressed steel radiator is:
   (a) nitrogen
   (b) carbon dioxide
   (c) hydrogen
   (d) oxygen.

10 Warm, moist air coming into contact with a cold water pipe in a kitchen will form:
   (a) surface tension
   (b) condensation
   (c) evaporation
   (d) cold draughts.

11 If a copper cold water pipe is subjected to freezing conditions the pipe is likely to burst when:
   (a)   the temperature rises above freezing point again
   (b)   the water changes into ice
   (c)   there are too many sharp bends in the installation
   (d)   non-manipulative compression fittings are used.

12 The main reason for sleeving a hot water pipe through a wall is to:
   (a)   protect it from corrosion
   (b)   allow it freedom of movement
   (c)   enable it to be removed easily for maintenance
   (d)   strengthen the wall after making good the hole.

13 A bi-metallic strip is used in a gas water heater to:
   (a)   act as a safety device
   (b)   ensure an even flow of gas
   (c)   regulate the flow of gas
   (d)   regulate the flow of water.

14 In which of the following is heat transfer by convection most important:
   (a)   joint wiping
   (b)   space heating
   (c)   capillary soldering
   (d)   refrigerated spaces.

15 Which one of the following joint types relies upon the principle of capillarity to make an effective joint:
   (a)   screw thread joint
   (b)   type A manipulative compression fitting
   (c)   solvent cement welded joint
   (d)   end feed solder joint.

16 Gravity circulation in a hot water system depends on:
   (a)   conduction currents being set up
   (b)   the pressure due to the head of water in the cistern
   (c)   residents drawing off hot water at the taps
   (d)   the difference in density of the flow and return columns of water.

17 Copper or gunmetal fittings should be used on underground copper pipelines in preference to brass in order to:
   (a)   improve the earth connection
   (b)   make a tighter joint
   (c)   avoid dezincification
   (d)   avoid desalination.

18 Fractures near a soldered dot fixing on a lead covered dormer cheek are caused by:
   (a)   restriction of thermal movement
   (b)   electrolytic corrosion between the solder and the lead
   (c)   the brass screw being too tight
   (d)   moisture movement of the timber.

19 Plumbo-solvency is a characteristic of some types of:
   (a)   metal
   (b)   water
   (c)   pipework
   (d)   plastics.

20 Galvanizing of a sheet steel cistern is achieved by:
   (a)   spraying the cistern with zinc oxide powder
   (b)   dipping in molten tin
   (c)   painting with a zinc-based paint
   (d)   dipping the cistern in molten zinc.

4 Cold water systems

After reading this chapter you should be able to:

1 Understand the requirements of the relevant Water Supply Regulations.

2 Define and classify water and its sources of supply.

3 Describe the method of mains connection for each of the different materials used.

4 Recognize and name each type of cold water system and draw its component parts.

5 State the functional requirements and working principles of each system.

6 Describe the properties of materials used in the manufacture of cisterns.

7 List the factors relevant to siting a cistern.

8 Describe methods of jointing all materials used for conveying cold water.

9 Describe the features of valves.

Water supply regulations

All the materials i.e. pipes, fittings and appliances as well as the manner in which they are installed, are covered by the *Water Regulations*. Before any work is carried out the plumber must be conversant with all aspects of the work and the relevant regulations. The Regulations are framed by various government bodies and give the guidance about materials and the method of installation. These are generally accepted throughout the whole country. Reference should also be made to all relevant British Standards and European Standards BS/EN.

To summarize the main reasons for the implementation of the regulations are:

1 Ensure a wholesome supply of drinking water

2 Prevent contamination of the water supply

3 Prevent waste and undue consumption of water

4 Endeavour to obtain uniformity of design, installation testing and maintenance.

Contamination

Submerged supply pipes, faulty float-operated valves, uncovered cisterns, use of incorrect materials, non-use of check valves.

Waste

Dripping taps, faulty float-operated valves, leaking pipes and cisterns.

Uniformity

Use of approved and tested materials and fittings to British Standards Certification and/or European Community Directive.

Definition of water

Water is a chemical compound of two gases – hydrogen and oxygen. It is formed when the gas hydrogen or any substance containing hydrogen is burned. One of the most important properties of water is its solvent power. It dissolves numerous gases and solids to form solutions.

The purest natural water is rainwater collected in the open country. This contains small amounts of dissolved solids, mainly sodium chloride (common salt) dissolved from the air, and also dissolved gases: nitrogen, oxygen, carbon dioxide, nitric acid and ammonia.

Rainwater collected in towns contains higher percentages of dissolved substances, and also soot, etc. In particular it may contain acids – sulphuric and carbonic – which may cause damage to certain stones and metals, particularly limestone and zinc.

Spring water usually contains more dissolved solids, the amount and kind depending on the type of soil and rocks through which it has passed.

River water is water which has passed through and over the ground and can vary considerably in hardness and purity.

Classification

Water can be classified as soft or hard.

Soft water

A water is said to be soft when it lathers readily. It is not very palatable as a drinking water, and has a detrimental effect on most metals. It can cause rapid corrosion particularly of those metals which contain organic solutions.

Hard water

A water is said to be hard if it is difficult to obtain a lather. A hard water can be temporarily hard, permanently hard or both, depending upon the type or earth strata through which the water has passed i.e. sulphates and carbonates of lime.

Measurement of hardness

This is known as the pH of water and was known as degrees of hardness and classified according to Clarks scale in which 1 degree of hardness represents 1 grain of calcium per 4.5 litres of water, it is now expressed in parts per million (weight/volume) or in milligrams per litre (Table 4.1).

Table 4.1

Water	Parts per million or mg/litre	pH
Soft	0 to 50	0
Moderately soft	50 to 100	
Slightly hard	100 to 150	7
Moderately hard	150 to 200	
Hard	200 to 300	
Very hard	over 300	14

A water is considered neutral if it has a pH 7, above this figure it would be getting harder, below this figure it would be softer. The scale runs from 1 (strongly acidic) to 14 (strongly alkaline).

Sources of water supply

Rainfall

Rainfall is the source of all natural fresh water. When rain falls on the surface of the earth part immediately runs off to ditches or natural water courses (streams and rivers). The remainder soaks into the ground, some to remain underground for the whole of its journey until it mingles with salt water in the seas, some to break forth as springs, and some to be artificially extracted from wells (see Figure 4.1).

Whatever source of water is used – lakes, rivers, springs or wells – the available supply depends on the nature and size of the catchment area and the amount of rain that falls on it. The greatest supply is usually obtained where moisture-laden winds from the ocean pass over the mountain range which deflects them into higher and cooler regions where the moisture is condensed.

The average rainfall in England is about 0.875 m per annum, in Wales and Scotland 1.25 m, in Ireland 1.17 m and in the British Isles as a whole 1.15 m.

For purposes of water supply, rain which falls in autumn, winter and spring is the most important. A large proportion of the summer rainfall is lost by evaporation. Summer supplies from wells are not reliable because summer storms are so short and intense that more runs off the surface,

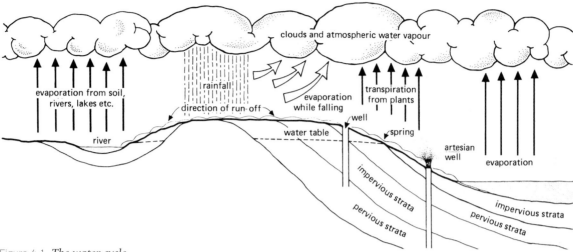

Figure 4.1 *The water cycle*

instead of collecting in the way that long steady winter rainfall does.

Sources of water supply may be grouped under two headings:

Surface sources
These include:

1 Run-off from natural catchments into natural or artificial lakes, reservoirs formed by damming valleys, and the run-off of small catchments into ponds (dew ponds).
2 Intake of water from streams and rivers.
3 Collection of rainwater directly from roofs or areas specially paved for the purpose.

Underground sources
These include:

1 Springs
2 Shallow wells
3 Artesian and deep wells.

Surface sources

Upland catchments
Water from upland catchments is usually excellent quality, being soft and free from sewage or animal contamination or suspended matter. It is liable to be acid in character if it contains water from peat areas, and corrosive to lead pipes (a condition known as plumbo-solvency). As lead is poisonous such water must not be conveyed in lead pipes in an untreated state.

Lakes
Supplies from natural mountain lakes are similar in nearly all respects to those from upland catchments with artificial storage. Pond water cannot usually be regarded as a fit source of domestic water supply.

Rivers and streams
The quality of river water varies. A moorland stream which is not unduly peaty will generally be wholesome. Downstream a river will usually collect drainage from manured fields, farmyards and roads and become progressively more contaminated. Most lowland rivers receive sewage from towns and industrial effluent from factories and have become heavily polluted.

Outside manufacturing districts most large rivers furnish water for the supply of one or more towns past which they flow, where the amount of impurity is not too great.

River water varies in hardness according to the proportion of water reaching it from springs and surface run-off. Water in the upland reaches of rivers is naturally soft, that in the lower reaches is generally hard.

Collection tanks
Rainwater can be collected from roofs of buildings into tanks, in order to be stored over dry periods.

Slate roofs are best for collecting rainwater: tile or galvanized iron roofs will also serve. Roofs covered with lead, copper or tarred material are unsuitable.

Rainwater is the softest natural water, but after collection on a roof it may contain leaves, insects and bird droppings. If required for domestic purposes it should always be filtered, and before drinking it should be subjected to a suitable purification process – boiling is sufficient where small quantities are involved.

Underground sources

Public water supplies from underground sources are usually drawn from water-bearing formations deep in the earth and covered by impervious strata. The water is often derived from gathering grounds several miles away, and the long journey underground provides a very thorough filtration.

Springs

Spring water is water that has travelled through the ground and come to the surface as a result of geological conditions. Its qualities are similar to those of a well in the same circumstances.

Spring water may vary considerably in quality. When it has travelled long distances through a stratum of rock it may be free from contamination, but hard. When a spring is fed by local rainfall organic pollution is possible, but the water may be soft.

Wells

Wells are classified as shallow or deep according to the water-bearing strata from which they derive their water. A well that is sunk into the first water-bearing strata is classified as shallow. A deep well is one sunk into the second water-bearing strata. Since the earth's strata vary in thickness, it is possible to have a shallow well deeper than a deep well.

Artesian wells

An artesian well or borehole is one which pierces an impervious stratum and enters a lower porous zone from which water rises as a 'gusher' above ground level. A similar borehole where water rises part of the way but does not reach the surface is called 'sub-artesian'.

Table 4.2 *River pollution commissioners' classification table*

Wholesome	1	Spring water	Very palatable
	2	Deep well water	Very palatable
	3	Uplands surface water	Moderately palatable
Suspicious	4	Stored rainwater	Moderately palatable
	5	Surface water from cultivated land	Moderately palatable
Dangerous	6	River water	Palatable
	7	Shallow well water	Palatable

Table 4.2 gives a summary of the quality of water from various sources.

Most of the waters we use come from rivers, the remainder comes from underground sources called 'aquifers'. The water is treated to make sure that it looks and tastes good and is safe to drink before piping it to consumers.

Water from aquifers is pumped up from under the ground and has already been filtered by nature as it soaks through the ground, and usually needs little treatment before it is safe to drink.

Water from rivers needs more treatment. First it passes through screens, these are large sieves that remove any floating matter, the water is then pumped into storage reservoirs where it is held for at least ten days. It is then aerated which circulates oxygen through it.

From the storage reservoirs the water is passed through sand filters which take out any remaining impurities in a process similar to that in nature, sometime mild chemicals are used to take out impurities. In some cases the water is passed through a layer of granular activated carbon which has thousands of tiny pores that take up traces of organic substances in the water. Ozone treatment is also used to improve the colour, smell and taste of the water.

Finally the water is disinfected with chlorine. This ensures the water stays fresh, clean and free from germs from the treatment works to the consumer's tap.

Connection to company main

Water authorities are able to run a water main from their reservoirs to supply wholesome water to most premises except those in very isolated areas. Water mains are constructed of PVC or cast iron and are laid to as convenient a place as possible. The method of making the connection to the main will differ according to the type of material and its thickness and strength. See Figures 4.2 and 4.3.

The connection is generally a screw-down gun-metal stop valve ferrule to which the communication pipe serving the house is connected (see Figure 4.4). Figure 4.5 shows an exploded view of a screw-down valved ferrule. The LA (local

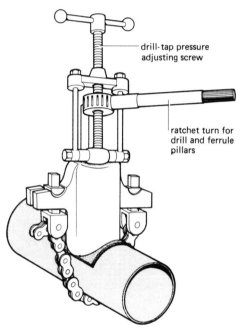

Figure 4.2 *Under-pressure tapping machine in place on main*

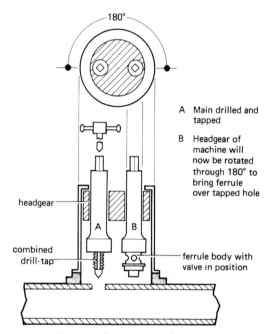

A Main drilled and tapped

B Headgear of machine will now be rotated through 180° to bring ferrule over tapped hole

Figure 4.3 *Diagram of tapping machine showing principle of design and operation*

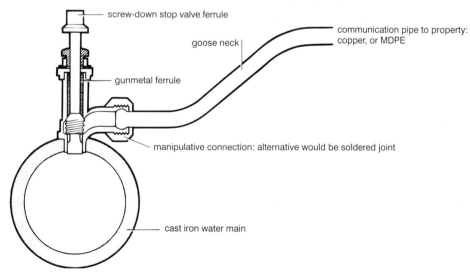

Figure 4.4
Connection to communication pipe

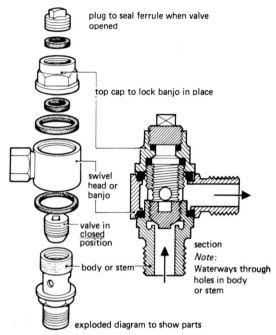

plug to seal ferrule when valve opened

top cap to lock banjo in place

swivel head or banjo

valve in closed position

body or stem

section

Note: Waterways through holes in body or stem

exploded diagram to show parts

Figure 4.5 *Exploded view of screw-down valved ferrule*

Materials

PVC and other plastics

Polyvinyl chloride is now becoming more widely used and in some instances has sufficient thickness and strength to make drilling and tapping a possibility.

Cast iron

Cast iron mains have been by far the most commonly encountered and time has proved them to be the most satisfactory. The only real problem is their susceptibility to deterioration in certain corrosive soils, but this can be overcome by taking protective measures. This material has now been superseded by plastics.

authority) mains are buried to great depths depending upon local conditions but they must have a minimum cover of 900 mm. The service pipe must have a minimum cover of 900 mm beneath roads and 750 mm elsewhere in order to prevent damage owing to settlement and/or frost.

Cold water supplies

At the boundary of the premises a screw-down stop-valve is provided. This control should be housed in a properly constructed box that is accessible from ground level. The box is usually covered with a hinged cover and frame.

It was common practice to use 150 mm 'seconds' glazed stoneware drain pipe these have now generally been superseded by a plastics pipe. The Water Authority provide control valves and meters if required just outside the boundary wall. See Figure 4.6.

As the service pipe continues and enters the premises, a combined stop-valve and drain-valve is fitted just above the finished floor level (see Figure 4.8). This permits the supply to be controlled easily from within the premises. It also provides for a complete draining down of the installation for alteration or repair work or when frost damage is likely during periods of absence, etc.

Types of system

Direct system

In districts where the mains supply is capable of delivering adequate quantities of water at good pressure, the water undertaking may permit a direct system of supply to all buildings.

All pipes to the cold draw-off points are taken directly from the rising main or service pipe and are subject to pressure from the main. There is therefore no risk of possible contamination that may occur when water is stored within the premises.

Figure 4.9 shows a suitable installation for the average dwelling, supplying a sink, WC, wash basin, bath and a cold feed cistern for the domestic hot water supplies.

A drain-off point should be provided on the service pipe where it enters the buildings immediately above the stop-valve and this is most conveniently catered for by a combined stop-valve and drain-valve.

Indirect system

In some areas the cold water supply is provided by use of the indirect system (see Figure 4.10).

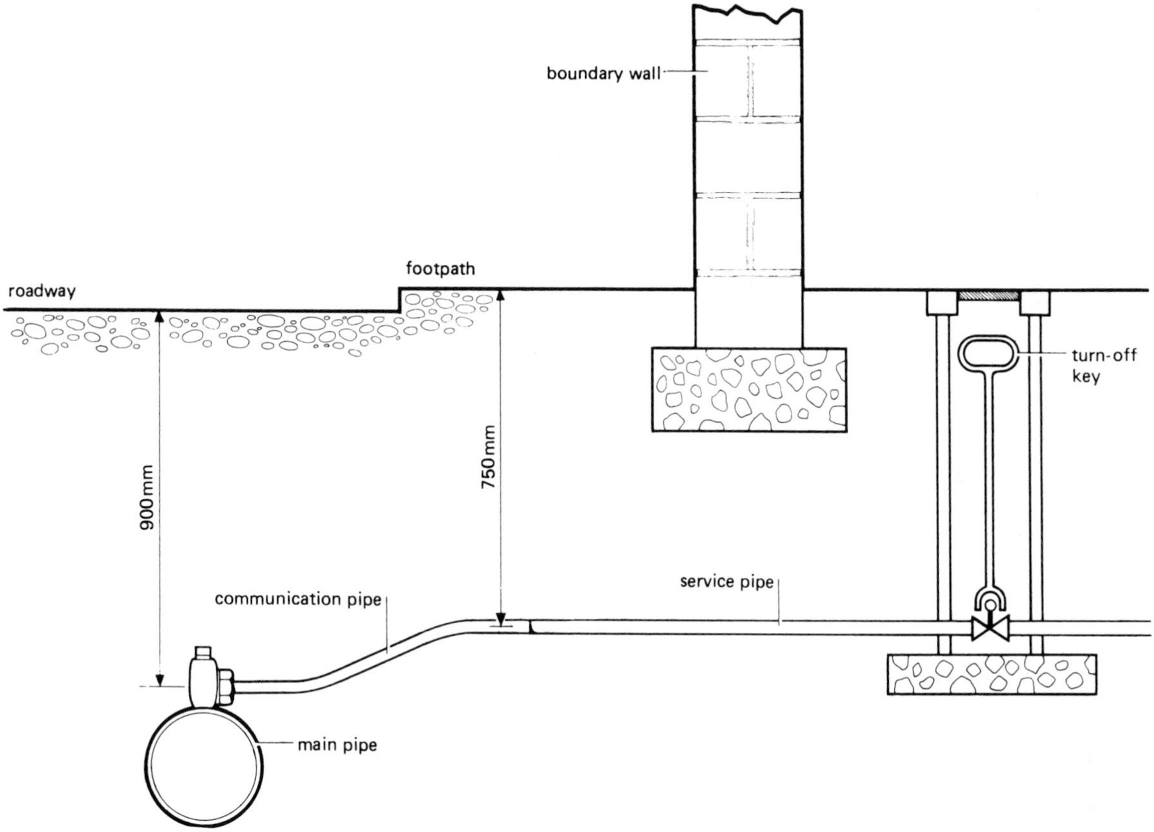

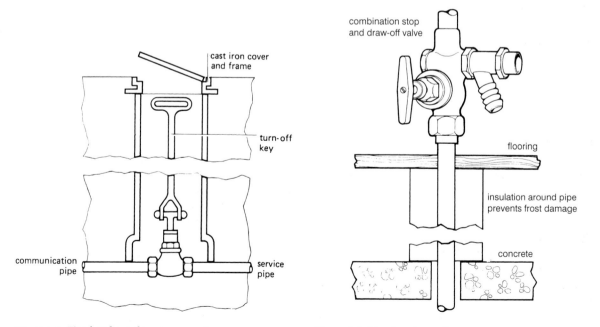

boundary wall

roadway

footpath

900 mm

750 mm

turn-off key

communication pipe

service pipe

main pipe

Figure 4.6

cast iron cover and frame

turn-off key

communication pipe

service pipe

Figure 4.7 *Fixed at boundary to property*

combination stop and draw-off valve

flooring

insulation around pipe prevents frost damage

concrete

Figure 4.8 *Fixed on immediate entry to property*

This means that the service pipe rises through the building to the cold water storage cistern and only one direct draw-off point for drinking purposes is permitted. The remaining cold water draw-off points are supplied from the storage cistern.

Storage capacity is normally based on a full twenty-four hour period, but some undertakings may only require storage for periods of six or twelve hours.

As in the direct system, a drain-off should be provided on the service pipe where it enters the building immediately above the stop tap.

Cisterns

A cistern is an open topped vessel designed to hold a supply of cold water, which will have a free surface subject only to the pressure of the atmosphere. It should be fixed as high as possible to give adequate pressure flow and fitted with a lid and to the requirements of regulations.

Storage cistern

This is designed to hold a reserve of water to supply cold water to the various appliances fitted to the system.

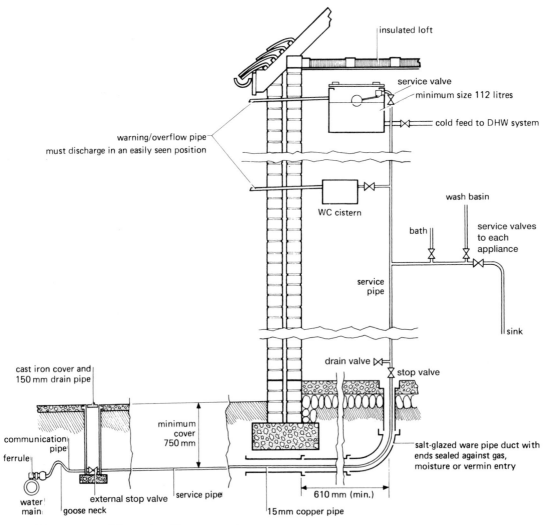

Figure 4.9 *Direct system of domestic cold water supply*

Feed cisterns

In districts where all cold water taps and sanitary appliances are supplied from the service pipe, it is not usual to provide a storage cistern as such although it is necessary to have a feed cistern to supply the domestic hot water system.

This may look exactly like a storage cistern, but it is meant only to hold a reserve of water for the hot water system, and is an essential part of a boiler-cylinder type of hot water system.

Combined storage and feed cistern

This should be large enough to combine both functions and will supply cold water both to appliances and the hot water system. This form of cistern is used where the water undertaking insist that only a drinking supply be taken direct off the company's main.

Capacities and connections

Minimum capacities for cisterns are prescribed by BS/EN Regulations.

Capacity is defined by the number of litres a cistern will hold, when filled to a level 25 mm, or the internal diameter of the overflow pipe (whichever is the greater) below the invert of the overflow pipe at its connection to the cistern. *Nominal capacity* indicates the number of litres a

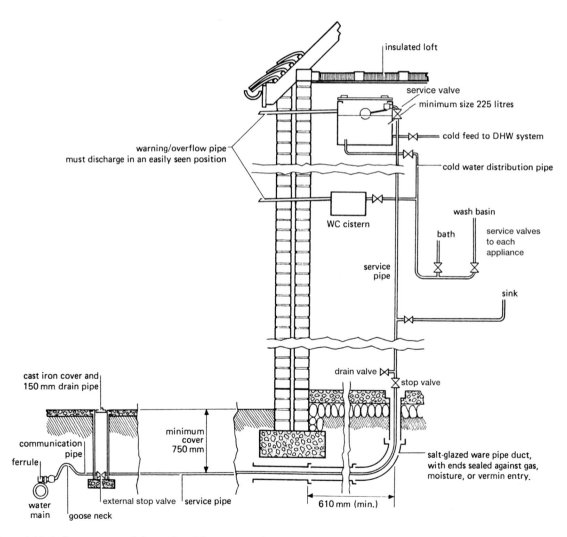

Figure 4.10 *Indirect system of domestic cold water supply*

cistern would hold if it were filled to its top edge.

Actual capacity means the number of litres a cistern would hold when filled to its working water line – that is, the level at which a properly adjusted float operated valve closes.

Regulation 16

Cisterns storing water for domestic purposes

1 Every storage cistern for water supplied for domestic purposes, shall:

(a) be installed in a place or position which will prevent the entry into that cistern of surface or ground water, foul water, or water which is otherwise unfit for human consumption; and

(b) comply with paragraph (2).

2 Every cistern of a kind mentioned in paragraph (1) shall:

(a) be insulated against heat and frost; and

(b) when it is made of a material which will, or is likely to, contaminate stored water, be lined or coated with an impermeable material designed to prevent such contamination.

(c) have a rigid, close fitting and securely fixed cover which:

  i   is not airtight,

  ii   excludes light and insects from the cistern,

  iii is made of a material or materials which do not shatter or fragment when broken and which will not contaminate any water which condenses on its underside,

  iv   in the case of a cistern storing more than 1000 litres of water, is constructed so that the cistern may be inspected and cleansed without having to be wholly uncovered, and

  v   is made to fit closely around any vent or expansion pipe installed to convey water into the cistern and

(d) be provided with warning and overflow pipes, as appropriate, which are so constructed and arranged as to exclude insects.

Figure 4.11 shows a cistern designed to meet the requirements.

Comment – Regulation 16

Where a cistern storing water for domestic purposes is being replaced, the replacement will be treated as a new and not as a replacement cistern.

Every water storage cistern from which cold water is drawn or may be drawn for domestic purposes must be insulated against heat and frost.

Cisterns must be fitted to the requirements of The Water Regulations:

(a) Sited well away from the roof and eaves (350 mm minimum).

(b) All to be suitably insulated.

(c) Cisterns made from rigid material supported on bearers.

(d) Plastics cisterns to be fully supported.

The adjusted float operated valve will shut off and water will cease to feed into the cistern. For cisterns of up to 455 litres capacity the working water level would be about 100 mm below the top edge.

Where a cistern is used only as a storage cistern, an appropriate minimum capacity could be 112 litres. Where used as a storage and feed cistern its minimum capacity could be 225 litres. If the storage is more than 4500 litres it is often advantageous to arrange it in a series of cisterns so interconnected that each cistern can be isolated for cleaning and inspection without interfering with flow of water to the fittings.

This can be done by the use of a header pipe of adequate size into which each cistern is connected and from which the distribution pipes are taken. Each branch into and out of the header pipe is provided with a control valve. Each cistern should have its oven float operated valve and overflow pipe, and a drain valve to facilitate cleaning out. In large storage cisterns, the outlet should be at the end opposite to the inlet to avoid stagnation of the water. If two or more cisterns are coupled together in series without header pipes, the inlet and outlet should be at opposite ends of the series.

Note: a service valve must be fitted in the pipeline before each float operated valve.

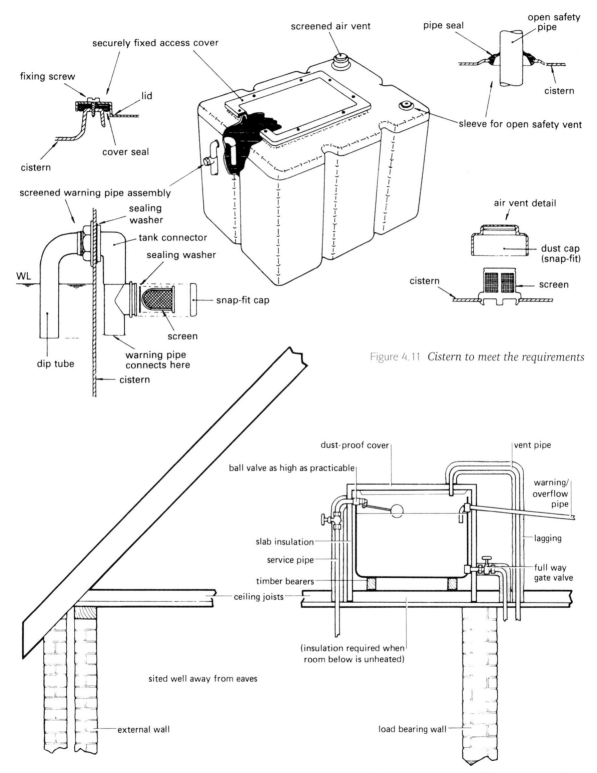

fixing screw

securely fixed access cover

screened air vent

pipe seal

open safety pipe

lid

cover seal

cistern

sleeve for open safety vent

cistern

screened warning pipe assembly

sealing washer

tank connector

sealing washer

WL

snap-fit cap

dip tube

screen

warning pipe connects here

cistern

air vent detail

cistern

dust cap (snap-fit)

screen

Figure 4.11 *Cistern to meet the requirements*

dust-proof cover

vent pipe

ball valve as high as practicable

warning/ overflow pipe

slab insulation

lagging

service pipe

timber bearers

ceiling joists

full way gate valve

(insulation required when room below is unheated)

sited well away from eaves

external wall

load bearing wall

Figure 4.12 *Siting a storage cistern*

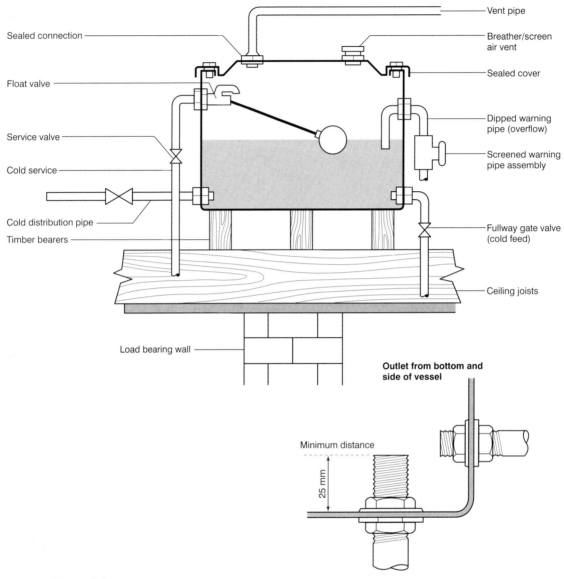

Figure 4.13 *Cistern fixing to meet requirements of water regulations*

Note: Bottom outlet not used on plastic cisterns

Storage cisterns

Storage cisterns and lids for domestic purposes should not impart taste, colour, odour or toxicity to the water, nor promote microbial growth.

BS 6700 follows the Model Water Byelaws in specifying that water cisterns for domestic purposes must be of the 'protected' type, on the grounds that water is likely to be drunk from all taps in the dwellings. This is a departure from past practice which should greatly improve cisterns hygienically.

All cisterns should be supported on a firm level base capable of withstanding the weight of the cistern when filled with water to the rim. Where cisterns are located in the roof space, the load should be spread over as many joists as possible.

Siting and fixing cisterns

Siting and fixing any cistern will normally be governed by the following factors:

Space available

The space available in a loft, roof space, or cistern housing must be adequate to accommodate a cistern.

Head

The cistern must be a certain height, or head, about the fittings, in order to provide sufficient water pressure.

Ease of access for maintenance

Periodic inspection, cleaning and float operated adjustment will be necessary.

Temperature

The cistern and its contents must not be subjected to extremes of temperature.

Structural tolerances

Water is heavy, 1 litre weighs 1 kg, 1 m^3 of water weighs 1000 kg.

Figure 4.13 shows a good siting of a storage cistern.

It has already been pointed out that the storage cistern, which is fed from the service pipe, must be placed in an elevated position in order to give sufficient pressure at the draw-off points to meet the demand rates. If, however, the storage cistern is placed on the same storey as the draw-offs, larger sized pipes must be installed to compensate for the lack of height.

The distribution pipe, conveying the water from cistern to the draw-off valve, etc., should be controlled by a control valve fitted close to the storage vessel. To minimize loss by friction, a full way gate valve is recommended.

In addition each appliance should be controlled by an independent control valve so each appliance is able to be isolated.

Connections for these distribution pipes should be located in the storage cistern in such a way that silt cannot be drawn into pipes. This means that the outlet should be taken from the side, and located at least 25 mm above the bottom of the cistern. In cases where the outlet is taken from the bottom of the vessel, a suitable connector, providing a 'stand up' above the bottom of the cistern of at least 25 mm, should be used. Figure 4.13 shows both methods.

All cisterns used for storing water should be provided with an over-flow pipe. This should be at least one size larger than the incoming supply pipe, and situated 25 mm below it (see Figure 4.13).

A dust-tight and air-tight cover should be placed over the cistern to prevent dust, animals and insects from gaining access and bringing about a risk of contamination

Commissioning and testing

The system should be flushed through to remove any foreign matter. It should then be subjected to a pressure test equal to 3 Bar for a period of 3 minutes or alternatively the test would be 1^1/$_2$ times the nominal working pressure of the area.

Float valves BS 1212

Every pipe supplying water connected to a storage cistern shall be fitted with an effective adjustable valve capable of shutting off the inflow of water at a suitable level below the overflow level of the cistern. These valves must conform to the BS parts 1, 2, 3 and 4.

Other types of valves such as delayed action valves or float switches controlling electrically operated valves may be used.

BS part 1 (copper alloy body, Portsmouth pattern)

If provision to adjust the water level does not exist then a double check valve must be fitted to the supply pipe (bending the lever arm is not permitted).

BS parts 2 and 3

These are both diaphragm types, part 2 being a copper alloy body, part 3 being a plastics body.

BS part 4

This valve is also a diaphragm type of a compact design and used mainly in WWPs (flushing cisterns).

Materials used in the manufacture of feed and storage cisterns

Galvanized steel

Cisterns manufactured in galvanized steel have been widely used for many years (see Figure

4.14). They are obtainable in many thicknesses and sizes. They are formed from black mild steel sheet, and then dipped into baths of acid to remove grease and scale. After this they are dipped into a bath of molten zinc and so coated with a corrosion resistant skin of zinc.

This protective hot-dip galvanizing treatment has been developed from experiments carried out by Galvani, a scientist after whom the process is named.

The cisterns are available in either riveted or welded construction. Their top edges are stiffened with an angled curb which is formed during manufacture and the open top corners are sometimes braced with corner plates which are riveted or welded depending upon the construction.

They are self-supporting and can be situated directly on timber bearers. The holes are cut by means of a tank cutter or expanding bit and the jointing done by means of couplings and nylon washers. It is good practice to ensure that no cuttings are left in the tank and that the inside is painted with an approved non-toxic paint prior to commissioning.

Plastics materials

Various plastics are now being extensively used for cold water storage cisterns: polyethylene, polypropylene and polyvinyl chloride (PVC), to mention just a few (see Figure 4.15). They are strong and resistant to corrosion of all types, virtually everlasting, very hygienic and light in weight. They do not cause, nor are they subject to, electrolytic corrosion, and they have low rates of thermal conductivity – so the stored water retains its heat longer in cold weather.

These cisterns are quieter in filling than a metal cistern, and this is a useful advantage, particularly when the cistern is sited near bedrooms. Furthermore, they are easy to squeeze through small openings, which is an advantage when the cistern is to be placed in an attic or loft (see Figure 4.16).

Plastic cisterns are manufactured square, rectangular, or circular in shape and are black to prevent algae growth. They must be fully supported by being placed on a solid decking (see Figure 4.17). Holes to enable pipe connections to be made are cut by circular saw cutters. The jointing is by means of plastic washers, but no oil-based paste of any description must be used, because this softens the material and causes it to break down.

Plastics materials are comparatively soft, so care must be taken in handling and fixing. Sharp

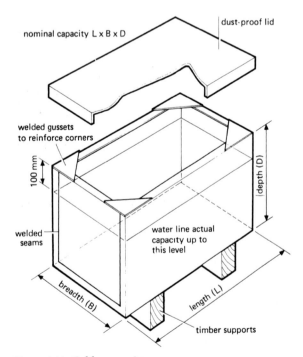

Figure 4.14 *Cold water cistern*

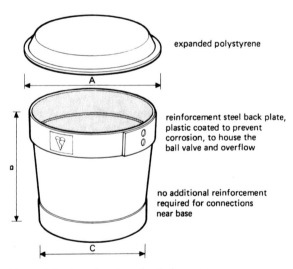

Figure 4.15 *Low density polyethylene cistern*

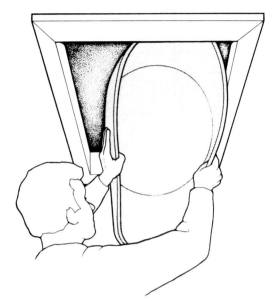

Figure 4.16 *Flexibility of material enables the cistern to be passed through small openings*

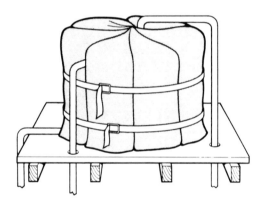

Figure 4.17 *Base must be fully supported*

instruments and tools can easily cut or puncture the cistern. Naked flames and excessive heat will also damage the material.

Joints for pipework

The following materials are used to make pipes to convey water in the domestic system.

Lead

This material is no longer used for new installations, and should be stripped out of houses wherever it is found.

Copper

This is by far the most popular metal used today for the manufacture of pipes for domestic use. It has many advantages:

1 It is neat in appearance
2 It is strong
3 It is easy to join
4 It is cheap to install.

Copper can be jointed with compression fittings, capillary fittings, silver soldering, brazing and bronze welding.

Plastics

These materials are now widely used in the industry. It is interesting to note that the methods of jointing are similar to those for copper tube – both manipulative and non-manipulative fittings are used. Fusion welding can also be used.

Galvanized low carbon steel tube

Steel pipes are not very popular for domestic work but are very extensively used for industrial work. They form an important part of the plumber's work.

Steel pipes are jointed by the threading (screwing) method and the use of purpose made fittings.

Stainless steel

This extremely strong and attractive looking metal has been able to claim only a very limited share of the market. The method of jointing is by means of compression and capillary fittings as described in the jointing of copper tube.

Jointing of copper tube

Copper tube can be jointed in several ways. The most common methods are:

1 Compression fittings
2 Capillary fittings
3 Silver solder
4 Brazing
5 Bronze welding
6 Copper welding
7 Push-fit fittings

Compression joints

These are divided into two groups:

1 Manipulative fittings
2 Non-manipulative fittings.

Manipulative fittings

These require the end of the tube to be cut square and to length. The nut is then slipped over the tube end and the tube opened to allow for a brass olive ring to be inserted or a ridge formed with a rolling tool. The nut is then tightened and the copper tube end trapped and squeezed between fitting or olive and nut. This manipulation of the tube end ensures the nut will not pull off. Many water authorities insist on this type of fitting being used on underground services. This type of joint is unaffected by vibration and withstands tensile and other stresses.

There are several types of manipulative fittings on the market. Their common factor is that some form of work (manipulation) is performed on the end of the tube before it is assembled with the fitting. The following examples show two types of these fittings.

1 In this case a special tool is inserted in the end of the tube which by means of a small ball bearing, forms a bead on the tube (see Figures 4.18 and 4.19).

2 In this case the end of the tube is flared out so that it can be compressed between the inner cone of the fitting and the shaped compensating ring.

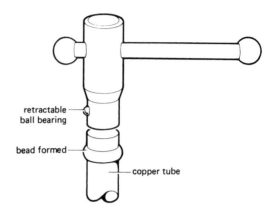

Figure 4.19 *Copper tube with special tool*

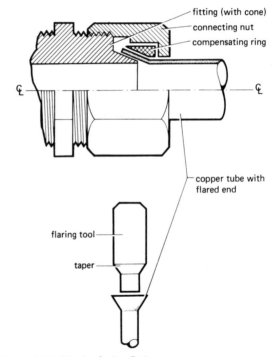

Figure 4.20 *Manipulative fitting*

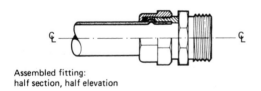

Assembled fitting:
half section, half elevation

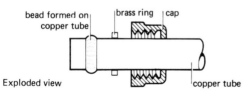

Figure 4.18 *Manipulative fitting*

Method of jointing with manipulative fitting

■ Ensure the tube is cut square and true.

■ Remove any burr from the pipe.

■ Place nut and compensating ring washer on to the pipe before any work is done in the preparation of the flared end of the tube.

■ Using the appropriate tool, i.e. flaring tool, form the end of the tube as required.

- Insert tube and assemble fitting. Slide ring or washer and nut into place.
- Tighten nut by hand, ensure that the nut runs freely and is not cross-threaded. Then with the aid of a correctly fitting spanner tighten nut half to one revolution.

Note: Do not over-tighten compression joints as this can cause distortion and subsequent leaks.

Non-manipulative fittings

These require only the end of the tube to be cut square and to length. The nut is then slipped over the tube end followed by a soft copper wedge shaped ring or brass wedding shaped ring. The tube is then inserted into the fitting socket, the nut tightened and the ring compressed between the inside of the fitting and the outside of the copper pipe (see Figure 4.21).

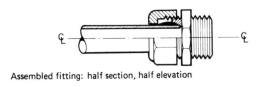

Assembled fitting: half section, half elevation

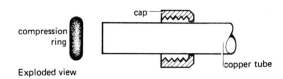

Figure 4.21 *Compression joint: non-manipulative*

This type of fitting is very popular due to its ease of application on both half hard and thin wall copper tube. They can also be used in certain cases on soft copper tubes. It is important to remember that this type of fitting should never be used on copper tube buried in the ground.

Method of jointing with non-manipulative fittings

- Cut the ends of the tube square (by means of a fine-toothed hacksaw).
- Remove any burr with a fine file.
- Place nut and compression ring on pipe and insert pipe in fitting up to the stop (see Figure 4.22).

- Tighten nut hand-tight. Ensure good fit and then with a correctly fitting spanner tighten half to one revolution.

Note: Do not over-tighten; experience will tell you the correct amount, as a guide hand tight plus a $1/4$–$1/2$ turn with a spanner.

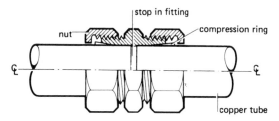

Figure 4.22 *Compression joint: non-manipulative*

Capillary joints

Capillary type fittings are also divided into two groups:

1 Fittings with integral lead free solder rings
2 End feed fittings.

Integral solder rings

This type of fitting can be used for both above and below ground work. It can be used extensively in construction, gas, refrigeration, marine and engineering pipelines conveying gas, air, water and oil. It is also without doubt an extremely attractive looking fitting and is also simple to make. The above points make this type of fitting the most popular, and the best known of all capillary fittings. Figure 4.23 shows an example of this type of fitting. The joint relies on the phenomenon of capillary attraction in the making of the joint.

Each fitting contains the correct amount of solder as an integral part. When heat is applied, the solder turns from a solid to a liquid and is then drawn by capillarity around the whole of the joint in the tight space between the outside of the pipe and the inside of the fitting, regardless of position.

Method of jointing with integral solder ring

- Cut the ends of the tube square, and remove burrs.
- Thoroughly clean the outside of the tube and the inside of fitting with steel wool or sandpaper.

- Lightly coat both cleaned surfaces with a suitable flux.
- Insert the tube into the fitting until it reaches the stop.
- Apply heat (torch) until the solder appears at the mouth of the fitting and forms a complete ring.
- Allow to cool without disturbance.
- Remove flux residue.

Owing to the fact that copper is an excellent conductor, the heat applied at one point is quickly transferred round the whole of the joint, particularly for the smaller size of pipe. It is still advisable to apply the heat wherever possible to all parts of the joint. In some cases where this is not possible, the use of a heat resistant mat around the back of the joint to reflect the heat is recommended. It is worth noting that *under-heating* results in more joint failures than over heating.

End feed fittings

This type of fitting is used in exactly the same circumstances as for the integral solder ring fittings and is identical in all ways except that the solder has to be added to the end of the fitting. The jointing surfaces inside the fitting and outside of the pipe are first cleaned, fluxed and assembled. Heat is applied until the temperature is high enough to melt the solder which is then drawn by capillary attraction into and around the whole joint. All surplus flux residue should be removed, otherwise it will continue to act on the surface of the copper and so leave staining and corrosion marks.

Method of jointing with end feed fittings
- Cut the ends of the tube square.
- Remove all burrs.
- Thoroughly clean outside of the tube and inside of the fitting with steel wool or sandpaper.
- Lightly coat both cleaned surfaces with a suitable flux.
- Assemble tube and fitting (tube to touch the stop).
- Apply heat until the temperature is high enough to melt the rod of fine solder.

- Add sufficient solder to fill the jointing space.
- Allow to cool without disturbance, wash off flux residue.

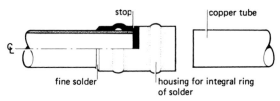

Figure 4.23 *Capillary joint*

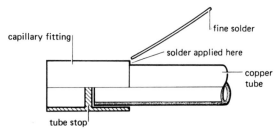

Figure 4.24 *Capillary soldered joint*

Push fit fittings

These are now widely used and are manufactured for both copper and plastics pipework systems.

Speedfit systems

Plastics pipes and fittings are becoming more popular, one of the fairly recent developments is the speedfit system which is suitable for:

1 Mains and indirect cold water systems
2 Vented and unvented hot water systems
3 Vented central heating systems
4 Sealed central heating systems when installed to comply with the relevant BS parts 1 and 3 class C.
5 Underfloor heating.

There is available a full range of fittings to cover most eventualities. The system is approved by BS/EN requirements, and should be installed to conform with good plumbing practice.

Note: This system must not be used for gas, oil or compressed air systems.

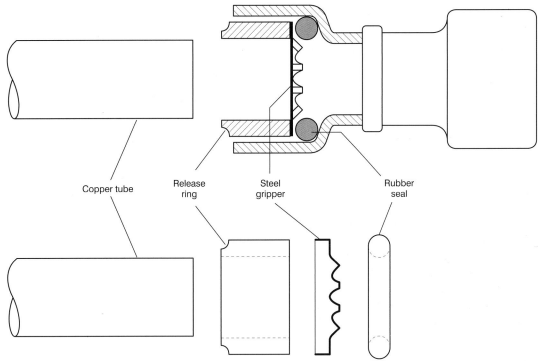

Figure 4.25 *The principle of the push fit fitting suitable for copper tube*

Table 4.3 *Working temperatures and pressures*

System	Temperature	Pressure
Cold water	20°C max	12 Bar
Hot water	65°C max	6 Bar
Heating	92°C max	3 Bar

Some of the advantages of this system are:
1 A permanent leak proof connection
2 No blowpipe work, easy to work
3 Corrosion free and scale built up
4 Lead free and non-toxic
5 Pipe elasticity reduces danger of frost damage
6 Less noise from flow of water and expansion/contraction.

Push-in fittings

This type of fitting has been in use for some considerable time and over this period they have been further improved.

This type of fitting illustrated has one great advantage in that it can be readily dismantled and the grab ring does not need replacing therefore the fittings are reusable.

Figures 4.26–4.29 illustrate the method of assembling, checking and dismantling push-in (speedfit) fittings.

1 Cut the pipe square along an insertion mark
2 Remove all burrs and/or sharp edges
3 Chamfer tube to assist entry
4 Push tube into fitting past 'O' ring until it reaches the stop

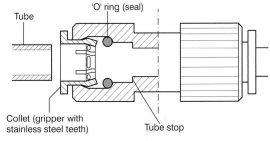

Figure 4.26 *Joint before assembly*

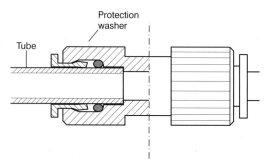

Figure 4.27 *Joint assembled*

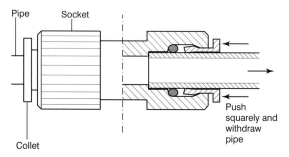

Figure 4.28 *Disconnecting joint*

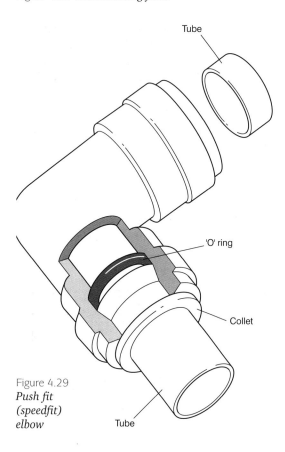

Figure 4.29
*Push fit
(speedfit)
elbow*

5 Pull pipe to check it is secure (impossible to pull out).

Method of disconnecting joint

1 Push in the collet (gripper) squarely.

2 Push inwards and at the same time withdraw the pipe (see Figure 4.28).

The tube is manufactured to BS/EN requirement and in sizes of 10 mm to 28 mm and in lengths of 2, 3 or 6 m, or in coils. The pipe can be bent cold, or alternatively, use elbows, i.e. by using the purpose made cold forming bend, a much tighter and neater bend can be made (Table 4.4).

As with all pipes proper adequate support is required but not closer than 60 mm from the end of the fitting (Table 4.5).

Pipes in roof spaces or ceiling clipped every 1 m.

Expansion is rated at 1 per cent of its length when the temperature is raised from 20°C to 82°C.

Example:
A pipe 3 m in length is raised from 20°C to 82°C.

$$3M = 3000 \text{ mm}$$
$$1\% = \frac{1}{100} \quad \text{increase}$$
$$\frac{3000}{100} = 30 \text{ mm}$$

Table 4.4 *Bending table*

Size of pipe	10	15	22	28
Min. radius with clips	100	175	225	300
Min. radius with cold forming bend	30	75	100	

Table 4.4 *Support table*

Pipe diameter	Clip spacing	
	Horizontal	Vertical
10–15 mm	300 mm	500 mm
22 mm	500 mm	800 mm
28 mm	800 mm	1000 mm

Jointing of polythene pipes

Compression fittings

These require only the end of the tube to be cut square. The nut is passed over the tube end followed by copper wedge ring or brass wedding shaped ring, into the open end of the tube a copper liner is inserted to strengthen the tube walls and then the tube end is inserted into the socket of the fitting. The nut is then tightened and the ring compressed into the polythene tube wall previously strengthened by the copper insert (see Figure 4.30). Liquid jointing compound must not be used.

Manipulative jointing

The ends of the tube are cut square. The nut is then slipped on to the pipe followed by a brass washer. The pipe end is then placed in a special tool which heats and forms a jointing flange which exactly marries up to the inside surface of the fitting. The newly formed flange is now inserted into the socket of the fitting. The brass ring and the nut are then secured into place, making a water-tight joint.

Fusion welding (solvent cement)

This term is used for the jointing of plastic by means of a special solution which has the property of melting the surface of the plastic. The tube and fitting must be thoroughly cleaned with spirit. The outside of the tube and the inside of the fitting is then quickly coated with the solution and assembled. The joint should be held together for a short time until the solution hardens. It takes approximately 24 hours for full maturity to take place. The two surfaces now form one homogeneous mass.

Jointing steel pipes

It is usual for low carbon steel pipes to be jointed by fittings such as elbows, bends, tees, couplings, flanges, etc. All these fittings are threaded with the British Standard Pipe Thread, thus making all fittings interchangeable irrespective of manufacturer. Steel tubing is usually supplied with a BSPT at each end plus a socket at one end (pipes without threads are also available). The tube is cut square and to the required length while held in a pipe vice. A thread is then cut on the end of the pipe with a tool called stocks and dies (see Figure 4.31). In one type there are four cutting dies housed in an adjustable stock head. This tool is set on the pipe end and rotated around the pipe. Long handles attached to the stocks head provide the necessary leverage. A lubricant such as oil or grease must be applied to prevent possible thread tearing.

Method of jointing steel pipes

- Cut the end of the tube square.
- Remove internal burr with the reamer, external burr with a file.
- Set stocks and dies to the correct pipe size (for large size pipes it may be necessary to make the thread in two or more cuttings, the required adjustments being made after each cut).
- Cut thread to correct length.
- Remove the stocks by releasing the dies with the release lever.
- Try a fitting to check that the thread is correctly cut.
- Wrap the threads with a special plastic tape.
- The fitting is now screwed on to the pipe, trapping the tape between the threads and so making a water-tight joint.

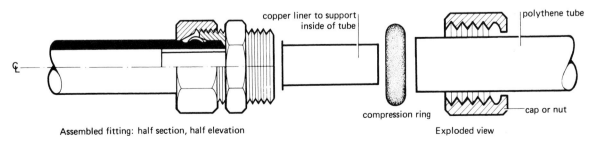

copper liner to support inside of tube

polythene tube

compression ring

cap or nut

Assembled fitting: half section, half elevation

Exploded view

Figure 4.30 *Non-manipulative compression fitting for polythene pipe*

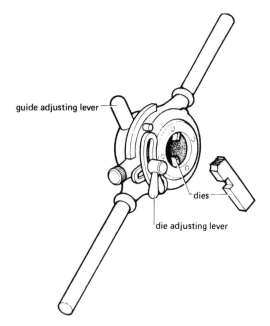

Figure 4.31 *Hand stocks and dies*

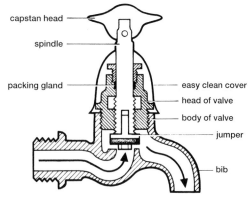

Figure 4.32 *Bib valve*

Note: Care must be taken not to over-tighten the joint. Fittings can become distorted and can even split if over-tightened, causing leakage.

Valves

The terms 'taps', 'valves' and 'cocks' are now classified as valves and are used to describe fittings whose purpose is to control the flow of water at draw-offs or outlet points or at intermediate positions in the system. Such fittings are usually made of brass or gunmetal with spindles or stems of phosphor bronze or manganese bronze. It is important to know and understand the differences in construction between the different types and to appreciate how these differences affect their use.

Screw down bib valve (Figure 4.32)

This is the best known of all water fittings and is in general use as a draw-off valve. Many types have a crutch head handle and an exposed body, but for use indoors a capstan head marked hot or cold, together with an easy-clean cover over the head of the valve gives a much better appearance. A loose jumper may be used if the mains pressure is sufficient to lift the jumper from the seating

when the spindle is unscrewed by the operation of the crutch head. In situations where mains pressure is not available as in cases of distribution supplies drawn from a storage cistern, the jumper must be fixed (pinned), i.e. secured to the spindle so that it is lifted off its seating as the valve is opened. Although the jumper is fixed in the worm (threaded spindle) it must be done in such a way to allow the jumper to rotate. This prevents the washer being worn away quickly by the turning action of the spindle. Washers are usually made of rubber, or nylon, which are suitable for both cold and hot water. There is a composition washer now available suitable for both hot and cold water. Figure 4.33 shows an exploded view of a bib valve.

Pillar valve (Figure 4.34)

This variation of the ordinary bib is fitted to baths, basins and sink units. Although it is fixed in the appliance, it is so designed that the nozzle outlet must always terminate above its flood level.

Disc valves

In outward appearance these valves appear the same as a normal valve. The difference is in the working parts of the valve instead of a jumper and washer there are simply two very highly polished discs, one disc is fixed and the other rotates. When the two holes coincide the water then passes through the valve and is stopped when the two holes are not in line. The advantage of these valves is that there is no maintenance or waste of water. Figure 4.35 shows an exploded view of one of these types of valve.

Spring loaded valves

This type of valve might be considered a special variety and would not normally be used on domestic work. Its use would be recommended for places where a large number of people are using the facilities, such as factories or public places. They might also find a use in hotels, hospitals and homes for children and mentally handicapped, where quite often valves are left running. This results in tremendous waste of both water and in the case of hot water, fuel also.

The valves are operated by exerting a downward pressure on the head of the valve which in turn depresses a spring holding the main valve on its seating. Water now flows through the valve, and will continue as long as this pressure is maintained. As soon as this pressure is released, the spring loaded valve is returned to its seating and the flow of water ceases. This type of valve can cause Water Hammer in areas of high pressure unless they are of the non-concussive type. They are also more prone to problems and more expensive than ordinary bib or pillar valves.

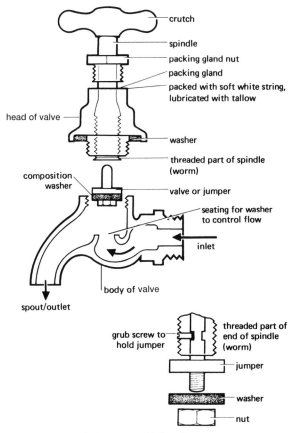

Figure 4.33 *Exploded view of bib valve*

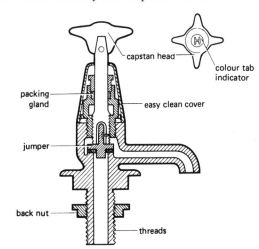

Figure 4.34 *Pillar valve*

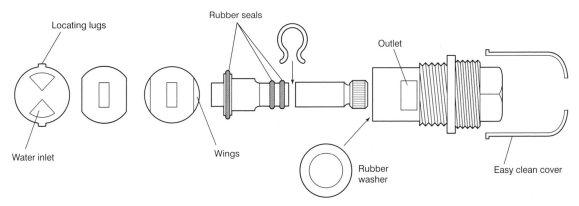

Figure 4.35 *Exploded view of disc valve*

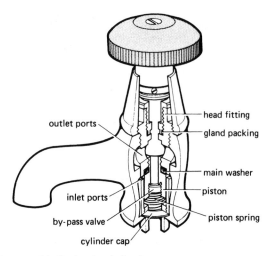

Figure 4.36 *Spring loaded valve*

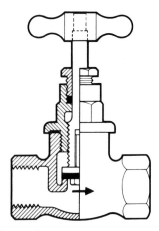

Figure 4.37 *Stop valve*

Stop valves (Figure 4.37)

The mechanism of this type of valve is the same as for that of the bib valve, but has a different body shape to suit its particular function which is to control the flow of water in a section of pipework. These valves have inlets and outlets to suit connection to steel, copper and plastics. The word inlet, or alternatively a direction arrow, must be stamped on the body of the valve. This is most important, for should the valve be fixed the wrong way round no water will be able to pass through. In all cases a stop valve must be provided just inside the boundary of the premises in a convenient and accessible position, usually in the drive or path and invariably the service pipe follows near to the line of drains. Because the stop valve will be at least 760 mm below the ground level, a loose key is advisable to facilitate its operation. Stop valves are recommended inside the property, the number being governed by the size and complexity of the building.

Stop and draw off valves (Figure 4.38)

Some stop valves have incorporated in the design an additional small valve to enable the householder to drain the system of water. This combination of stop and draw off should be fitted at the entry of the service pipe into the dwelling and is particularly useful in frosty conditions or when the occupants are going away. The drain valve is usually in the form shown with a hose pipe connection.

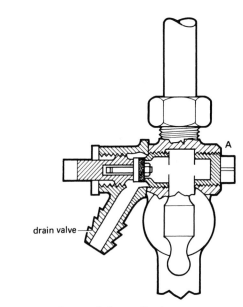

Figure 4.38 *Stop and draw off valve*

Mixer valves

Many combination hot and cold mixer sets are available for all sanitary appliances, in which the hot and cold supplies are separately controlled. The outlet appears to be just one pipe but in fact it contains two separate pipes. Figures 4.39, 4.40 and 4.41 show three different types of mixer valve, while Figure 4.42 shows an internal view.

Drain valve (Figure 4.43)

These draw off valves are generally fitted to drain off the water from boilers, cylinders and even sections of pipework. They are available with either

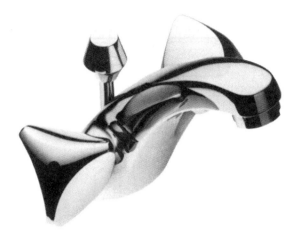

Figure 4.39 *Basin mixer*

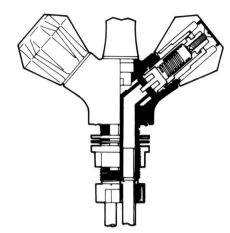

Figure 4.42 *Alternative type of mixer: sectional view*

Figure 4.40 *Sink mixer*

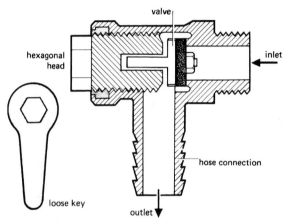

Figure 4.43 *Drain valve*

Figure 4.41 *Pillar bath mixer*

loose jumpers and/or fixed jumpers according to the requirement. They are operated by a purpose made loose key which fits on the hexagonal head.

The key should not be fixed to the valve permanently for general use, only sited near for use in emergency or for repair work.

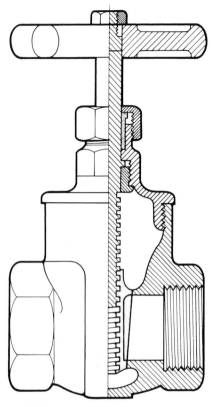

Figure 4.44 *Fullway gate valve*

Gate valves

These are known as fullway valves due to the fact that when they are fully opened there is no restriction to the flow of water through them. These types of valve are recommended where it is important that there should be the least possible restriction to the flow, as with low pressure distribution from storage cisterns. A gate valve is operated by a screw spindle generally of the non-rising type. In this, instead of rising out of the fitting when the valve is opened, it is rotated without rising and the wedge shaped gate climbs up the spindle screw thread.

Water regulations

Backflow prevention

The subject of the prevention of backflow into the water undertaker's mains has been a requirement of the Byelaws for some time, but because there

have been a significant enlargement on this subject in the Water Regulations, this subject matter may be considered new.

The Water Regulations has adopted an international method of categorizing fluids (water quality), which is based on its perceived risk hazard to health, these fluid categories are numerically listed from category 1, 'no risk' to category 5, 'serious risk'.

Fluid Category 1

Wholesome water supplied by the water undertaker and complying with that required by the Water Industry Act 1991. *No health hazard or impairment to quality.*

Fluid Category 2

Water in Category 1 whose quality is impaired by temperature or appearance. *Aesthetic quality impaired.*

Fluid Category 3

Fluid which represents a slight health hazard due to concentrations of substances of low toxicity. *Slight health hazard.*

Fluid Category 4

Fluid which represents a significant health hazard due to concentrations of toxic substances. *Significant health hazard.*

Fluid Category 5

Fluid representing a serious health hazard due to concentrations of pathogenic organisms, radio active or very toxic substances. *Serious health hazard.*

The Water Regulations require that for each given application, the fluid hazard risk category is identified and precautions taken to prevent backflow into the water undertaker's main by either an air gap, backflow prevention device, pipework arrangement, or combination of any of them. The choice of the devices and/or pipework arrangements used must comply with the Water Regulations.

The former Water Byelaws listed two categories of air gap, type A and Type B, these have both been retained but have been re-classified and joined by a further eight non-mechanical devices making a total of ten, in which all but type DC are forms of air gaps. Type DC is a pipe interrupter having a permanent atmospheric vent. The Water Regulations have chosen a form of nomenclature to classify the air gap types or mechanical backflow devices, whereby the first letter denotes the family type such as A for air gap, B for backflow preventer, C for disconnector, D for pipe interrupter, E for non-return type check valves, H for hose union connecter and L for pressurized air inlet valve. The second letter denotes the types within that family group.

From this it can be seen that a wash hand basin and bath that are classified as fluid category 3, may require an air gap protection that is greater than that required for a domestic kitchen sink which is classified as fluid category 5. This is explained in the guidance document by the fact that the sink type pillar valves and bi-flow mixers are of a high neck pattern that will permit a container such as a bucket to be filled under it, and therefore the air gap is likely to be around 150 mm. However, there is no requirement for high neck pattern pillar valves to be fixed to kitchen sinks.

Notes on Reduced Pressure Zone valves (RPZ)

It is said that the RPZ valve is relatively new to UK practice, but it has been available in the UK for many years and has had limited use in certain industrial applications.

The valve is one device or method that is permitted by the Water Regulations to safeguard against both backpressure and backsiphonage for a fluid category 4, *Significant health hazard*.

The Water Regulations requires them to be installed by *Approved Contractors* only, who in turn are not required to seek permission from the water undertakers before installing them, but are required to certify them both to the property owner and the water undertaker advising them of its existence together with the details of the RPZ valve and its installation.

Once the RPZ valve has been installed and commissioned, it must be inspected, tested and if required maintained, on a twelve monthly basis by an *Approved Contractor*, who is then required to certificate this valve for a period of another twelve months, and send copies to the property owner and water undertaker. This is one method

that the water undertakers have of keeping record of each RPZ valve installed legally in their supply area.

The RPZ valve is an automatic form of mechanical valve that incorporates a series of individual valves within its main body. Although the detail of the RPZ valves will vary between manufacturers, these valves comprise two spring assisted check valves, one located upstream, and one located downstream of a central pressure sensitive zone that incorporates a third spring assisted valve. This third valve incorporates a diaphragm that is operated by a change in pressure from equilibrium pressures, two unequal pressures thereby activating the three valve assembly to shut off the downstream side, and drain the pressure sensitive zone area, thus protecting the incoming water supply main.

The RPZ valve will also incorporate a series of test point bleed valves to facilitate the testing procedure, allowing unequal pressures to be simulated causing the activation of the valve assembly, and therefore checking the correct functioning of the valve.

The RPZ valve should be installed in accordance with the manufacturer's instructions and the requirements of the Water Regulations. The valve should be installed with an isolating valve both upstream and downstream of it to facilitate both maintenance and testing. The RPZ valve should also have been installed between the upstream isolating valve and the inlet connection of the RPZ valve, an inline strainer, having a mesh or micron rating in compliance to the manufacturer's recommendations. A combined test and drain valve should also be installed between the inlet connection of the RPZ valve and the upstream isolating valve to assist with maintenance and testing of the entire valve assembly.

As well as the ten types of non-mechanical backflow prevention devices, there are fourteen mechanical arrangements classified in the Water Regulations.

These all consist of some form of valve arrangement ranging from non-return type check valves, reduced pressure zone (RPZ) valves, vacuum breaker type valves, pipe interrupter valves, diverters and pressurized air inlet valves.

All of these valve types are designed to prevent reversal of flow of a fluid in a pipe, some incorporating an integral method of verifying the correct functioning of the valve.

The twenty-four mechanical and non-mechanical devices and methods listed in the Water Regulations are required to prevent backflow into the water undertaker's mains, with backflow being defined as being caused by backpressure or backsiphonage.

Backpressure may be described as a situation whereby the pressure upstream becomes lower than the pressure downstream, causing a reversal of flow to occur. Normally the mechanical family group of devices such as B, C, E and L are selected to prevent or check this occurrence happening.

Backsiphonage can be described as a situation where the downstream end of the pipe or device is subjected to atmospheric pressure that is greater than the upstream pressure which has become negative, this situation will create a siphon causing water or fluid to be siphoned back down the pipe. Typical backflow prevention devices or arrangements to check backsiphonage occurring are mainly those in family group A, (air gap) or family groups D and H.

Notes on air gaps

It should be noted that the air gap classifications listed in the Water Regulations, such as types AA, AB, AC, AD, AF, AG, AUK1 and AUK3 all require a minimum air gap of 20 mm, or twice the *internal* pipe diameter of the incoming water supply pipe, the only exception is a type AUK2 air gap.

AUK2 type air gaps differ in accordance to the Water Regulations as follows:

Size of incoming water supply pipe	
(pillar valve) (½)	20 mm
Exceeding (½) but not exceeding (¾)	25 mm
Exceeding (¾)	70 mm

It should be noted that domestic wash hand basins, baths, domestic clothes washing machines, domestic dishwashing machines and shower trays are classified as **Fluid Category** 3, which require a **AUK2** type air gap. Over the rim supply bidets are also classified as this fluid category.

Domestic kitchen sinks, WC pans, urinals and bidets with submerged inlets are all classified as **Fluid Category** 5, which require a type **AUK3** air gap.

The Water Regulations also requires all RPZ valves to be installed in a lockable cabinet to reduce the risk of unauthorized interference, protected from frost, not a hazard to electrical equipment and clearance in compliance with the regulations.

The discharge drain from the RPZ valve should be at least 300 mm above floor level and incorporate a type AA air gap with a tundish inlet to the discharge drain pipe.

Note no drinking water is permitted to be drawn off downstream of these valves, and caution should be exercised when installing RPZ valves on low-pressure applications, or high pressure boosted supplies. In these situations the pressures involved may upset the balance of the valve resulting in surge conditions occurring, or valve failure due to high-pressure drop of the flow.

Note any pipeline strainer may be required to be oversized so as to reduce the pressure drop through the basket of the strainer.

Notes on hose union bibvalve

The practice of installing hose union bibvalves that incorporate a double check valve, now classified as **HUK1**, is no longer permitted on new installations under the Water Regulations, but they are permitted to be installed as replacement valves to existing hose union bibvalves that do not incorporate any backflow device. New installations should be arranged to have a separate double check valve on the pipeline, prior to the hose union bibvalve, and located internal of the building to protect it from frost.

Bib (hose) valve

Regulations now require that all service pipes must be protected from backflow by the fixing of check valves as part of the fitting as shown in Figure 4.45 or as a separate device as illustrated in Figure 4.46 when fixing an outside hose valve.

Servicing valves

These valves comprise of a highly polished ball with a waterway through the centre. It is operated by turning the ball with a screwdriver or

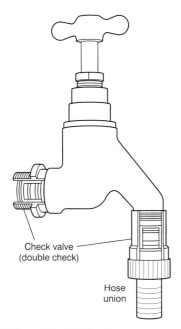

Figure 4.45 *Hose (check) bib valve*

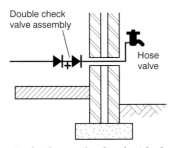

Figure 4.46 *Garden hose valve fitted with check valves*

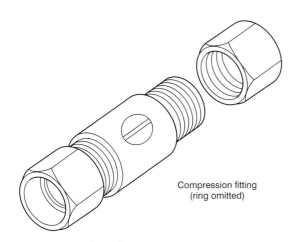

Figure 4.47 *Service valve*

even a coin. Only a quarter turn is required to fully close or open the orifice of the valve. The purpose of this type of valve is to isolate appliances in the hot or cold system so as to allow repairs or replacements to be carried out without affecting the rest of the system or having to run off to waste large volumes of water.

The valve is fully opened or closed by a quarter turn of the ball using a screwdriver or a coin.

Float valves

This valve is simply a control actuated by a lever arm and a float which closes off the water supply when a predetermined level of water has been reached. There are many different types and it is important to be able to recognize each to know when and where each should be fitted.

All valves must be manufactured from non-corrosive material. The two used today are brass and plastics. The brass valve is made from a hot pressing which is then drilled, threaded and dressed to the dimensions required. The nozzle seating is a replaceable nylon unit and the washer must be a good quality rubber 3 mm thick. The float which is attached to a 6 mm brass rod is generally made of plastics which has taken over from copper which for many years was the accepted material.

Float valves are classified as follows:

High pressure valves These must be capable of closing against pressures of *1380 kN/m^2*.

Medium pressure valves These must be capable of closing against pressures of *690 kN/m^2*.

Low pressure valves These must be capable of closing against pressures of *276 kN/m^2*.

The visual difference between the three types of valve is in the size of the orifice (the hole through the nozzle). For high pressure water supply the orifice must be small, for medium pressure the size of the orifice will be a little larger, while for low pressure the orifice is larger still, almost full way.

Croydon float valve (Figure 4.48)

This was widely used at one time but has been generally superseded by the others. The movement of the piston is in a vertical direction and tends to have a rather jerky and sluggish action.

Portsmouth-float valve BS 1212 (Figure 4.49)

This has the advantage of being easily dismantled or removed from the cistern in the case of malfunction, without having to undo the back-nuts to remove the body. The parts are renewable and readily accessible. This is particularly important with regard to the nylon seating which can be suitable for low, medium or high pressure, thus making this type of valve universal.

Nylon is a chemically inert plastics material and so has a high resistance to mechanical wear and is now extensively used in place of metal seatings.

The movement of the piston in this type of valve is in a horizontal direction and has a much smoother and better action.

Figure 4.50 is an exploded view showing all the component parts. It will be noted that the outlet is not threaded to receive a silencing pipe. It is now against regulations to fit these, as they may cause back syphonage. This type is being superseded by valves with a top outlet.

Diaphragm valve (Figure 4.51)

This type of valve is similar in some respects to the BS 1212 Portsmouth. It resembles the British Research Station's diaphragm valve in some features, but differs from it in that the design of the outlet completely overcomes the problem of back syphonage.

Although small holes in the outlets were at one time accepted as a prevention to back syphonage, this is no longer the case. Several methods have been tried, including a very thin plastic tube which collapsed when the supply was subjected to negative pressure. There are now several variations to the above with regard to the outlet. The one illustrated delivers the water through a number of fine streams, which is comparatively quiet and completely overcomes back syphonage. This type of valve is covered by BS 858.

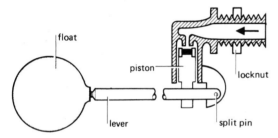

Figure 4.48 *Croydon pattern valve*

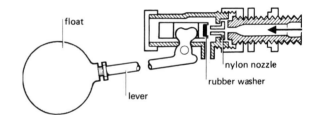

Figure 4.49 *BS 1212 Portsmouth valve*

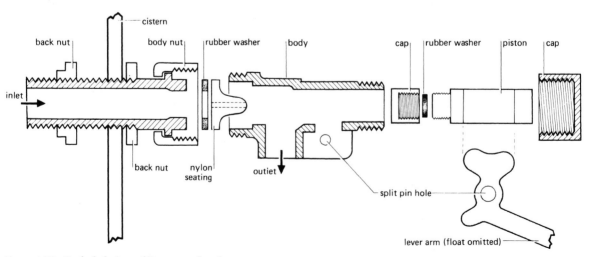

Figure 4.50 *Exploded view of Portsmouth valve*

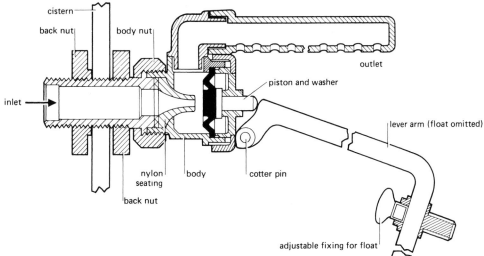

Figure 4.51 *Diaphragm valve*

Building Research Station diaphragm valve (Figure 4.52)

The Government's Building Research Station (BRS) design team set about designing a valve free from corrosion problems and from the nuisance caused by lime deposits, both of which interfere with the proper working of the valve. The advantages of this valve are:

1 No water touches the working part, thus overcomes the corrosion and lime problems.

2 The large inlet chamber breaks the speed with which the water enters and so reduces wear and noise.

3 A simple and convenient method of adjustment for the float.

Rewashering is a simple operation and is achieved by unscrewing the large nut which secures the ball valve head to the body (see exploded view in Figure 4.53). In this respect the BRS diaphragm valve resembles the BS 1212 valve.

The equilibrium float valve

The diagram illustrates the principle of the equilibrium float valve. This type of valve is fitted where the water pressure is very high which could result in the valve malfunctioning. It could also cause what is known as water hammer (a knocking noise in the pipe as the water recoils)

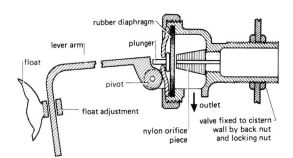

Figure 4.52 *British Research Station diaphragm ball valve*

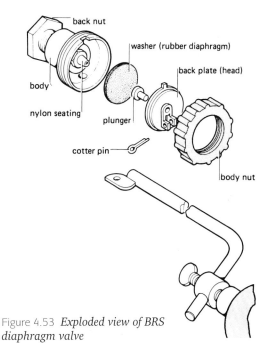

Figure 4.53 *Exploded view of BRS diaphragm valve*

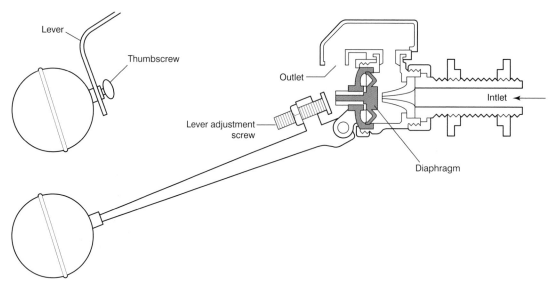

Figure 4.54 *Typical details of a float-operated valve*

apart from the noise factor damage could also be a result. It will be readily seen that there is a water-way right through the valve so allowing the water to act with equal pressure on each end of the valve hence the name equilibrium. In all other aspects it is very much the same.

The levers of float valves should be comparatively strong and fitted with some easily adjustable device for setting the water level, bending the float arm is not an accepted method and does not meet the requirements of the regulations.

Fluidmaster valve (no float)

This is a recently developed type of valve. This valve is of American manufacture and is approved by WRAS. It is manufactured from a plastics material. The joint therefore to the appliance must not be made by plumber's putty, only by means of the sealing washers supplied. As is the case with all plastics the connecting nuts must not be overtightened usually hand tight then approximately 1/4 turn with a spanner. It should be noted that chlorine (bleach) cleaner or related products

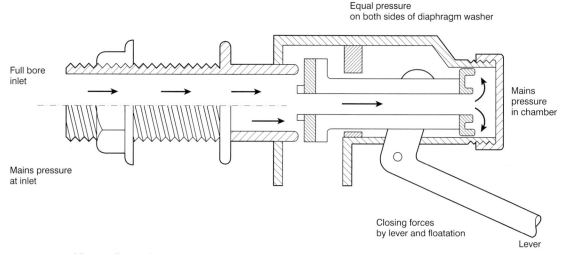

Figure 4.55 *Equilibrium float valve*

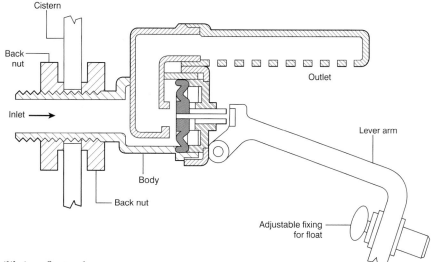

Figure 4.56 *Equilibrium float valve*

in the cistern may have a detrimental effect on the valve.

Before the cistern is put into operation always flush out any debris that may be in the system. This is illustrated in the diagrams. Should the water level require adjusting this is achieved by the simple adjustment of the clip attached to the float.

See Figure 4.58, A, B, C, D.
Clear debris from valve.

1 Turn off the water.

2 Unscrew valve top ⅛ turn anti-clockwise pressing down lightly on cap.

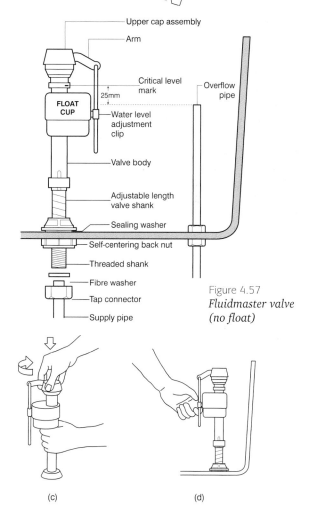

Figure 4.57
*Fluidmaster valve
(no float)*

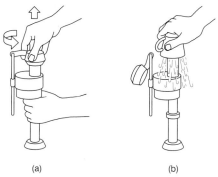

|(a)|(b)|(c)|(d)|

Figure 4.58 *Adjusting water level*

3 Hold a vessel over open valve and turn on the water momentarily to remove debris.

4 Turn off water again and replace valve top ensure the top is turned to the locked position.

5 Turn on the water supply.

Legionnaires' Disease

This is a type of Pneumonia which affects only a small number of people and only those who are susceptible to it.

The most at risk are those who are:

1 Aged people

2 Respiratory sufferers

3 People with underlying illness

4 Smokers

5 Males.

The bacteria which causes this condition cannot be passed from one person to another, it is passed on in the air as droplets of water which must be deeply inhaled into the lungs by the susceptible individual and in significant amounts. The bacteria which is from the *Legionellaceae* family can be found in small numbers in fresh water but probably do no harm. Once in the distribution system if the conditions are favourable, i.e. water temperature sediment, etc. the bacteria multiplies and can become very serious.

Prevention

1 Do not store the water below 20°C or above 60°C.

2 Avoid stagnation of water.

3 Pipe lengths to be as short as possible.

4 Cleanliness of the system is vitally important.

5 Avoid aerosol (spray) conditions.

Good practice

1 Minimize storage volumes.

2 Eliminate stagnation and short circuiting.

3 Use WRC approved fittings and material.

4 Use copper or plastics material.

5 Fit single cisterns.

6 Inspect, regularly, clean as necessary.

When dealing with Health Care Premises the following should be done *annually*.

1 Drained

2 Inspected

3 Cleaned

4 Repainted as necessary (metal cistern)

5 Disinfect with 50 mg/litre chlorine for one hour.

Note

Further information on this subject can be obtained from HSE books Tel. 01787 881165.

In conclusion the code of practice places responsibility on employers and persons responsible for buildings.

Domestic systems

Due to the frequent change of the water in the domestic system there is very little risk of the Legionella multiplying and causing a problem.

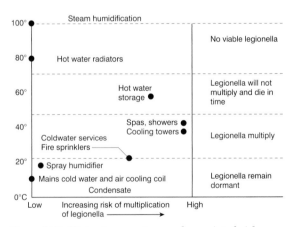

Figure 4.59 *Water temperature and associated risk*

Flushing and disinfecting of water systems

The water regulations state: Water fittings shall be flushed and where necessary disinfected prior to use.

The disinfection should always be carried out:

1 In all new hot and cold installations, appliances, and components, except single dwellings and other similar instances.

2 In all systems that have been modified or altered.

3 Below ground pipe work except for minor repairs or insertions of 'T' junctions (of which any material used should be immersed in disinfectant before insertion into pipeline).

4 On pipe work left for 30 days unused (whether new or existing).

5 On any pipe work installation if contamination is suspected.

The BS requires that all systems other than single private domestic dwellings will need to be disinfected before being put into use. The BS lays down two methods, using either chlorine or a product chosen from the 'Drinking Water Inspectorates List' and published in the **Water Fittings and Materials Directory**.

Disinfection should start at the water mains, then work through the system supply pipes, then the cisterns and the distribution pipe work.

Supply pipes, cisterns and distribution pipes

The cistern and the pipe work should be flushed twice, all the draw off points and service valves closed. The system refilled and chlorine added to get a sufficient concentration of 50 ppm. When the system is full, check at each draw off point that there is a strong smell of chlorine when the valve is opened. Top up the cistern. Leave for one hour then ascertain that there is 30 ppm concentration of chlorine at the draw off points, if below this level repeat the disinfection process. Drain and flush until at an acceptable drinking water level.

Single domestic dwellings

Single domestic dwellings should only need to be flushed, unless it is known to be contaminated in which case the system should be disinfected.

Measuring and checking flow rate

This is achieved with the aid of a simple device known as a *Halstead Combi-cup*.

Method of measuring flow

1 Ensure all other valves, washing machines, etc. are turned off.

2 Turn on the valve to be measured to maximum flow rate.

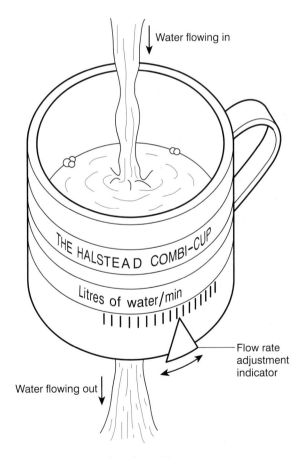

Figure 4.60 *The Halstead Combi-cup*

3 Place the Combi-cup below the water flow so that the cup is filling with water.

4 Adjust the indicator on the Combi-cup until the water level inside the cup is constant.

5 Read the *Flow-rate* from the indicator on the cup measured in litres/mm.

How to set the flow rate

1 As above.

2 As above.

3 Set the indicator on the Combi-cup to the desired rate on the scale.

4 Adjust the variable restrictor until the water level within the Combi-cup is constant.

Whilst every effort is made to ensure the accuracy of the Combi-cup it is inevitable that with such a simple device there will be some variance in the flow rate readings and as such it should be used only as a good guide.

Self-assessment questions

1 The minimum depth of cover required for a cold water service pipe is:
  (a)   600 mm
  (b)   750 mm
  (c)   900 mm
  (d)   975 mm.

2 The type of control valve fitted to a cold distribution pipe from a cold water cistern to allow a full-bore flow of water is:
  (a)   ball valve
  (b)   gate valve
  (c)   pillar valve
  (d)   safety valve.

3 Upon freezing water expands by approximately:
  (a)   2%
  (b)   10%
  (c)   15%
  (d)   20%

4 Water with a pH of 7 is classified as:
  (a)   Hard
  (b)   Acidic
  (c)   Alkaline
  (d)   Neutral.

5 Which type of ball valve has a large circular washer which also acts as a water seal to prevent corrosion of the working parts of the ball valve?
  (a)   Croydon
  (b)   diaphragm
  (c)   Portsmouth
  (d)   Equilibrium.

6 Which of the following are suitable materials for connecting a pipe to a polythene cold water storage cistern?
  (a)   lead washers and putty
  (b)   hemp and oil based jointing compound
  (c)   metal washers and jointing paste
  (d)   plastic washers.

7 To avoid damage to the ferrule in the water main due to settlement the communication pipe is:
  (a)   given a swan neck bend
  (b)   wrapped in greased tape
  (c)   surrounded in concrete
  (d)   carried through an earthenware drain.

8 Which one of the following types of joint relies upon the principle of capillarity to make an effective joint?
   (a) Screw thread joint
   (b) Type A manipulative compression joint
   (c) Solvent cement welded joint
   (d) End feed solder joint.

9 Corrosion to a galvanized cold water storage cistern before fixing can be minimized by:
   (a) applying two coats of non-toxic solution
   (b) lining it with copper
   (c) applying two coats of white lead paint
   (d) softening the water.

10 Underground copper service pipes are required by water authorities to be joined by means of:
   (a) bronze welding
   (b) wiped soldered joints
   (c) manipulative compression fittings
   (d) non-manipulative compression fittings.

11 A pipe supplying water under mains pressure to fitments is termed the:
   (a) communication pipe
   (b) cold feed
   (c) service pipe
   (d) distribution pipe.

12 The overflow pipe from a cistern is required to discharge:
   (a) into a gutter if possible
   (b) in a conspicuous position
   (c) over a drain
   (d) where it will cause least nuisance.

13 Potable waters are permanently hard if they contain:
   (a) magnesium bicarbonate
   (b) calcium sulphates
   (c) ferric salts
   (d) calcium bicarbonates.

14 Which one of the following groups gives essential requirements of an efficient cold water installation?
   (a) high pressure, full flow float valves, constant water temperature
   (b) low pressure float valves, restriction on flow at bath and basin, reduction of night pressure
   (c) adequate flow at taps, no contamination, even pressure
   (d) one days storage capacity, full bore draw-off filters on inlets.

15 At normal atmospheric pressure expansion of water will take place:
   (a) only when the temperature reaches 0°C
   (b) when there is a change of temperature from 4°C
   (c) only when the temperature goes outside the fixed points
   (d) when the temperature of the water increases from 0°C to 4°C.

16 Stagnation of water in large cold water storage cisterns is best avoided by arranging the inlet and outlet at:
   (a) opposite ends of the cistern
   (b) one end of the cistern
   (c) the bottom of the cistern
   (d) the top of the cistern.

17 A plastics cold water storage cistern installed in a roof space should be placed on:
   (a) two wooden bearers
   (b) two iron bearers across the joists
   (c) the ceiling joists
   (d) a flat boarded base.

18 The minimum internal diameter for an overflow pipe from a WC flushing cistern is:
   (a) 32 mm
   (b) 25 mm
   (c) 19 mm
   (d) 13 mm.

19 A common cause of frozen water in waste discharge pipes is:
- (a) sharp bends
- (b) dripping taps
- (c) excessive fall
- (d) use of shallow seal traps.

20 The main advantage of the diaphragm float valve is that:
- (a) it prevents back siphonage
- (b) the float is adjustable
- (c) the working parts are out of the water
- (d) it is easy to install.

5 Domestic hot water and heating systems

After reading this chapter you should be able to:

1 Identify and name types of domestic hot water and heating system.
2 Recognize different types of hot water storage vessel.
3 Demonstrate knowledge of types of boiler and their operation.
4 Identify and name the component parts of hot water and heating systems.
5 State the function of the control components of hot water and heating systems.
6 Understand the methods of heat transfer related to different forms of heat emitter and radiator.
7 Identify and name different types of thermal insulating material.
8 Describe the fitting and application of thermal insulating materials.
9 Understand and describe the basic methods of water heating by electricity.
10 Demonstrate knowledge of installation procedures related to hot water and heating systems.

Before we discuss domestic central heating systems, it may be advantageous to clarify what is meant by:

1 Direct hot water supply systems
2 Indirect hot water supply systems (conventional)
3 Indirect hot water supply systems (single-feed cylinder).

Direct hot water supply system

This system comprises three main components:

1 The feed cistern from which the cold water is supplied
2 The hot storage vessel, usually a copper cylinder
3 The boiler where the water is heated.

The three components are connected together by the appropriate pipework to form a simple and popular common system.

The direct system is defined as one in which the water that is heated in the boiler rises by convectional currents into the hot storage vessel to be stored until required. As that heated water is drawn off at the taps, the system is refilled by fresh raw water from the feed cistern.

Advantages:
1 Simple system
2 Inexpensive to install
3 Fairly quick recovery period.

Disadvantages:
1 Not suitable where domestic space heating is required from same boiler
2 Not suitable where the water is temporarily hard

3 Precautions to be taken where the water is very soft.

Domestic hot water and heating combined

Where a hot water system is combined with a heating system, several different metals are used to construct that system, with the result that corrosion takes place.

The different materials used may include:

1 Piping and fittings: copper, brass, gunmetal
2 Cylinder: copper, steel
3 Cisterns: galvanized steel, plastics
4 Boiler: cast iron, steel, copper
5 Radiators: steel, cast iron
6 Pump: cast iron, steel.

The use of dissimilar metals may cause corrosion by 'electrolytic action', and this action is quite common in most domestic systems. There may also be a problem of de-zincification.

The greatest problem is the formation of 'iron oxide'. This is the result of an attack on iron and steel components. A black oxide sludge forms, which can lead to:

1 Pump malfunction (seizure)
2 Partial blockage in radiators causing poor circulation and possible cold areas
3 Perforation of the mild steel sheet used to construct radiators or boilers
4 Excessive gas formation affecting the efficiency of radiators.

Each of the above-mentioned terms will be dealt with later. To keep the discoloured and contaminated water in the radiators separate from the clean water required for domestic purposes, some form of indirect hot water system must be used.

Indirect hot water supply (conventional)

In this system there are two cold water cisterns, a calorifier and a boiler, plus all the appropriate pipework. Essentially there are two separate systems in one (see Figure 5.1).

One cistern supplies cold water to the boiler and the calorifier. A calorifier is a hot storage vessel (cylinder) with another vessel (either cylinder or coil) inside. This part of the system is also connected to the space heaters (i.e. radiators).

The second cistern supplies water to the outer cylinder which in turn supplies the water to the hot discharge points (i.e. appliances). The water in the outer cylinder is heated *indirectly* by means of hot water circulating between the boiler and inner cylinder or coil. Therefore the water that is heated in the boiler is used only to heat *indirectly* the water that is drawn off at the appliances plus any space heating. Only a small amount of fresh water is added to the heating circuit to replenish that which may be boiled away, or lost by evaporation from the feed and expansion cistern. The indirect hot water system is used:

1 Where the water is of a temporarily hard nature
2 Where the heating medium is steam (not covered in this book)
3 Where both domestic hot water and heating systems are fed from the same boiler.

Indirect hot water supply (single-feed cylinder)

Mention has already been made of the fact that where domestic hot water and domestic space heating are required from the same boiler, some form of indirect system must be installed. Where a dwelling with an ordinary hot water supply system is to have a space heating system added, a saving in cost can be achieved by the adaptation of the existing system to incorporate a single-feed cylinder (see Figures 5.2 and 5.3). The single-feed system is exactly the same as the ordinary direct hot water system except that, in the place of the direct cylinder, a special (patented) self-venting cylinder is fitted.

There are several types of self-venting cylinders on the market: although they may differ slightly in appearance, basically they operate on the same principle. You may recall that in an indirect system there are two separate systems, and that the waters

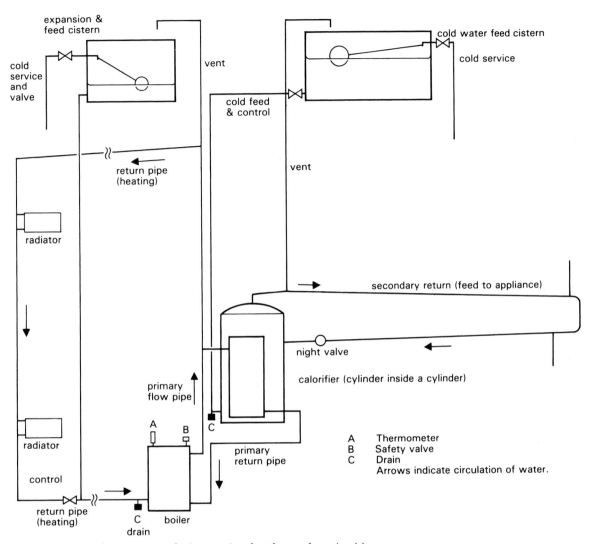

Figure 5.1 *Indirect hot water supply (conventional and secondary circuit)*

in the two systems must be kept separate. In the single-feed system, the waters are prevented from mixing by an airlock in the heat exchange unit, between the heating water in the primary circuit and the domestic supply to the appliances. Although the single-feed system operates as an indirect system it is cost effective because it does not require the additional feed and expansion cistern and appropriate pipework, i.e. the additional cold feed, vent pipes, cold supply and overflow.

Advantages:
1 Cost effective

2 Simple to install

3 Minimum of work needed to alter existing hot water system to include heating.

Figures 5.2 and 5.3 illustrate how, by means of a self-venting cylinder, an ordinary direct domestic hot water system (DHWS) can be adapted to incorporate a heating system, so converting it into an indirect system, a requirement when radiators are fed from the domestic hot water system boiler. There are many different types of self-venting cylinders available which work on the principle of forming an airlock in the

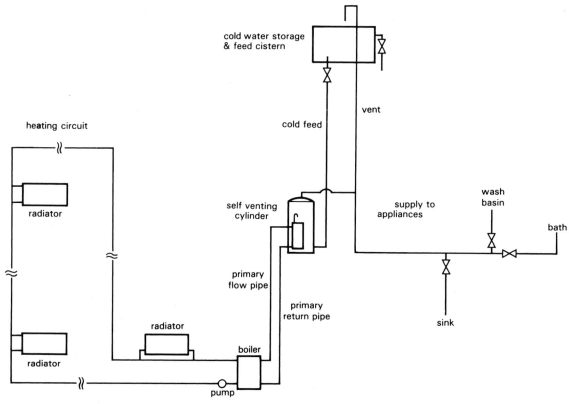

Figure 5.2 *Domestic hot water supply and heating system (indirect single feed cylinder)*

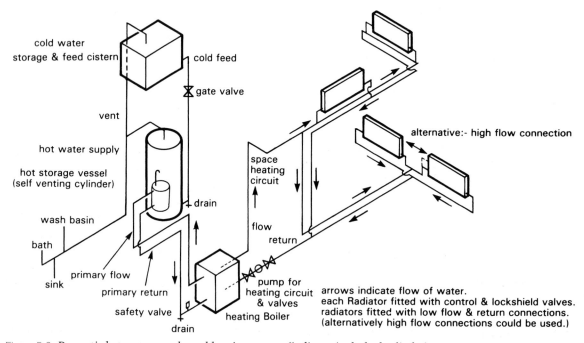

Figure 5.3 *Domestic hot water supply and heating system (indirect single feed cylinder)*

inner cylinder to prevent mixing of the waters. It is advisable to check the manufacturer's literature when selecting and installing these self-venting cylinders.

Self-venting cylinders

The primatic cylinder illustrated in Figure 5.4 is one of these self-venting cylinders.

Primary circuit

This consists of the flow and return pipes and heat exchange unit, and contains a small quantity of water which is not consumed but re-circulates.

Filling

(This should be done slowly.) Figure 5.5 illustrates the flow of water from the cistern. The water enters the primary circuit via a number of holes at the top of the vertical pipe immediately under the upper dome. As the primary circuit is filling, the system is self-venting by means of the air vent pipe which is incorporated. When the primary circuit is filled, the next stage is to fill the secondary supply, that which is drawn off at the appliances. When this is complete two air seals are automatically formed and permanently maintained, being self-charging during operation by air liberated from the heated water. When the water in the primary circuit is heated through the normal temperature range, the increase in volume of about 4% is accommodated in the unit by displacing air from the upper dome into the lower one.

Draining

The normal provision of fitting a drain valve to the boiler should be carried out. In addition to this, a further drain valve must be provided on the cylinder or cold feed to enable the whole system to drain.

Dead legs

Many dwellings have hot water systems in which some appliances are fitted a long way from the hot storage vessel. This results in long lengths of pipe, known as 'dead legs', being necessary to convey the water to the appliances, Figure 5.6. Regulations state that any pipe conveying hot water to an appliance must not exceed a set limit, see Table 5.1.

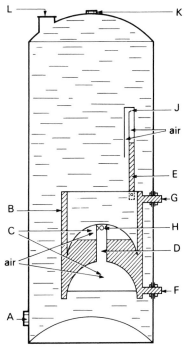

Figure 5.4 *The primatic self-ventilating indirect hot water cylinder*

Key
A   *Cold feed connection*
B   *Heat exchange*
C   *Air locks*
D   *Tube connecting air lock spaces*
E   *Vent pipe*
F   *Primary return pipe (to boiler)*
G   *Primary flow pipe (from boiler)*
H   *Holes to allow initial filling and later venting*
J   *Inverted air trap*
K   *Hot water to appliances*
L   *Immersion heater connection*

The reason for this regulation is to prevent the unnecessary waste of water which occurs when hot water is required at a distant point and a considerable amount of cooled water must be drawn and run to waste before hot water is delivered at the appliance. When the required quantity of hot water is obtained and the tap closed, the pipe remains full of hot water which, in time, loses its heat. The cycle repeats itself when hot water is again required.

From the householder's point of view this waste of water means a waste of money. Therefore the grouping of appliances, and short pipe

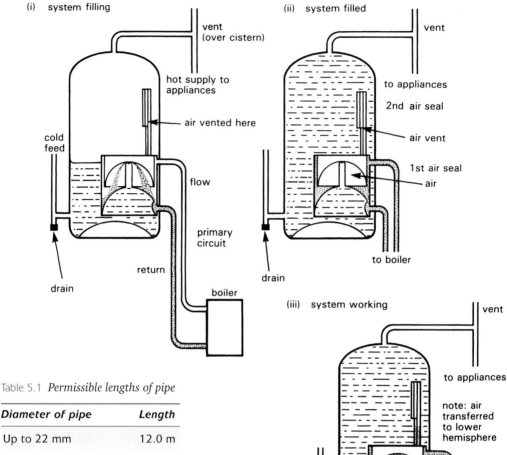

Table 5.1 *Permissible lengths of pipe*

Diameter of pipe	Length
Up to 22 mm	12.0 m
22 mm to 28 mm	8.0 m
Over 28 mm	3.0 m

Figure 5.5 *Filling of system*

runs, are factors of good design. Where such considerations are not possible, the problem can be overcome by the introduction of secondary circulation, see Figure 5.7.

Secondary circulation

Secondary circulation is the circulation of hot water from the top of the hot storage vessel or vent pipe to or near to the appliances, reconnecting into the hot storage vessel. This return connection should be in the top third of the vessel.

It must be appreciated, that water will constantly be circulating around this circuit so long as the water is hot. There will, therefore, be a cooling of the hot water and a subsequent and expensive loss of heat. This loss of heat can be curtailed at night time, when the water would not normally be required, by the fixing of a night valve. The night valve is simply a valve fitted in the circuit which can be manually or automatically operated to stop the circulation when demand is low or nil, as shown in Figures 5.7 and 5.8.

Although it is desirable that the secondary return should be re-connected to the hot water system in the top third of the hot storage vessel

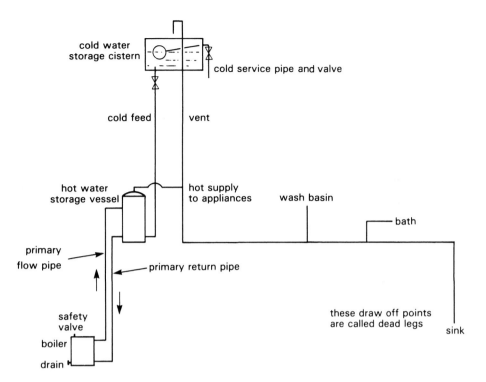

Figure 5.6 *Direct hot water supply with dead legs*

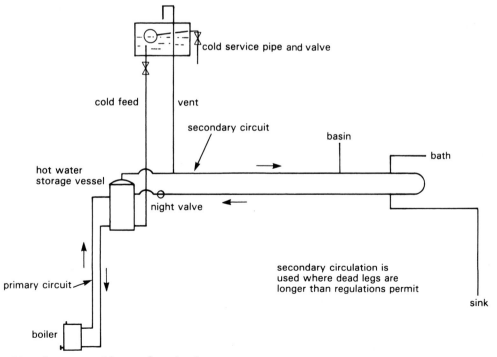

Figure 5.7 *Direct hot water with secondary circuit*

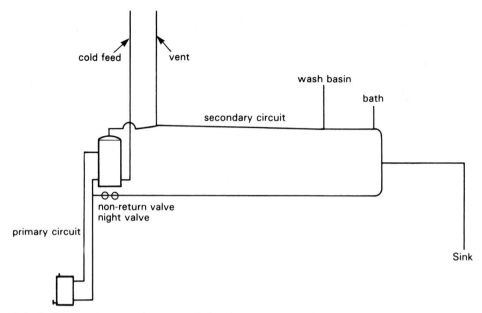

Figure 5.8 *Method of connecting secondary return below hot storage vessel*

(cylinder), there are some occasions in practice where this is not possible. Figure 5.8 clearly shows how this problem is solved by the fitting of a non-return valve in the secondary circuit at a point between the *last* draw off and the point where the secondary circuit rejoins the primary circuit.

Stratification

This is the term given to the formation of layers of water in a hot storage vessel from the hottest water at the top down through the temperature range to the coldest water at the bottom. To enable the system to function to maximum efficiency, this stratification must be encouraged and fostered. Important factors include the shape and size of the hot storage vessel. The best shape is cylindrical. The taller the vessel the better and it should be fitted in a vertical position: horizontally fitted vessels have poor stratification properties. The next important point is that the entry of the cold feed should always be in a horizontal direction, so that the incoming cold water does not disturb or destroy the existing stratification. In the normal cylindrical vessel the cold feed connection ideally is placed with a horizontal connection near the base. This is not the case with

cylinders fixed in a horizontal position, therefore some modification is necessary. The modification takes the form of a spreader tee fitted inside the vessel, which diverts the flow of water from a vertical to a horizontal direction. Figures 5.9 and 5.10 clearly indicate what is meant by stratification and how the design and connections play such a vital part in obtaining and retaining the supply of hot water.

Hot storage vessel

This is an important part of the hot water system. It must be:

1 Large enough to hold the maximum estimated quantity required

2 Designed and fitted so that only the hottest water is delivered to the taps (stratification)

3 Strong enough to withstand the internal pressures exerted by the pressure of water in the system.

Materials

The choice of material can be influenced by the type of water in your area, which may be either hard or soft. Should the water be temporarily hard

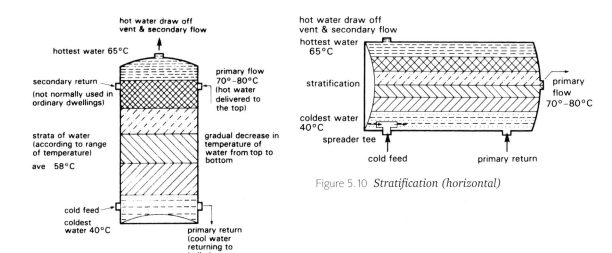

Figure 5.9 *Stratification (vertical)*

Figure 5.10 *Stratification (horizontal)*

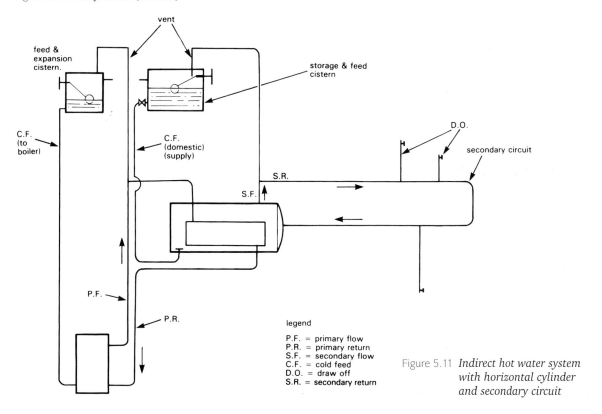

legend

P.F. = primary flow
P.R. = primary return
S.F. = secondary flow
C.F. = cold feed
D.O. = draw off
S.R. = secondary return

Figure 5.11 *Indirect hot water system with horizontal cylinder and secondary circuit*

then a lime deposit known as furring will take place when the water is heated above approximately 60°C. For this reason, copper and steel storage vessels and pipes are generally fitted. Adequate provisions should be made to facilitate the cleaning out of the component parts of the system, such as access covers (hand or manhole covers) and tees with cleaning access at change of direction. There would be no adverse affect if copper was used although it may suffer some damage from frequent cleaning due to not being as strong as low carbon steel. In districts where the water is of a soft nature it would be advisable to fit copper pipes and hot storage vessels.

Design

The size and shape of the vessel is extremely important. The size must be large enough to hold the estimated maximum requirement of water, plus an additional amount for extra demand. For ordinary domestic work the size of vessel is not calculated but is selected using the knowledge and experience of the installer.

When calculating the size of hot storage vessels, the following facts must be known:

1 Number of occupants
2 Number and type of appliances
3 Temperature of water required
4 Frequency of use.

Table 5.2 *Hot water requirements for domestic appliances*

Bath	70 litres
Shower	25 litres
Wash basin	9 litres
Bidet	4 litres
Sink (with washing-up bowl)	7 litres

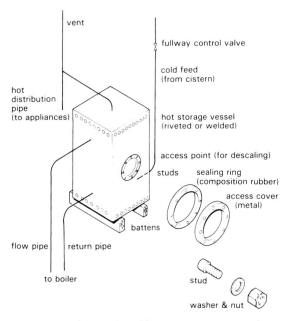

Figure 5.12 *Galvanized steel hot water storage vessel*

It will be appreciated that the amount of hot water required, and consequently the size of the storage vessel, can vary considerably from one household to another, even if each has the same number and type of appliances. Therefore only an approximation can be made. It is always good practice to err on the large side.

Experience has proved that the best shape of vessel is cylindrical. This gives great strength and enables the vessel to be manufactured from thin material, which leads to an economical product. An added advantage of the cylindrical vessel is that it aids stratification, as described earlier.

Cold water is heavier per unit volume than hot water therefore the cold feed supply pipe must always be connected at the base and the hot water distributing pipe must always be connected at the top of the hot storage vessel. The best location for the vessel is as near to the boiler as is practical, so that the primary circulation pipes are kept short and heat losses reduced to a minimum. These can be reduced still further by insulation.

Unvented hot water systems

The supply of hot water in the United Kingdom has for many years been of the boiler/cylinder/cistern type system as described and illustrated previously.

Changes in the model water bye-laws (1986) now permit the installation of unvented hot water systems. This type of system allows the hot water system to be connected directly to the main cold water service. The whole system is then under a controlled pressure, and a pressure reducing valve is fitted.

The advantages of this type of system over the traditional system are:

1 No cold water storage cistern, together with the relevant pipework, is required. This overcomes many problems regarding the use of storage cisterns, e.g. space, head, frost damage, etc. There is also a reduction of on-site work, with an inevitable saving of cost.

2 Because both hot and cold water supplies are at the same (equal) pressure, the use of mixer units – and in particular showers – is

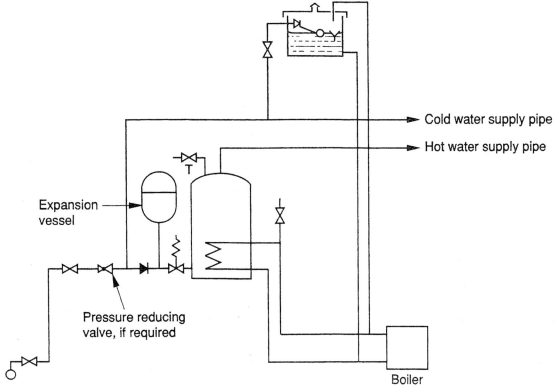

Figure 5.13 *Example of an indirect unvented (vented primary) system*

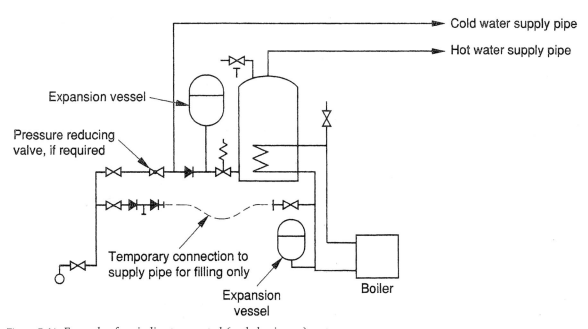

Figure 5.14 *Example of an indirect unvented (sealed primary) system*

improved. The increased pressure also gives a more invigorating shower.

3 Due to the pressure available, smaller diameter pipes can be used, also giving a saving in the cost of installation, water and fuel.

4 It is a much quieter system in operation than the traditional one.

5 Less risk of reduced flow with simultaneous demand.

6 Less risk to health because there is no stored cold water which could become contaminated.

7 Greater flexibility is gained in siting the unit.

The component parts and their functions

Figures 5.15 and 5.16 illustrate the component parts of a typical unvented system. It is important

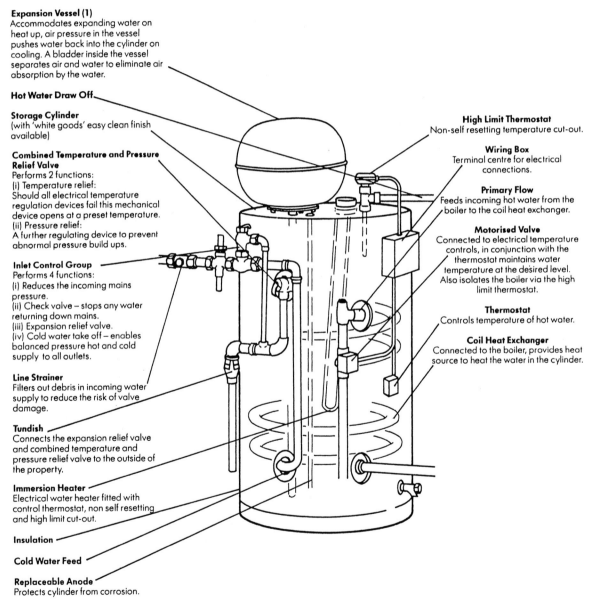

Expansion Vessel (1)
Accommodates expanding water on heat up, air pressure in the vessel pushes water back into the cylinder on cooling. A bladder inside the vessel separates air and water to eliminate air absorption by the water.

Hot Water Draw Off

Storage Cylinder
(with 'white goods' easy clean finish available)

Combined Temperature and Pressure Relief Valve
Performs 2 functions:
(i) Temperature relief:
Should all electrical temperature regulation devices fail this mechanical device opens at a preset temperature.
(ii) Pressure relief:
A further regulating device to prevent abnormal pressure build ups.

Inlet Control Group
Performs 4 functions:
(i) Reduces the incoming mains pressure.
(ii) Check valve – stops any water returning down mains.
(iii) Expansion relief valve.
(iv) Cold water take off – enables balanced pressure hot and cold supply to all outlets.

Line Strainer
Filters out debris in incoming water supply to reduce the risk of valve damage.

Tundish
Connects the expansion relief valve and combined temperature and pressure relief valve to the outside of the property.

Immersion Heater
Electrical water heater fitted with control thermostat, non self resetting and high limit cut-out.

Insulation

Cold Water Feed

Replaceable Anode
Protects cylinder from corrosion.

High Limit Thermostat
Non-self resetting temperature cut-out.

Wiring Box
Terminal centre for electrical connections.

Primary Flow
Feeds incoming hot water from the boiler to the coil heat exchanger.

Motorised Valve
Connected to electrical temperature controls, in conjunction with the thermostat maintains water temperature at the desired level. Also isolates the boiler via the high limit thermostat.

Thermostat
Controls temperature of hot water.

Coil Heat Exchanger
Connected to the boiler, provides heat source to heat the water in the cylinder.

Figure 5.15 *Indirect system*

to note that there are many unvented system designs available, the system illustrated is merely a typical system and does not claim to be the universal standard.

Hot water ADLI

The Approved Document L of the revised Building Regulations for England and Wales requires hot water to be provided consistent with certain basic rules relating to energy efficiency.

First the hot water cylinder must comply with BS 1566, BS 3198, BS 7206 or equivalent approval that confirms compliance with the regulations. What the regulations demand from the hot water cylinder is sufficient insulation and a large enough heat exchanger, or coil, which prevents frequent boiler cycling.

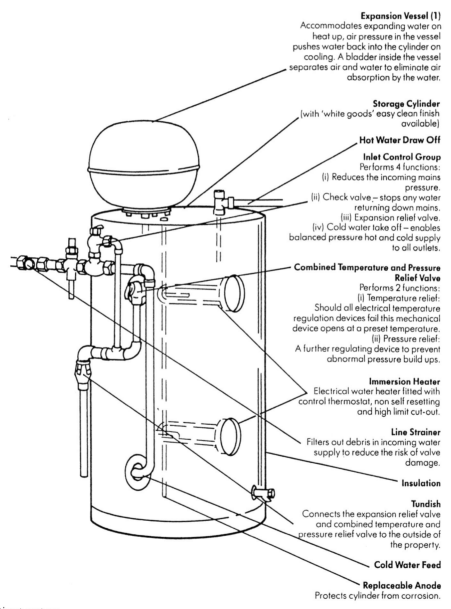

Expansion Vessel (1)
Accommodates expanding water on heat up, air pressure in the vessel pushes water back into the cylinder on cooling. A bladder inside the vessel separates air and water to eliminate air absorption by the water.

Storage Cylinder
(with 'white goods' easy clean finish available)

Hot Water Draw Off

Inlet Control Group
Performs 4 functions:
(i) Reduces the incoming mains pressure.
(ii) Check valve – stops any water returning down mains.
(iii) Expansion relief valve.
(iv) Cold water take off – enables balanced pressure hot and cold supply to all outlets.

Combined Temperature and Pressure Relief Valve
Performs 2 functions:
(i) Temperature relief:
Should all electrical temperature regulation devices fail this mechanical device opens at a preset temperature.
(ii) Pressure relief:
A further regulating device to prevent abnormal pressure build ups.

Immersion Heater
Electrical water heater fitted with control thermostat, non self resetting and high limit cut-out.

Line Strainer
Filters out debris in incoming water supply to reduce the risk of valve damage.

Insulation

Tundish
Connects the expansion relief valve and combined temperature and pressure relief valve to the outside of the property.

Cold Water Feed

Replaceable Anode
Protects cylinder from corrosion.

Figure 5.16 *Direct system*

The Water Manufacturers Association (WMA) has helped identification by marking all compliant vented cylinders with a large green Part L label, and also by providing a Benchmark label on which can be recorded the installation and commissioning details. Unvented cylinders and thermal stores produced by WMA members all complied with the new regulations before they were published.

The next element of compliance is connected with controls, and energy conservation. The regulations demand that the primary flow to the indirect cylinder is pumped in new installations, and that existing systems be upgraded where possible. The reason for this is obvious, increased flow rates reduce re-heat times. They also increase the temperature drop on the return to the boiler, which is vital if a condensing boiler is to operate at peak efficiency. It is possible that the SEDBUK rating of the boiler will be affected if the primary circuit is not pumped.

The hot water system should be a separate zone from domestic heating. Again this is obvious, as hot water is needed all year round, whereas heating is only needed at certain times. Choosing whether this should be a two, or more zone system – S plan – or flow share or hot water priority will depend upon boiler size and existing pipework, if there is a replacement. The cylinder must also have a thermostat to prevent excessively hot water.

Directly heated electric storage systems are not dealt with specifically in the regulations. However, some basic thought at the time of selecting the cylinder will help minimize energy costs. Make sure that the store is large enough. Low tariff electricity is cheap and spreads the power generation day. If you are going to rely upon this as an energy source, remember that the hot water generated at night has to last until the following evening, unless you are going to use more expensive daytime electricity.

Immersion heaters should include a safety cut-out independent of the main thermostatic control. This cut-out must break the electrical supply to the immersion heater to prevent the water in the storage vessel from exceeding 98°C. It must not reset automatically.

This brings immersion heaters for vented systems into line with the existing requirements for unvented water heaters and other domestic appliances. It is intended to ensure that the equipment is protected against overheating, even if the main thermostatic control fails in the ON condition. From 1 April 2004, manufacturers ceased production of immersion heaters to the old British Standard BS 3456. Instead all immersion heaters must comply with the new standard **BS EN 60335–2–73: Fixed Immersion Heaters**.

The requirement for an independent safety cut-out can be satisfied in two ways:

■ A one-shot device, which necessitates replacement of the thermostat or thermal cut-out if the latter operates, or

■ A manually re-settable cut-out that will not reset automatically but can be reset manually.

Installers, distributors and specifiers have a special responsibility to advise their customers of the new requirement and to begin changing over to the new devices. Failed immersion heaters should be replaced by units that comply with BS EN 60335–2–73 and all new vented cylinders should be fitted with immersion heaters to the new standard.

Boilers

Most homes today are fitted with domestic hot water systems, and many of them also have either full or part space heating systems. The householder of today has a wide choice of heating boiler and heating fuel. The type of boiler chosen will be governed by personal preference, whether free standing or wall mounted, cost, and availability of a particular fuel.

The choice of fuel would be between:

1 solid fuel

2 gas

3 oil

4 electricity.

System choice

Open or sealed systems

The most common type of system used with a regular boiler is the open-vented system with an indirect hot-water cylinder. It is termed 'open vented' because it includes a separate vent pipe, which is open to atmosphere. It also includes a feed and expansion cistern, which will allow for changes in the system water volume resulting from fluctuations in water temperature. The cistern must be at the highest point in the system, usually in the loft space where it must be protected from freezing.

An increasingly popular arrangement is the 'sealed' system, in which the expansion cistern is replaced by an expansion vessel that incorporates a diaphragm to accommodate the changes in water volume. The system is not open to atmosphere and the pressure within the system increases as the temperature rises. As the system is not open to atmosphere there is little possibility of oxygen being absorbed into the water, and therefore reduced risk of corrosion occurring within the system. These systems also require additional safety controls (often incorporated into the boiler) since there is no open vent, nor is there a permanent connection to a water supply. The system will include a relief valve, which will need connection to a suitable external discharge point. These systems may remove the need to install pipes and cisterns in the roofspace and so reduce the risk of freezing.

Replacement systems

Most boilers installed are replacements for older units. Many of the older boilers were installed with gravity circulation to the hot water cylinder. This provides a relatively poor hot water service and it will not usually have a boiler interlock which can give rise to excessive cycling, i.e. the boiler fires to keep itself hot even though there may not be a real heating or hot water demand.

When boilers are replaced the systems should always be upgraded to full pumping for both space heating and hot water circuits and new controls installed as this has a significant impact on efficiency. This will give the system an improved response and more effective control of room and domestic hot water temperature. Additional controls which give further enhanced features may also be considered.

When converting from gravity to fully pumped operation, it should be noted that the pump may need to be repositioned, motorized valve(s) installed, and additional piping and wiring will be required between the boiler and the hot water cylinder. These alterations are best done before kitchens and bathrooms are modernized.

Simple size-for-size boiler replacement is not recommended. The dwelling heating and hot water requirements should be checked before a new boiler is selected, since insulation levels may have been improved or the original sizing may have been incorrect. Oversizing will lead to less efficient operation as well as unnecessarily increased capital cost.

Regular boiler systems will often employ a vented indirect storage hot water cylinder. For small dwellings with a single bathroom this is typically of 120 litres capacity. Larger dwellings with more than one bathroom will require a larger cylinder capacity. Unvented cylinders are also available which operate at mains pressure with either an internal expansion facility or a dedicated external expansion vessel.

High-performance cylinders are now available containing a rapid heating coil, which reduces the time taken for the water to be heated, and may reduce boiler cycling. This helps to increase the system efficiency, especially with older boilers. Most hot water cylinders and thermal stores are now supplied with factory-applied insulation. Hot water cylinders should meet British Standards Requirements. Pre-coated cylinders should always be used in preference to cylinders with separate jackets.

Thermal stores are also available where the high temperature water from the boiler is stored directly. These systems are available either for 'hot water only' or 'hot water and space heating'.

Mains-fed water systems such as combi boilers, unvented cylinders, thermal stores and CPSU units are available which usually provide higher water pressures. These are particularly, beneficial where multiple draw-offs are required. However, it is particularly important to ensure that the incoming water supply pressure and flow to the dwelling are adequate before these units are fitted.

SEDBUK (Seasonal Efficiency of Domestic Boilers in the UK)

This is the preferred measure of the seasonal efficiency of a boiler installed in typical domestic conditions in the UK and is used in SAP Assessments (Standard Assessment Procedures for calculating the energy efficiency performance of dwellings) and the Building Regulations.

The SEDBUK efficiency is expressed as a percentage. An A to G Scale has also been defined as Table 5.3.

SEDBUK efficiency figures are based on test results submitted by manufacturers and certified by an independent body accredited for the testing of boilers to European Standards.

The boiler must have sufficient output to meet its maximum load, which includes the radiators, the domestic hot water cylinder and the heat losses from the distribution pipework. It should also have sufficient extra capacity to warm the house up in a reasonable time when the system is switched on from cold. It should not be oversized, however, as that will increase its capital cost. Oversizing can also adversely affect efficiency and hence running cost, although most modern boilers are capable of operating efficiently under part load conditions.

Boiler types

Boilers are defined as follows:

Regular boilers

This is the name given to boilers, which are not combination boilers. Historically they were the most commonly specified boiler and referred to as conventional or traditional units. Regular boilers are available for wall-mounting or floorstanding. A Back Boiler unit is a regular boiler designed specifically for installation within a fireplace. All Regular boilers are capable of providing space heating directly but require connection to a separate hot water storage system since they do not have the capacity to provide domestic hot water directly.

Regular boilers for sealed systems, which have components such as pumps, expansion vessels, etc. within the boiler casing are known as 'system boilers'.

Combination boilers

Combination boilers provide both space heating and direct domestic hot water. The most common type of unit is the instantaneous 'Combi' boiler, which heats water on demand without maintaining an internal store of water already heated. The units are capable of providing hot water continuously, but at a lower flow rate than could be expected from typical hot water storage systems. Therefore, these appliances may be less suitable for dwellings where multiple simultaneous draw-offs from separate taps are likely, i.e. multi-bathroom/shower room dwellings. Combi boilers will save space because

- They are fed directly from the water mains supply, and there is no need for a hot water storage cylinder or cistern to feed it.

- They are usually intended for use in a sealed system and so do not need a feed and expansion cistern, giving the opportunity to have 'dry' roof space.

Before selecting a Combi boiler, it is important to ensure that the dwelling has satisfactory water pressure and an adequate water supply pipe size to prevent the possibility of inadequate hot water performance.

Condensing boilers

Condensing boilers are becoming an increasingly important choice in the UK when boilers are being specified. In some European countries, they are already well established and have a major share of the market. Gas fired condensing boilers offer significantly higher efficiencies than can be achieved from non-condensing boilers. They may have a higher capital cost than non-condensing

Table 5.3 *SEDBUK*

SEDBUK range	Band
90% and above	A
86%–90%	B
82%–86%	C
78%–82%	D
74%–78%	E
70%–74%	F
Below 70%	G

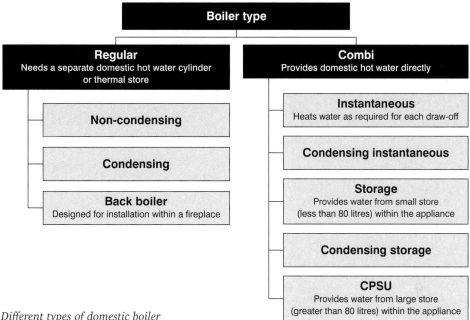

Figure 5.17 *Different types of domestic boiler*

boilers but are usually cost-effective for larger dwellings (houses with three or more bedrooms).

Size and efficiency

Boilers are described by their heat output. Heat output is the amount of useful heat extracted from the respective fuel, and this figure will be governed by the boiler's efficiency. For example, a 30 kW boiler is capable of producing 30 kW of heat when the boiler is run for one hour.

The efficiency of a boiler is expressed by its output as a percentage of its input. Example: A boiler with an efficiency of 85 per cent means that only 85 per cent of the heat contained in the respective fuel is being satisfactorily used.

Heat required for dwelling	15 kW
Boiler efficiency	85%
Heat lost	15%
Heat required	$\dfrac{15 \times 115}{100}$
	17.25 kW

Solid fuel boilers

These come under two headings:

1 Back boilers

2 Independent or free standing boilers.

Back boilers

These are very popular in many parts of the country and are still the most economic means of providing space heating and water heating during the winter, although some alternative, usually an electric immersion heater, may be advantageous in summer.

Although the most basic, solid fuel boilers are limited in their use, they will produce most of the domestic hot water required for an average household at very little extra cost. It forms part of the fireplace, and derives its heat from the fire which would be normally heating only that particular room. Larger and more intricate back boilers are available (Figure 5.18), some with the flue forming an integral part of the boiler enabling the maximum amount of heat to be extracted from the hot flue gas as it passes up to the chimney. Although these boilers are able to impart a much greater heat value to the water, this would still not be sufficient satisfactorily to heat even a small domestic dwelling, but could supply what is known as 'background heating' in addition to the domestic hot water requirement.

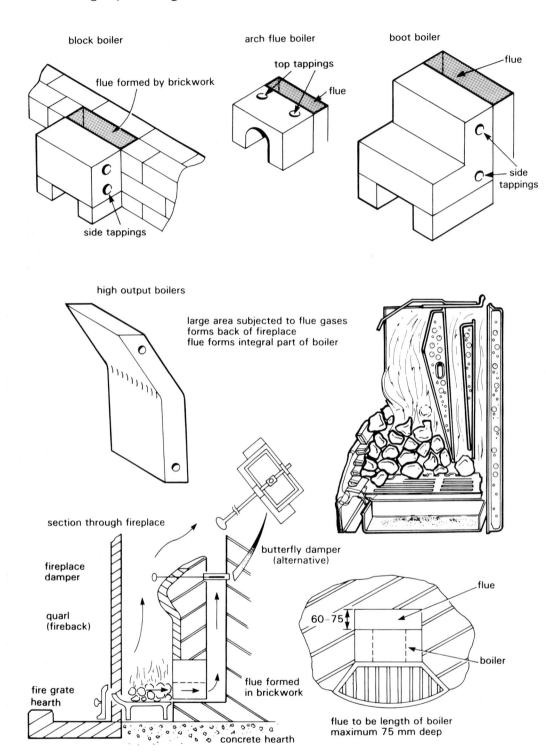

block boiler

flue formed by brickwork

side tappings

arch flue boiler

top tappings

flue

boot boiler

flue

side tappings

high output boilers

large area subjected to flue gases
forms back of fireplace
flue forms integral part of boiler

section through fireplace

fireplace
damper

quarl
(fireback)

fire grate
hearth

butterfly damper
(alternative)

flue formed
in brickwork

concrete hearth

flue

60–75

boiler

flue to be length of boiler
maximum 75 mm deep

Figure 5.18 *Solid fuel boilers*

Link-up system

Solid fuel boilers can be used as the sole heating system or in conjunction with another heating system. An existing gas or oil fired system can be linked up to an open fire or solid fuel room heater, thereby giving all the advantages of a real fire.

There are two types of 'Link-up system', both of which mean a saving can be made on oil or gas bills.

1 The solid fuel boiler is connected to the hot water storage vessel to provide domestic hot water while the gas or oil boiler continues to provide the space heating via the radiators.

2 The system is connected so that you have the choice of using either or both systems to provide heating and hot water. It must be remembered that solid fuel open fires or room heaters have a lower heat output than gas or oil boilers, therefore a less efficient system would result if using only the solid fuel appliance. It may be necessary either to reduce the temperature in some rooms by automatically controlled thermostatic valves or to turn off selected radiators when using only the solid fuel appliance.

Not all existing central heating systems can be successfully linked up to an additional boiler. Some of the factors governing this are:

1 Location and distance from solid fuel appliance to point of link-up into existing system

2 Location of hot water storage vessel

3 The need for each appliance to have a separate chimney.

Independent or free standing solid fuel boilers

The natural progression from the back boiler with its limited use is to a more sophisticated and larger boiler capable of providing both the domestic hot water and space heating of the dwelling. Independent boilers can be fired by solid fuel, gas or oil and modern boilers are now operated automatically, being fitted with electrical and thermostatic controls.

There are two main types of independent solid fuel boilers: (i) sectional boilers; (ii) gravity feed boilers. Both come in sizes large enough to provide all the domestic hot water and satisfactorily to space-heat domestic dwellings.

These boilers are suitable for use with gravity central heating, small bore and micro bore systems, and the systems can be controlled by thermostat, time clock and individual thermostatic radiator valves for economic and automatic control.

Sectional solid fuel independent boilers

The sectional solid fuel independent boilers on the market today are capable of providing the heat required for the small domestic household, for both water and space heating. The boiler flue outlet must be connected to a chimney, independent of any other appliance. The flue and chimney must be regularly cleaned depending upon its use: it is recommended that this should be carried out at least once a year.

Once the fire is lit and the thermostat set, the boiler requires only a minimum of attention. Refuelling and ash removal is required from time to time, possibly two or three times a day when the boiler is working at its maximum. The fuel is fed through either a front or a top door. The door incorporates a fireproof gasket, ensuring an airtight seal when closed, which results in controlled burning.

Gravity feed solid fuel independent boilers

One of the disadvantages of solid fuel boilers is the necessity of having to load the fire regularly with fuel. Irregular loading will result in fluctuation of the heat imparted to the water for heating. This problem is overcome by the use of gravity feed hopper type boilers which also incorporate fan assisted burning controlled by thermostat. This type of boiler only requires loading once each day, when working at its maximum. The fire burns continuously, being automatically fed with fuel from the hopper. The sensing thermostat governs the fan when heat is called for and boosts combustion. This ensures a high burning temperature, which in turn fuses the ash into a solid clinker, which may be automatically ejected from the fire-bed by a simple lever operated de-clinkering mechanism, once or twice a week depending on its load.

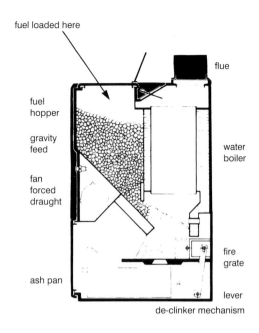

Figure 5.19 *Solid fuel independent boiler*

Oil-fired boilers

In their outward appearance, solid fuel and oil fired boilers appear very similar. It is only upon opening the boiler casing and carrying out an inspection that it becomes evident which fuel is being used. Oil-fired boiler flues are usually connected to chimneys to discharge the products of combustion into the atmosphere at a safe height. It is now possible to purchase boilers which burn *kerosene* with low level, balanced flues, so overcoming the necessity of being tied to a chimney location or the expense of building one.

Note: It cannot be over-emphasized that adequate air is essential for correct combustion and ventilation sources should be as close to the boiler as practicable.

Burners

The function of the burner is to convert the oil into a state in which it can be satisfactorily mixed with air for combustion, and to intimately mix the oil and the air in the correct proportions to obtain maximum efficiency from the combustion process. This is achieved by one of the following processes:

1 Vaporization (by heating the oil)

2 Atomization (mechanically breaking up the oil into fine particles).

Atomizing burners

The pressure jet burner was developed primarily for very large domestic, commercial and industrial boiler applications and it has proved to be extremely efficient and reliable although somewhat noisy in operation. Fairly recent development has seen the introduction of smaller output burners of approximately 13 kW, with even greater reliability and quieter action: these boilers are the ones most commonly installed in the domestic market today. The oil is pumped under pressure through a specially designed nozzle, where it is emitted as a mist of tiny droplets. The oil mist combines with the air delivered by the fan and is then ignited within the combustion chamber of the boiler. For pressure jet boilers a gravity head for the oil supply is desirable (because of cost) but the fuel pump of the burner can be adapted to enable the oil storage tank to be below the level of the burner.

Fuel

All fuel oils consist mainly of combinations of carbon and hydrogen in varying proportions together with small quantities of sulphur.

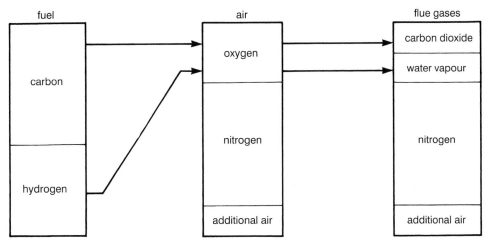

Figure 5.20 *The combustion process*

Combustion

This is a form of oxidation. When the oil is burned in the air, the carbon and hydrogen (oil) combine with the oxygen (air) to form carbon dioxide and water (see Figure 5.20).

Preparation of oil for combustion

Oil in bulk will not burn, in fact a lighted match would be extinguished if dropped into a tank of domestic oil: therefore some preparation of the oil for burning must first be carried out. Important considerations are:

1 Large surface area for contact with oxygen
2 Sufficient supply of oxygen (air)
3 Intimate mixing of oil and oxygen
4 Sufficient heat to promote combustion.

The installation of oil storage and supply pipes is covered by building regulations and by environmental legislation.

The following points should be considered.

Quantity to be stored

Larger storage capacity reduces the number of deliveries and fuel cost. For domestic use 2500 litres is recommended if there is space.

Fire protection

If the tank is closer than 1.8 metres to a building or 760 mm to a boundary, some simple fire protection measures are required. All tanks must be installed over a fireproof base. Dimensions and other details are given in BS 5410 Part I: 1997.

Environmental protection

Domestic oil tanks up to 2500 litres capacity do not have to be bunded unless their installation fails the risk assessment in OFTEC Technical Information Note TI/133. The TI/133 risk assessment must be completed for every tank installation.

Tank construction

Steel tanks should comply with OFS T200. Plastic tanks should comply with OFS T100.

Oil supply pipes

Oil supply pipes must be installed in accordance with the requirements of BS 5410: Pai-tl: 1997; remote acting fire valves to OFS E101 are required for all installations. Underground pipework should comply with OFTEC Technical Information Note TI/134.

Tank location

Storage tanks should be located on firm foundations in accordance with BS 5410: Part 1: 1997, with good access for delivery, inspection and maintenance.

Tank gauges

Oil storage tanks must be provided with an easily readable gauge, either mounted on the tank or remotely located in a convenient position. Gauges should comply with OFS E103.

The oil supply lines to the boiler may be constructed from either:

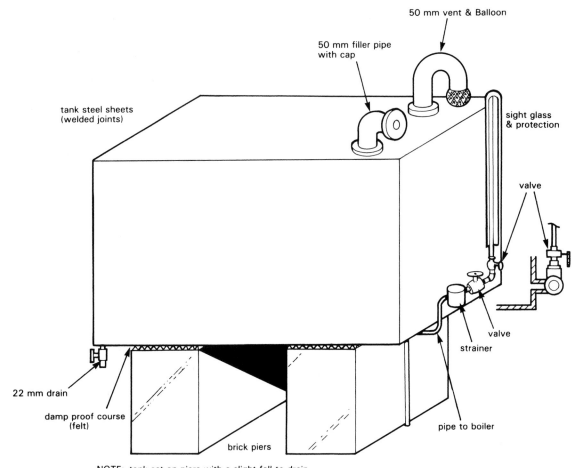

Figure 5.21 *Typical oil storage tank layout*

1 *Copper.* This should be soft temper or half hard with flared fittings (manipulative joints), no soft-soldered joints or jointing compounds to be used. Alternative brazed or silver soldering may be used; or

2 *Steel.* Black low carbon steel (not galvanized) with tapered threads only, sealed with oil resistant (shellac based) jointing compound. PTFE tape may be used.

Valves

These should be of the glandless fullway pattern. *Fire valves* to shut off the oil supply automatically on an excessive rise in temperature should be fitted, and may be of either fusible link, hydraulic or electric pattern.

Filters

To be of 5 meshes per millimetre and to be made from stainless steel, monel metal or phosphor bronze. Alternatively the more modern paper element filters which are of a throw-away pattern may be used.

General considerations

The layout should be as simple and economical as possible. with a minimum number of joints. Soft copper oil pipeline should be well supported, clipped at 500 mm spacing. Pipe should be laid to a steady fall (automatic venting) avoiding dips.

The oil supply should be controlled by an automatic fire valve fitted next to the point of entry to the building, the sensing element being near the boiler for safety reasons. The boiler must be set on a strong fire resistant surface unaffected by heat or oil, such as ceramic tiles or concrete. A good chimney is essential to ensure correct operation of vaporizing and rotary burners, and it should develop a draught sufficient to induce the correct quantity of air for combustion.

Electric boilers

The development of electric boilers now makes them an attractive alternative to other forms of energy. They can be fixed as a replacement for an existing boiler, or as part of new systems.

Figure 5.22 shows an electric boiler which outwardly appears similar to a vertical chest freezer. Inside is a well-insulated water storage unit heated by thermostatically controlled heating elements.

The hot water is pumped around the heating circuit by a pump as in the previously described systems. There is one important difference between electric water heating and other systems: the water is heated in the electric boiler to a much higher temperature, almost to boiling point, and is too hot to be used without being passed through the mixing valve (blender). The function of the blender is to mix the near boiling water from the boiler with cooler water returning from the heating circuit, giving a blended water at the usual 70°C for circulation to the heat emitters. The system is controlled in the normal manner with automatic control timers, room thermostats, etc.

The advantages of this type of boiler are:

1 Clean, quiet and reliable
2 No storage of fuel required
3 No flue required
4 Little or no maintenance required, maintains its efficiency
5 Pollution free
6 Can be sited almost anywhere (under cover with electric supply)
7 Easily installed
8 Factory assembled and tested
9 Suitable for new or existing installations.

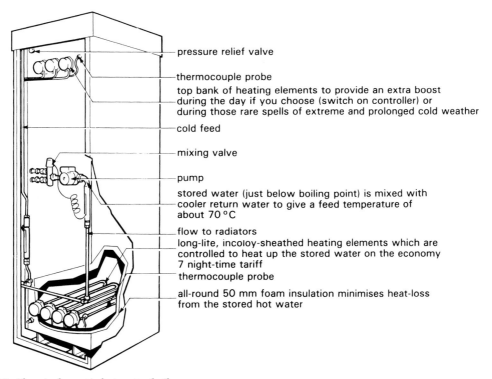

pressure relief valve

thermocouple probe

top bank of heating elements to provide an extra boost during the day if you choose (switch on controller) or during those rare spells of extreme and prolonged cold weather

cold feed

mixing valve

pump

stored water (just below boiling point) is mixed with cooler return water to give a feed temperature of about 70 °C

flow to radiators

long-lite, incoloy-sheathed heating elements which are controlled to heat up the stored water on the economy 7 night-time tariff

thermocouple probe

all-round 50 mm foam insulation minimises heat-loss from the stored hot water

Figure 5.22 *Electric domestic hot water boiler*

Figure 5.23 illustrates how an existing system of hot water and space heating can be adapted to incorporate an electric boiler and dual immersion heater in hot storage. The cylinder should be well insulated and isolated from the heating system. Figure 5.24 illustrates a typical arrangement of a domestic hot water and space heating system.

All boilers

- The flue must be correctly designed and sized, use suitable materials and be provided with a suitable terminal.
- Back boilers make use of an existing chimney, which must have a suitable flue liner and terminal. Flue liners deteriorate with age and consideration should be given to replacing them at the same time as boiler replacement.
- Special consideration should be given to the siting of flue terminals for condensing boilers

due to the possibility of pluming at the terminal. This can cause a nuisance in some situations since the 'plume' will be visible for much of the time the boiler is in operation.

- All open-flue boilers require a purpose-made air vent to ensure there is sufficient air for combustion.
- Room-sealed boilers do not require special provision for combustion air in the room they are installed.
- Boilers installed in a compartment (whether open-flue or room-sealed) may need provision to supply additional air for cooling.
- Where an extractor fan is fitted in a room containing an open-flue appliance, additional ventilation may be required to prevent the fan affecting the boiler flue performance. By providing an air vent, colder outside air will

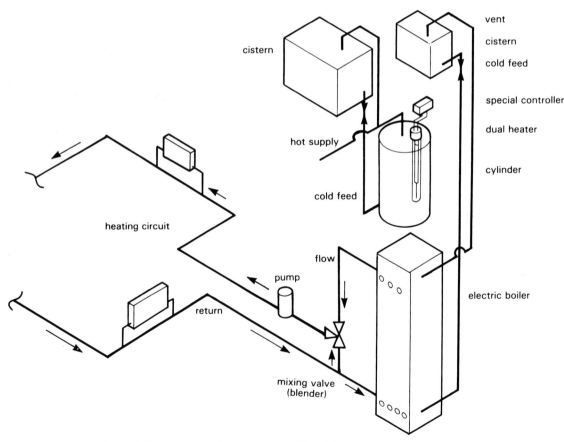

Figure 5.23 *Separate domestic hot water and heating system (electric)*

enter the room and increase the ventilation heat loss, and so will slightly increase running costs for space heating.

Gas boilers

■ Boilers with fan-assisted flues are often more energy efficient than those without fans, since they are usually more compact, have a smaller flue diameter which reduces heat losses when the boiler goes off and are more likely to include automatic ignition. The electrical energy input of the fan is very small relative to the overall gas energy input for space heating and hot water.

■ Regular and instantaneous combi boilers are available in all flue options, but for most other types of boiler the range is smaller. Boilers with fan-assisted flues are likely to have the fewest restrictions when siting the flue terminal. New, more stringent requirements for the positioning of natural draft room-sealed flue terminals are expected to be introduced which make them more difficult to install near to windows and doors. This makes the selection of a boiler with a fan-assisted flue more attractive since it can be installed in a much wider range of positions.

■ A room-sealed boiler should be chosen where possible. Room-sealed boilers are inherently safer than open-flue boilers since in these appliances there is not a direct path for combustion products to spill into the room. However, all new open-flue boilers must incorporate a safety device that, in abnormal draught conditions or flue blockage, is designed to turn off the boiler and limit the release of combustion products into the room.

■ Extended flues are now available for a wide variety, of appliances and an increasing number of boilers have separate connections for the air inlet and the flue pipe. In some cases these allow total flue lengths of over 8 metres with a number of bends.

■ Open-flue gas boilers of input greater than 7 kW require a purpose-made non-closable vent in the room to ensure there is sufficient air for combustion.

■ Some gas boilers are available to fit to 'shared flues'. Expert advice must be obtained before this option is considered.

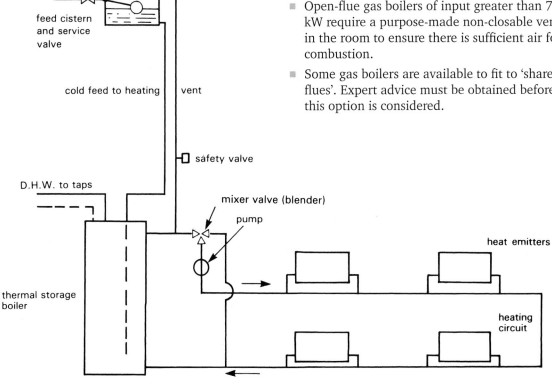

Figure 5.24 *Electric boiler combined system*

Oil boilers

▪ Oil-fired boilers have open or room-sealed balanced flues. All new boilers now have fan-assisted pressure jet burners, but some other oil-fired appliances use vaporizing burners.

▪ The efficiency of new oil-fired boilers is high and they operate with comparatively low flue-gas temperatures. A correctly sized, well-constructed lined flue is essential for satisfactory operation of open-flue models.

▪ A wide variety of flues are now available for oil-fired boilers which can be extended horizontally or vertically from the boiler.

▪ The requirements for flue terminal siting are different from those that apply to gas boilers.

▪ Open-flue oil boilers of output greater than 5 kW require a purpose-made non-closable vent in the room to ensure there is sufficient air for combustion.

System layouts

Gravity circulation systems

This is the system by which domestic hot water systems circulate the heated water and which depends on the movement of a column of hot water (lighter) being displaced by a column of cold water (heavier). This movement is known as convection. The very early space heating systems worked on the gravity circulation system and required careful and accurate pipe sizing, the use of large diameter pipes and large radius bends (Figure 5.29).

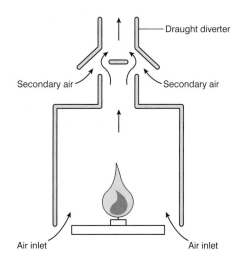

Figure 5.25 *Open flue*

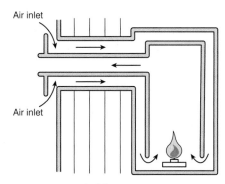

Figure 5.26 *Room-sealed flue*

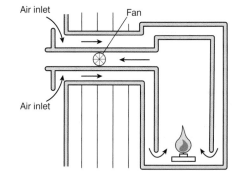

Figure 5.27 *Fanned flue*

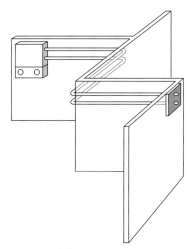

Figure 5.28 *Extended flue*

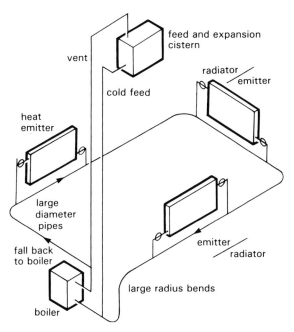

Figure 5.29 *Single ring main system (isometric projection)*

Forced circulation

This is the system in which the heated water is forced or drawn around the pipework system by the operation of an impeller. This type of system has many advantages over the gravity system. including:

1 The use of small diameter pipes
2 Frictional resistance in systems easily overcome
3 Isolated radiators present no problem (design)
4 Cheaper to install
5 Cheaper to operate
6 Due to pipes being small they can easily be hidden
7 Heat losses smaller (small diameter pipes used)
8 Rapid response to a demand for heat.

Gravity heating systems

Single ring main system (gravity)
In this system there is one single pipe carried around the building either above or beneath the floor. From this pipe branches are taken to feed

the radiators which are fixed above the main circulation pipe. The branch flow connection to the radiator can be connected either to the top of the radiator or alternatively to the bottom connection. The return pipe is connected from the bottom of the radiator at the opposite side to the flow connection and connected directly back to the ring main. This is a very simple system and serves very well for one large building. It is not so suitable for a building divided into rooms (ideal for community halls, etc.).

The main disadvantage of this system is that the cooled water from each radiator is returned to mix with the hot water in the main, so cooling it down. This means that each successive radiator along the main is cooler than the previous one, and in a large system the last radiator will be appreciably cooler. This can to some extent be minimized by accurate pipe sizing, and adjustment of lock-shield or thermostatic radiator valves.

Note: Today all modern systems are forced circulation systems.

By forced circulation we mean the introduction of a pump into the system the effect of the pump is to create sufficient energy to force the water around the circuit. The pump must be able to generate sufficient energy to overcome the frictional resistance between the water and the pipework.

The location of the pump in the system is always a matter of debate. Some people suggest that it should be fitted on the return pipe near to the boiler, while others advocate that it should be fitted on the flow pipe adjacent to the boiler.

Figure 5.30 shows a typical fully pumped system. It uses a three-way motorized valve to provide water to the heating system or domestic hot water cylinder as and when required. Similar operation can be obtained by using two port valves.

Heating system pipework must be sized so that each part of the circuit has sufficient circulation to deliver its rated heat output. The flow required for each heat emitter is directly related to its output and may be calculated using the design temperature difference. The pressure required to achieve circulation depends upon the resistance to flow in the circuit, which is affected by pipe

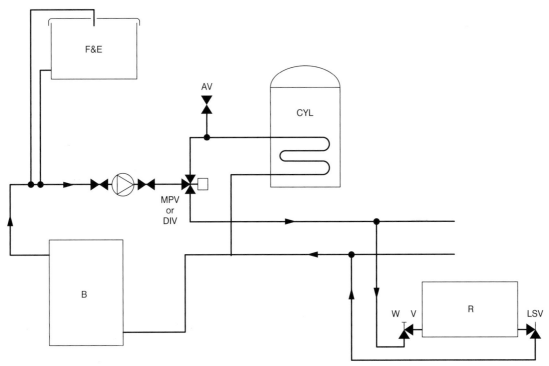

Figure 5.30 *Heating and domestic hot water using a mid-position or diverter valve*

length and diameter, the number and type of fittings and the resistance of components including the boiler and heat exchangers. The designer's task is to ensure that the circuit resistance is low enough for the circulator (pump) to achieve the required flow to all points in the circuit, without undue noise or a tendency to collect air in parts of the system.

A **reverse return** layout is shown in Figure 5.31. This is a variation of the conventional two-pipe circuit, which has the advantage of equalizing the pressure loss to all parts of the system. This layout is particularly useful when installing two boilers, as it helps to ensure both receive equal amounts of the water being circulated. It can be seen from the figure that the total length of the flow and return pipework from the branch tees is equal to both boilers.

Sealed heating systems

A sealed heating system eliminates the need for a feed and expansion cistern and its associated pipework, and virtually eliminates all corrosion risks since there is no possibility of ingress of air during normal operation of the system. When installed together with an unvented domestic hot water cylinder, there is no need for cisterns or pipework in the roof space. This considerably reduces the risk of frost damage and condensation in the roof space. Sealed systems are particularly advantageous in flats and bungalows where it may not be possible to obtain adequate static pressure from a cistern.

Figure 5.32 shows a typical sealed system. The system must be provided with a diaphragm expansion vessel complying with BS 4814, a pressure gauge, a means for filling, make-up and venting, and a non-adjustable safety valve. Boilers used in sealed systems should be approved for the purpose by their manufacturer and must incorporate a high limit thermostat. It is of particular importance that manufacturers' instructions are followed when installing the components of a sealed system.

The main component of a sealed heating system is the expansion vessel. It performs the

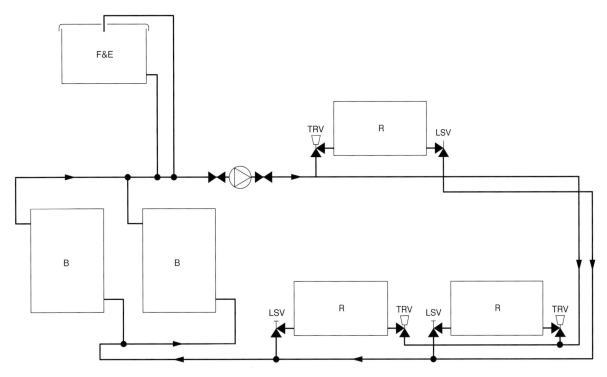

Figure 5.31 *Reverse return heating circuit*

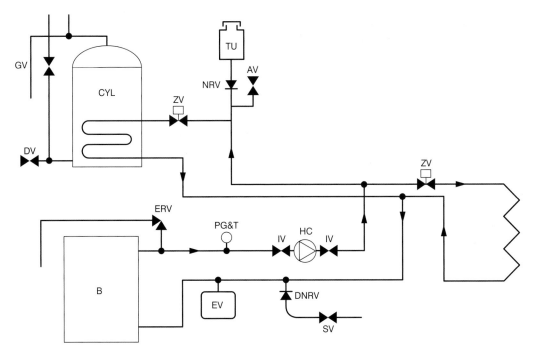

Figure 5.32 *Sealed heating system components*

same task as the feed and expansion cistern in an open vented system: to receive the increased water volume when expansion takes place as the system heats up and to maintain a positive pressure in the system. The expansion vessel contains a flexible diaphragm, which is charged on one side initially with nitrogen but is then topped up as required with air.

The expansion vessel should be located close to the suction side of the circulator to ensure that there is positive pressure in all parts of the system pipework. This will eliminate the possibility of air ingress through valve glands, etc. It should be connected in such a manner as to minimize natural convection currents in order to maintain the lowest possible temperature at the diaphragm. The pipe connecting the pressure vessel to the system should have a diameter of not less than 15 mm and must not contain any restriction such as an isolating valve.

It is important that the expansion vessel has sufficient acceptance volume to accommodate the expansion that would occur if the system water were heated from 10°C to 110°C. Consequently, it is important that the system volume is estimated with reasonable accuracy.

The initial charge pressure in the expansion vessel should be in accordance with manufacturers' instructions and must always exceed the static pressure of the heating system at the level of the vessel. Prior to connecting the expansion vessel to the system, the pipework should be flushed and tested. Following connection of the vessel, the system when cold should be pressurized to above the initial nitrogen pressure in the vessel (typically by 0.2 bar). This will result in a small displacement of the diaphragm as illustrated in Figure 5.33. When the system is operational the expansion water will move into the vessel, compressing the nitrogen so that when the operating temperature is reached the system pressure will rise and the diaphragm will be displaced to accommodate the additional volume (Figure 5.34). BS 5449 defines the practical acceptance volume of the vessel as what it will accept when the gauge pressure rises to 0.35 bar below the safety valve setting.

If the initial system water pressure is too high or the nitrogen fill pressure in the vessel is too low, the diaphragm will be displaced too far into the vessel, which will then be unable to accommodate the volume of expansion water. This will result in an increase in the system pressure and the safety valve will lift.

All sealed systems must have a non-adjustable safety valve set to lift at a gauge pressure not exceeding 3 bar (300 kPa). Safety valves must also have a manual testing device, valve seating materials that will prevent sticking in a closed position and provision for connecting a full bore

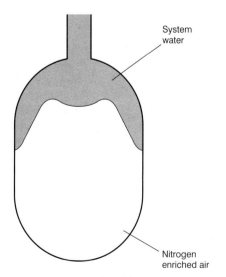

Figure 5.33 *Expansion vessel (cold)*

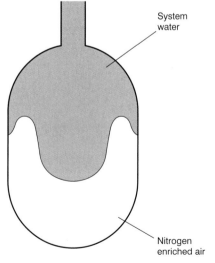

Figure 5.34 *Expansion vessel (hot)*

discharge pipe. The valve should be connected to the flow pipe close to the boiler with a metal discharge pipe installed to an open tundish, which should then discharge in a safe visible external low level location.

A pressure gauge is provided so that the pressure may be checked and the system charged to the correct pressure when commissioning or topping up. The pressure gauge must be readable from the system filling position. A thermometer which may be combined with the pressure gauge, shows the temperature of the flow water from the boiler. Care should be taken to ensure that the thermometer is fitted to the boiler or the flow pipe, not to a non-circulating pipe, and that the thermometer pocket does not restrict the bore of the pipework.

Filling and pressurizing the system is normally achieved by a direct connection from the cold water supply main through a special filling loop. This is an arrangement of fittings which incorporates a BS 1010 stop valve, double check valve with test point, and a flexible pipe which should be disconnected and removed after use and protective caps fitted over the ends of the stop and check valves.

Sealed systems do not lose water due to evaporation and so, provided there are no leaks, make up water should not usually be required except during the initial period of operation when air in the fill water is being removed from the system.

Lost water can be replaced by any of the following methods.

1 Refilling and pressurizing manually through the filling loop
2 Refilling automatically through a top-up unit
3 Refilling automatically through a make-up cistern.

The top-up unit is connected to the highest part of the system and is fitted with a double check valve assembly and automatic air eliminator. The unit, which has a water capacity of about 3 litres, is topped up manually when the water level drops. It is also useful as an indicator of leaks in the system. It should be connected either to the return side of the radiator distribution pipework or to the return side of the primary domestic hot water circuit.

The commissioning of a system is greatly simplified when a top-up unit is installed as initial pressurization is unnecessary.

Provision must be made for venting air from sealed systems, using either automatic or manual air vents. Automatic air vents should be fitted at the highest points of the system and should be float operated. Hygroscopic types of automatic air vent should not be used because they allow continuous evaporation of small quantities of water. The automatic vent incorporated into some boilers is sufficient when combined with a manual air vent fitted at the highest point in the system. Automatic air vents can be obtained with an integral shut-off valve, which allows cleaning to be carried out without draining down the system.

An air separator with an automatic air eliminator is an optional item to be considered on larger domestic systems to assist commissioning and for the easy removal of air. It is usually fitted on the flow pipe between the boiler and the circulator.

The installation of a make-up cistern for filling and make-up water should be considered for larger installations. This is typically a conventional 45 litre plastic cistern with lid, overflow and ball valve connected to the mains water supply.

The cistern should be sited at least 300 mm above the highest point of the system and the outlet fitted with a double check valve assembly and a stop valve. The cistern must not be used for any other purpose. The filling of the system will be simplified but the main advantage of a manual top-up unit, that of knowing water is being lost, will not be available because of the automatic replenishment of water.

Boilers with sealed system components already fitted within the casing, or with casing extensions to hide the components from view, are readily available and provide a compact alternative to installing separate components.

The circulator should be installed to maintain a positive pressure at all points around the system and, when fitted to an open vented system, be subject to at least the minimum static pressure specified by the manufacturer. This is defined in BS 5449 as being one-third of the maximum pressure developed by the circulator under no water flow conditions. For most domestic circulators this is between 1.7 and 2.0 metres of water.

The circulator should not be located at the lowest part of a system, where it is possible for sediment to collect, nor at the highest point, where air can be a problem.

Circulator installation should observe manufacturer's instructions about the orientation of the installation. This normally requires that the circulator should be located with its motor shaft in a horizontal position so that there is no undue load on the bearings, and should either be fixed into a vertical pipe so that the water is flowing upward, or in a horizontal position. It is often difficult to remove trapped air from the impeller casing and for this reason a circulator should not be installed facing vertically downwards.

System design faults or abuse can often be diagnosed from the sediment found in the circulator. They are usually either black encrustation (ferric oxide) or a red sediment (ferrous oxide).

The black ferric oxide is often as a result of the system not having been flushed out correctly on completion of the installation, leaving extraneous matter in the water and causing the residue to form hydrogen which has to be vented out of the system.

The red ferrous oxide is caused by air entering the system, possibly being entrained on the suction side of the circulator or from the feed and expansion cistern on an open vented system. This is commonly the result of incorrect positioning of the feed and venting pipework in relation to the circulator, resulting in failure to maintain a positive pressure at all points in the system. In such circumstances, the circulator is likely to fail, often followed by leaking radiators and heat exchangers.

Noise in a heating system can often be difficult to trace, especially if emanating from the circulator, since it can be transmitted and amplified along the pipework to remote parts of the installation.

Pipework noise also commonly occurs as a result of the expansion of pipework over joists or where the pipe has been left touching other pipes or a part of the building structure. Care must always be taken to ensure the pipework is correctly bracketed, is not in tension or compres-

sion and does not carry the weight of components such as the circulator. Room must be left for pipework to expand and contract without coming into contact with other pipes or the building structure.

Boilers with high efficiency usually have a low water content and require a minimum flow rate to be maintained through the heat exchanger to ensure the heat can be removed quickly enough. Small systems, which do not require a high circulator duty, can occasionally cause noise problems if the boiler requirement for a minimum flow rate is overlooked. In such cases, an automatic bypass valve should be fitted so that the minimum water quantity can flow through the boiler without relying on circulation through the radiators. In these circumstances the circulator must be capable of satisfying the duty of both the system and the bypass.

A boiler operating with inadequate flow through it is said to 'kettle', a description derived from the characteristic noise it makes. If allowed to operate in this manner too long before remedial action is taken, it will probably have a deposit of calcium in the heat transfer coil which will be extremely difficult, if not impractical, to remove.

Air trapped in the boiler heat exchanger can be another source of noise. This can sometimes be avoided by using an eccentric reducing bush for smaller pipes attached to horizontal boiler flow tappings.

A number of boilers, particularly combination and sealed system types, are supplied complete with circulators already piped and electrically wired as part of the package. In such cases the circulator is usually an important part of the operating sequence of the boiler which may also incorporate an automatic bypass valve and have an overrun requirement as part of the control function.

Boilers of this type tend to use a high proportion of the available circulator pressure to circulate water through the heat exchanger and will, for this reason, specify the pressure available to be used to circulate water through the rest of the system. Refer to the manufacturer's instructions for particular boilers to ensure that the correct temperature drop across the system is used when sizing heat emitters and pipework. If the integral circulator has insufficient capacity for the system, use a mixing header and a second circulator for radiator circuits; on no account add a second circulator in series with the integral circulator.

Controls for domestic central heating systems

Time switch

A switch operated by a clock to control either space heating or hot water, but not both. The user chooses one or more 'on' periods, usually in a daily or weekly cycle.

Programmer

Two switches operated by a clock to control both space heating and hot water. The user chooses one or more 'on' periods, usually in a daily or weekly cycle. A mini-programmer allows space heating and hot water to be on together, or hot water alone. A standard programmer uses the same time settings for space heating and hot water. A full programmer allows the time settings for space heating and hot water to be fully independent.

Room thermostat

A sensing device to measure the air temperature within the building and switch on and off the space heating. A single target temperature may be set by the user.

Night setback

A feature of a room thermostat that allows a lower temperature to be maintained outside the period during which the normal room temperature is required.

Programmable room thermostat

A combined time switch and room thermostat which allows the user to set different periods with different target temperatures for space heating, usually in a daily or weekly cycle.

Delayed start

A device, or feature within a device, to delay the chosen starting time for space heating according to the temperature measured inside or outside the building.

Optimum start

A device, or feature within a device, to adjust the starting time for space heating according to the temperature measured inside or outside the building, aiming to heat the building to the required temperature by a chosen time.

Optimum stop

A device, or feature within a device, to adjust the stop time for space heating according to the temperature measured inside (and possibly outside) the building, aiming to prevent the required temperature of the building being maintained beyond a chosen time.

Cylinder thermostat

A sensing device to measure the temperature of the hot water cylinder and switch on and off the water heating. A single target temperature may be set by the user.

Programmable cylinder thermostat

A combined time switch and cylinder thermostat which allows the user to set different periods with different target temperature for stored hot water, usually in a daily or weekly cycle.

Weather compensator

A device or feature within a device, which adjusts the temperature of water circulating through the heating system according to the temperature measured outside the building.

Load compensator

A device, or feature within a device, which adjusts the temperature of the water circulating through the heating system according to the temperature measured inside the building.

Boiler energy manager

No agreed definition, but typically a device intended to improve boiler control using a selection of features such as weather compensation, load compensation, optimum start control, night setback, frost protection, anti-cycling control and hot water over-ride.

Boiler anti-cycling device

A device to introduce a time delay between boiler firing. Any energy saving is due to a reduction in performance of the heating system. The device does not provide boiler interlock.

Boiler thermostat

A thermostat within the boiler casing to limit the temperature of water passing through the boiler by switching off the boiler. The target temperature may either be fixed or set by the user.

Boiler auto-ignition

An electrically controlled device to ignite the boiler at the start of each firing, avoiding use of a permanent pilot flame.

Boiler modulator (water temperature)

A device, or feature within a device, to vary the fuel burning rate of a boiler according to measured water temperature. It is often fitted within the boiler casing. The boiler under control must have modulating capability.

Boiler modulator (air temperature)

A device, or feature within a device, to vary the fuel burning rate of a boiler according to measured room temperature. The boiler under control must have modulating capability and a suitable interface for connection.

Pump over-run

A timing device to run the heating system pump for a short period after the boiler stops firing to discharge very hot water from the boiler heat exchanger.

Pump modulator

A device to reduce pump power when not needed, determined by hydraulic or temperature conditions or firing status of the boiler.

Motorized valve

A valve to control water flow, operated electrically. A 2-port motorized valve controls water flow to a single destination. A 3-port motorized valve controls water flow to two destinations (usually for space heating and hot water), and may be either a diverter valve (only one outlet open at a time) or a mid-position valve (either one or both outlets open at a time). The valve movement may also open or close switches, which are used to control the boiler and pump.

Automatic bypass valve

A valve to control water flow, operated by the water pressure across it. It is commonly used to maintain a minimum flow rate through a boiler and to limit circulation pressure when alternative water paths are closed (particularly in systems with thermostatic radiator valves).

Thermostatic radiator valve

A radiator valve with an air temperature sensor, used to control the heat output from the radiator by adjusting water flow.

Frost thermostat

A device to detect low air temperature and switch on heating to avoid frost damage, arranged to override other controls.

Pipe thermostat

A switch governed by a sensor measuring pipe temperature, normally used in conjunction with other controls such as a frost thermostat.

Boiler interlock

This is not a physical device but an arrangement of the system controls so as to ensure that the boiler does not fire when there is no demand for heat. In a system with a Combi boiler it can be achieved by fitting a room thermostat. In a system with a regular boiler it can be achieved by correct wiring interconnections between the room thermostat, cylinder thermostat and motorized valve(s). It may also be achieved by a suitable boiler energy manager.

Zone control

A control scheme in which it is possible to select different temperatures in two (or more) different zones.

Controls

Hot water and heating systems, regardless of the fuel used, can be fitted with various types of controls, some manually operated, others automatic. The function of these controls is to switch the boiler on and off automatically whenever heat for the space heating or domestic water heating is required, so effecting a saving of fuel by controlling the temperature of the water. The use of pumps enables mini-bore and small-bore systems of heating to be installed in most domestic dwellings almost regardless of design. To install an effective central heating and domestic hot water system in the home and to economize on the running costs it is important to have an effectively and efficiently controlled system.

The following are some of the controls available to assist in obtaining maximum efficiency at minimum operating cost.

1 Drain valve
2 Thermostat
3 Programmer
4 Pump
5 Safety valve
6 Diverter valve
7 Non-return or check valve
8 Economy valve
9 Thermostatic and lockshield valves.

Drain valve

This is perhaps the most simple and basic control in the whole system, yet it is one of the most important. The purpose of the drain valve is to facilitate the draining of all the water from the system, i.e. boiler, storage vessel and pipework. Drain valves are usually made of brass with washers of fibre or nylon.

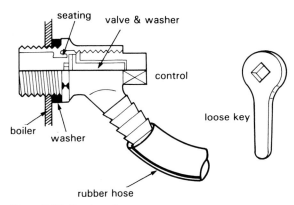

Figure 5.35 *Drain valve*

Thermostat

This is a very cost-effective control, its function being to close down the appliance when the temperature of the water has reached a predetermined level set on the thermostat. The setting of the thermostat should be in the region of 60°C for domestic installations. This is governed by the nature of the water and the requirements of the user.

Operation of thermostat

Sensing element

The thermostat sensing element comprises two metal plates, welded together at the rims, encapsulating a liquid whose pressure changes greatly in response to small variations in temperature. This dual diaphragm forms a bellows which expands and contracts with the ambient temperature changes. This movement serves to operate a snap-action micro switch rated to control the heating circuit.

The action of this type of thermostat is brought about by the fact that metals have varying degrees of expansion over the same range of temperature. The two metals in this case being brass with a high rate of expansion and invar steel with a low rate of expansion. When the system is cold the metals will have contracted to their shortest length and the electrical contact will be made. When the water is heated and the temperature has reached that set on the control the brass outer casing will have increased in length and, due to the fact that the invar steel rod is secured to the brass at the end, the rod is taken with it and so the electrical contact is broken. When the water is cooling the brass tube will also cool and contract in length until the electrical contacts touch again, whereupon the boiler fuel is ignited and the whole cycle is repeated.

Room thermostat

This type of thermostat is manufactured in many different sizes, shapes and colours to fit in with most decorations. Irrespective of appearance they all perform the same function, which is to automatically close down or alternatively, operate the boiler at a pre-set temperature.

The thermostat should be fixed on an inside wall approximately 1.5–2 m above the floor. The

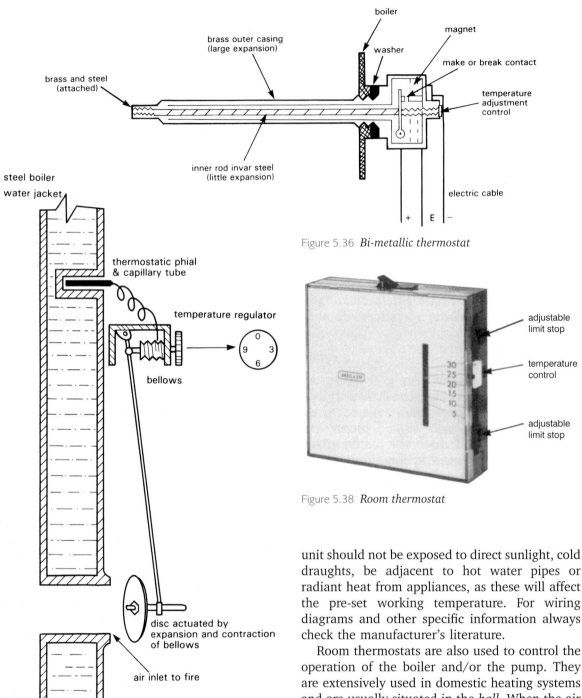

Figure 5.36 *Bi-metallic thermostat*

Figure 5.38 *Room thermostat*

Figure 5.37 *Automatic air control thermostat*

unit should not be exposed to direct sunlight, cold draughts, be adjacent to hot water pipes or radiant heat from appliances, as these will affect the pre-set working temperature. For wiring diagrams and other specific information always check the manufacturer's literature.

Room thermostats are also used to control the operation of the boiler and/or the pump. They are extensively used in domestic heating systems and are usually situated in the *hall*. When the air temperature of the hall has reached that at which the thermostat has been set, the electrical current to the appliance is broken and the boiler flame extinguished. Ignition of the flame is only restarted by a fall in temperature or adjustment of the thermostat setting. The only disadvantage of this control is that the whole building is heated to that setting and, due to the number of air changes that take place in the hail, it is quite possible to overheat the other rooms, resulting in a waste of heat and incurring unnecessary expense. This problem can be overcome by the fixing of individual thermostatic valves in each room, which is explained later in this chapter.

Cylinder thermostat

These thermostats are available as clip-on units to the outside of the hot water vessel, their function being to regulate the temperature of the water by automatically shutting down the boiler, pump or motorized valve when the water has reached the predetermined temperature (see Figure 5.40).

The unit is operated by a bi-metallic sensing element; some companies offer a unique remote sensor which is connected to the unit by cable. This allows flexibility in the placing of the unit.

Programmer

In addition to being able to control the temperature of the water as described, i.e. with the thermostat, it is also necessary to be able to set the boiler to start and shut off automatically at predetermined times of the day or night.

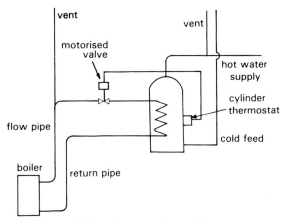

Figure 5.40 *Cylinder thermostat located in system*

Heating timeswitch

It may be that the system only requires a control to switch the heating circuit on or off. In this case a *timeswitch* has been designed which can be pre-set to operate the circuit once or twice a day. The selected programme can be overridden by the operation of the override selector on the switch. Some timeswitches automatically revert back to the set programme at the start of the next period, while others may be manually reset.

Pumps

The pump is a very important part of the modern central heating system. It enables the use of smaller diameter pipes and boilers than would be the case if a conventional gravity system was used. The function of the pump is to provide pressure inside the system which in turn will force water to circulate throughout the whole system of pipework and heat emitters.

The location of the pump within the system can be on the *flow pipe* (now generally accepted as the best position) or on the *return pipe*. In some instances it is connected adjacent to the boiler inside the boiler casing. One popular position is in the horizontal line with isolating valves at each side. These are to facilitate the pump removal for repair or replacement (see Figure 5.41).

Some pumps are designed to operate in either horizontal, inclined or vertical positions. Always check the manufacturers' information sheet as regards this point as damage could result if the

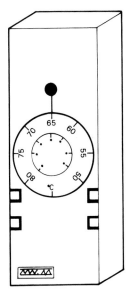

Figure 5.39 *Cylinder thermostat*

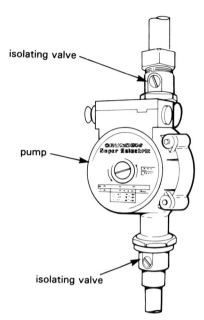

Figure 5.41 *Pump and isolating valves*

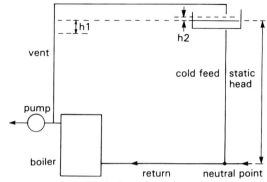

Figure 5.42 *Pump located on flow*

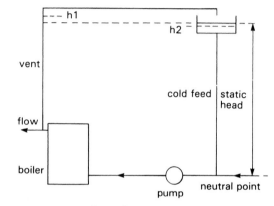

Figure 5.43 *Pump located on return*

pump is incorrectly located. Although the system may function satisfactorily with the pump fitted into the flow or return, the pressure should not be such as to force water out of the vent or create sub-atmospheric pressures in the system. This would result in an inefficient heating system and excessive corrosion caused by oxygen entering the system (see Figures 5.42 and 5.43).

Figure 5.42 illustrates the difference in levels which occurs when the pump fitted to the flow pipe is operating. The level in the cistern will increase by a small margin, h2, while the level in the vent will show a decrease equal to the pump pressure, h1. Figure 5.43 illustrates the difference in levels which occurs when the pump fitted in the return pipe is operating. A small quantity of water is drawn from the cistern showing a reduced level, h2. The level of water due to the pump pressure will show an increase, h1, in the vent pipe.

Safety valves

Although all systems are not fitted with safety valves, they are still an important consideration when dealing with boiler controls and should be fitted wherever possible. The most commonly used safety valve is of the type illustrated in Figure 5.44 which is of a spring-loaded control.

In addition to the valve shown in Figure 5.44 there are two other patterns: one is known as the *dead-weight valve*, the other as the *lever valve*. Both these types have been superseded by the *spring-loaded valve*, although you may still encounter all three types in site work. The working principle of each valve is shown in Figure 5.45. The function of the safety valve is to facilitate the escape of excess pressure in the system, should the pressure exceed that for which the system has been designed. The valve is set at the pressure of the *head of water*, plus an approved safety factor.

The recommended fixing position of the valve always causes some debate. If the water is *hard*,

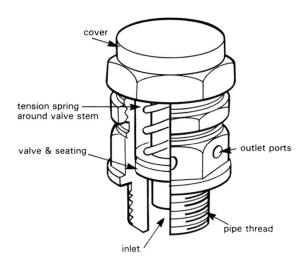

Figure 5.44 *Spring loaded safety valve*

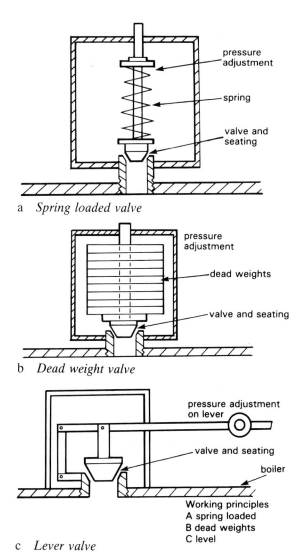

a *Spring loaded valve*

b *Dead weight valve*

c *Lever valve*

Figure 5.45 *Working principles of safety valves*

the type of hardness and the degree of hardness may have a bearing on the position chosen. In the case of temporary hard water, with the possibility of furring taking place in both the boiler and the flow pipe, serious consideration must be given to the position of the safety valve so that it is protected as far as possible from the risk of becoming sealed with the lime incrustation and so becoming useless. In this situation it may be advantageous to fix the valve in the return pipe where the least amount of furring will take place. It is of course argued that the valve in this position is now subject to fouling by sediment and possible corrosion but this is the lesser of the two problems. It is also believed that the pressure is greater at the top of the boiler, but this is not in fact the case as the pressure will be transmitted equally in all directions.

To sum up, if water is neutral or permanently hard, fix the valve either on or near the top of the boiler or flow pipe; if the water is temporarily hard (lime deposit), fix the valve on the return pipe.

Regular checking and maintenance of all safety valves are strongly recommended.

Motorized valves

These valves are manufactured as two port and/or three port controls with a motorized unit as an integral part. Both types are suitable for use in the fully pumped combined central heating and domestic hot water systems. The function of the valve is to control the flow of water through the system. To achieve full temperature control, room and cylinder thermostats should be used in conjunction with these motorized valves.

Two port valves

Figure 5.46 shows how both the domestic hot water and the central heating circuits can be controlled by the two port valves.

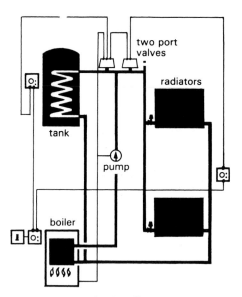

Figure 5.46 *Two port valve installation*

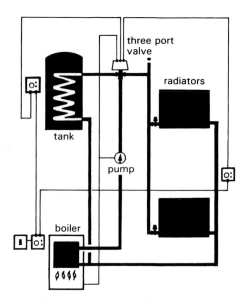

Figure 5.48 *Three port valve installation*

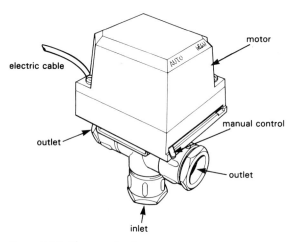

Figure 5.47 *Three port valve*

Figure 5.48 shows the positioning of a three port valve, also known as a diverter or mid-position valve. This valve will control both the domestic hot water and the central heating system simultaneously.

Check or non-return valve

These valves are manufactured in various shapes and sizes. Some are made from bronze with flanged plates; some with internal threads as shown in Figure 5.49, while others are manufactured from brass as illustrated in Figure 5.50 and

connected to the pipeline by normal compression joints. The function of the valve is to prevent a reverse flow taking place in a pipeline, as the valve will automatically close on to its seating in the event of such a mishap. Care must be taken to fit the valve as indicated by the direction arrow on the body of the valve. Some valves are suitable for horizontal fixing, others for vertical fixing, while others (see Figure 5.49) can be fixed in either position. Careful selection is essential. This type of valve is also used as an anti-gravity flow valve to prevent circulation of water in the heating system when the pump is switched off.

Economy valve

This is a form of three-way valve fitted into the return pipe from the cylinder to the boiler or heater. The function of this type of valve is to facilitate the heating of only part of the contents of the hot storage vessel when this is required.

Location of economy valve in system

Figure 5.52 shows the location of the valve in the system. Under normal working conditions with the valve open as in Figure 5.51 the whole of the contents of the hot storage vessel will be circulating through the heater. On occasions when only a small amount of hot water is required then by a simple quarter-turn of the valve head the

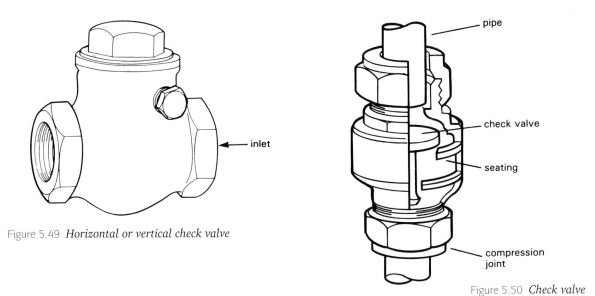

Figure 5.49 *Horizontal or vertical check valve*

Figure 5.50 *Check valve*

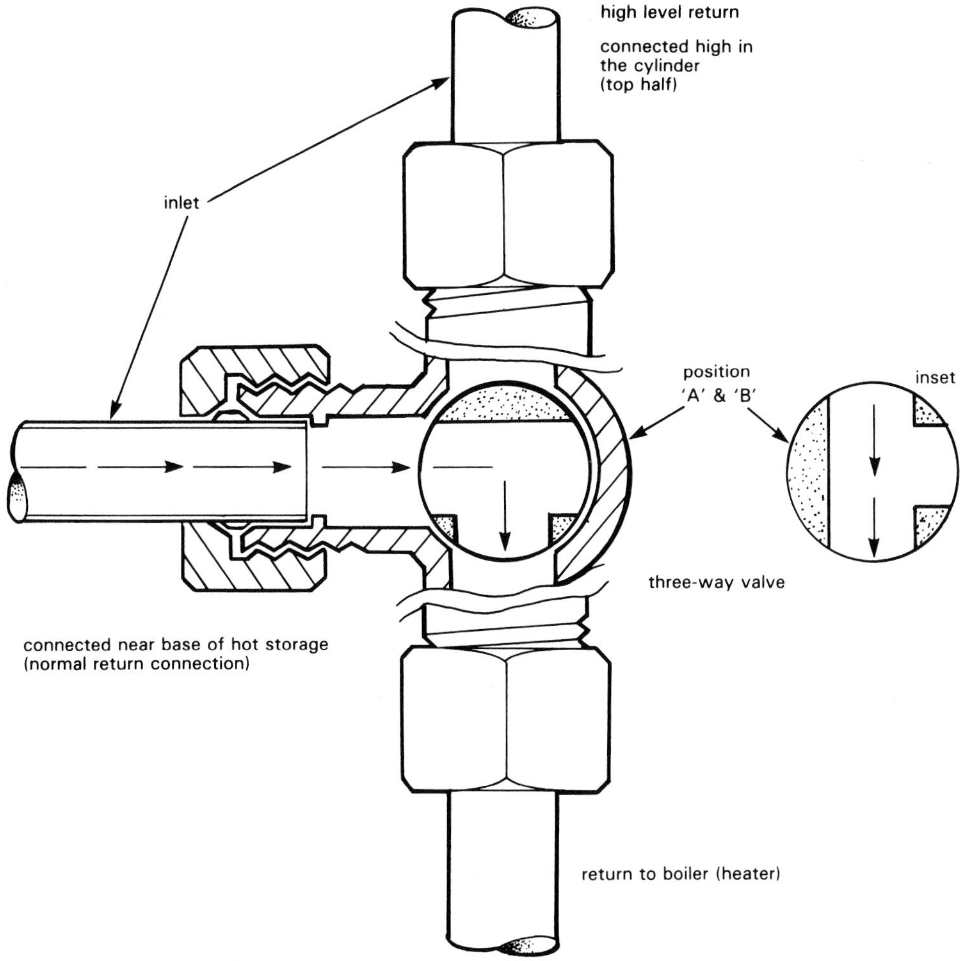

Figure 5.51 *Economy valve*

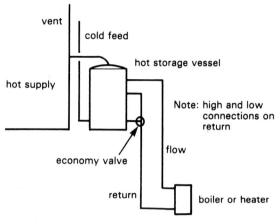

Figure 5.52 *Economy valve located in system*

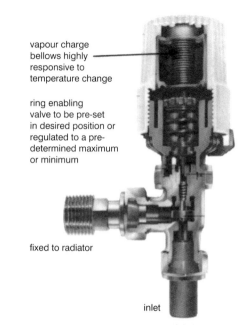

Figure 5.53 *Thermostatic valve*

waterways would be positioned as *inset*. This would then allow circulation of the water in the top half of the cylinder only, effecting a considerable saving in heating: hence its name, *economy valve*.

Thermostatic radiator valves

These valves are fitted to each heat emitter and control the temperature of rooms independently of each other. The fitting instructions supplied by the manufacturer must be carefully observed to ensure the valve's correct operation.

Thermostatic valves contain a heat sensitive fluid, wax or paste, which expands when the required air temperature is achieved – thus closing the valve: as the temperature in the room falls below that required, the sensitive compound contracts allowing the valve to open and hot water to flow into the heat emitter.

In systems where all the heat emitters are fitted with TRVs, it is essential that one, or a part of the circuit is kept open, so that in the event of all valves being closed and the pump starts, it is not operating against a static pressure.

TRVs are best fitted on the heating flow pipe leading into the heat emitter.

Lockshield valves

This type of valve is fitted to the outlet end of the radiator, its function being:

1 By adjustment of the regulator the flow of water through a radiator can be accurately

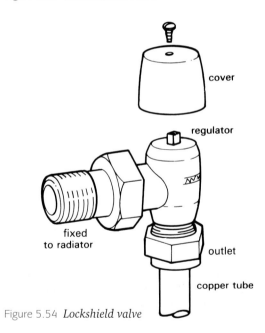

Figure 5.54 *Lockshield valve*

controlled. This may be necessary in the case of some radiators being starved of hot water. This is known as *balancing* the system.

2 The valve can be used as an 'off' control and, in conjunction with the other control valve, it is possible to remove the radiator for repairs or renewal or to allow re-decoration of the

wall behind the radiator to be carried out without draining down the whole system.

Heat emitters

There are several different types of heat emitters, they can be classified under two basic headings:

1 Radiators
2 Natural or fan assisted convectors.

Radiators (emitters)

The name given to this appliance would indicate that the heat given off is *radiant heat*, but this is not the case. Only a small amount (5–10%) is radiated (heat in straight lines from the source without heating the intervening space), the bulk of the heat being given off by conduction (the air is heated by touching the radiator) and circulating in the room by convection (the heated air moving). All three forms of heat movement are therefore being used.

It is difficult nowadays to know what name should be applied to this form of heat emitter. In our opinion the most correct term would be a 'heat conductor' with 'heat convector' taking second place. Because only approximately 10% of the heat is radiated heat the term 'radiator' would come a very poor third. It can be readily seen that, as in so many cases, the names given to some of our everyday commodities are sometimes misnomers, but perhaps the all-embracing term 'heat emitter' will best fit the bill.

There are many different types of emitters (radiators) available on the market today from the traditional cast-iron column type (see Figure 5.55) through to the pressed steel panel emitters, skirting panels and the more sophisticated thermal panels, all of which will be explained in some detail in this chapter.

Column (radiator) emitters

The open column types are generally more efficient as they present the maximum heating surface area to the air in the room. This type of emitter is the one used commercially, its main disadvantage being that because of its open construction it contains many dust traps. A variation of the column emitter is shown in Figure 5.56. The columns are of a solid or plain elevation, still presenting a fairly large surface area for heating purposes but not having the problem of the dust traps. This type of emitter is very hygienic and is therefore known as the hospital pattern. It is recommended where hygiene is listed on the requirements.

Figure 5.55 illustrates the column type emitter: the flow pipe can be connected either at the top or bottom of the emitter by means of a control valve. The return pipe is connected at the opposite end of the emitter to the bottom connection by means of a lockshield valve. Open columns present maximum surface area to the surrounding air which is heated by conduction (particles touching) and circulating by convection (particles moving), together with a small amount of radiated heat (direct rays). This type of emitter was originally made from cast iron. Today it is available in both cast iron or steel, the latter being more prone to corrosion than the former.

Figure 5.56 illustrates the similarity between the two types of heat emitters. The hospital pattern emitter would be fixed and connected to the heating circuit in the same manner. By virtue of the closed or plain columns dust traps are obviated.

There are on the market today a multiplicity of heat emitters (radiators) from the basic single panel (see Figure 5.57) to the double-panelled type as shown in Figure 5.58, and various

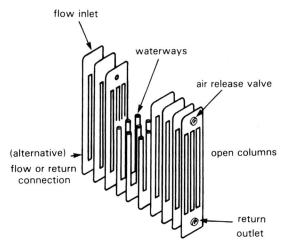

Figure 5.55 *Cast iron column emitter*

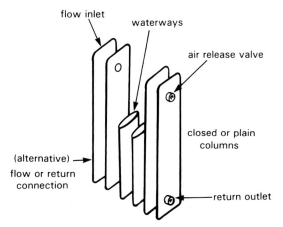

Figure 5.56 *Hospital pattern emitter*

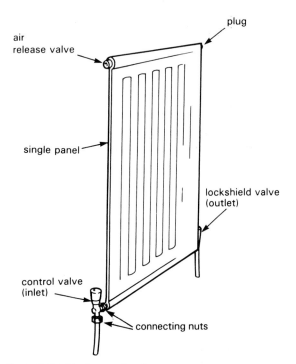

Figure 5.57 *Single panel emitter*

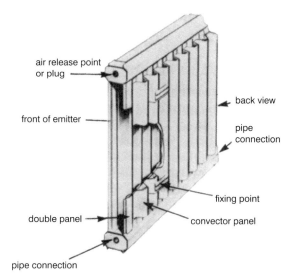

Figure 5.58 *Double panel convector emitter*

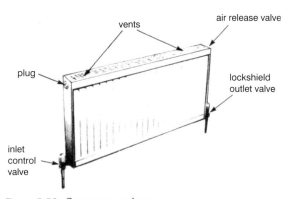

Figure 5.59 *Convector emitter*

modifications designed to increase the warmed surface area of the emitter as in Figure 5.58, which in turn will transfer more heat to the air surrounding it.

Siting of heat emitters

This is a very important factor and requires much consideration. Most draughts should as far as possible be eliminated but it is both impossible

and undesirable to seal a room completely: all rooms must have some ventilation and air change. The siting of the heat emitter should be as near as possible to the point of entry of the fresh air, i.e. near the door or at the most important place beneath the *window*. In the case of large areas it may be advisable to fit two emitters, one at each end of the room. To enable the emitters to function properly air movement (convectional currents) must not be restricted: the emitters should be fitted approximately 100 mm from the floor and 40 mm from the wall, which will allow cleaning access as well as air circulation. It is good practice to fit the emitter along the full length of the window and, in the case of a bow-window, at an extra cost the emitter

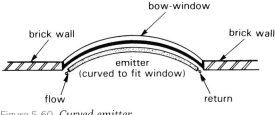

Figure 5.60 *Curved emitter*

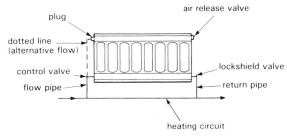

Figure 5.61 *Emitter controls and connections*

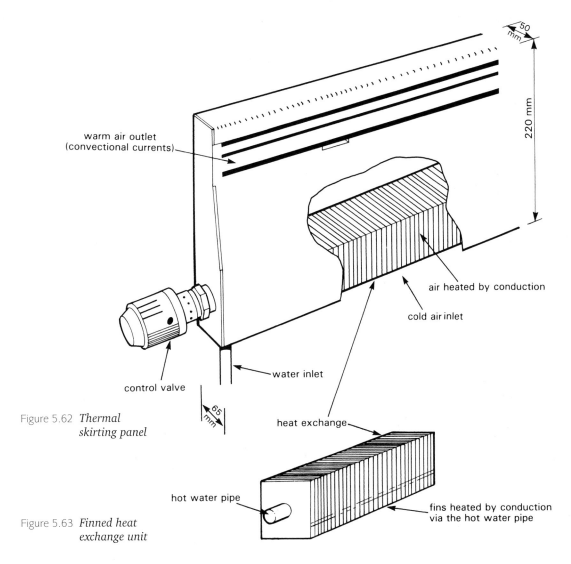

Figure 5.62 *Thermal
skirting panel*

Figure 5.63 *Finned heat
exchange unit*

can be curved to fit neatly into the curvature of the window.

The emitters should not be covered by curtains or fitted behind furniture as both would have an effect on efficiency. In addition, the heat and airborne dust can have a detrimental effect on both fabrics and the furniture. The efficiency of the system is also affected by the colour and type of paint used: for maximum radiated heat a black matt colour is the best, but no-one would suggest that this colour be used in any situation where appearance is to be a consideration.

In general the same points apply to all emitters, including the convector type, although it must be pointed out that in the case of the fan-assisted convector the situation is not so critical.

The emitters (radiators) normally have four tappings, two at the top and two at the bottom. It is possible to fit the flow connection either at the top or the bottom, with the return fitted at the bottom at the opposite end.

The flow pipe is connected to the emitter by means of a control valve (sometimes thermostatically controlled); the return pipe is connected by means of a lockshield valve (sometimes known as a balancing valve). An air release valve is fitted in one of the two top connections, its function being to bleed the system of air during the filling operation and it is sometimes required for the same purpose during the working of the system. The remaining spare tapping is simply plugged off.

Skirting pattern emitter

The cold air is surrounding and touching the heat exchanger which, when hot, will transfer its heat to the air by *conduction* (transfer of heat by bodies touching). Upon being heated the air becomes lighter and so *convectional currents* are set up: the warm air rises, passing out through the top vent into the room and being displaced by colder, heavier air (natural circulation of air).

This type of emitter is fitted, as the name implies, at skirting-board level or immediately above the skirting board; it is recommended that there is a space of 60 mm between the floor and the underside of the emitter.

Thermal panels

These are yet another variation of the space heating of a dwelling and in principle are similar to that of the skirting unit. The cold air in this case passes through the front panel where it is heated by being in contact with the hot-water pipes. The heated air then passes out of the top of the unit into the room by convectional currents (Figure 5.64).

Fan convectors

Fan-assisted convector heaters are perhaps the next progression from the skirting and thermal panel emitters described above. In this case the warm air, heated as previously described, is

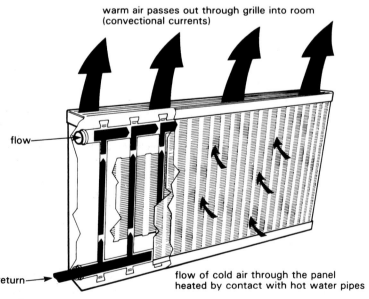

warm air passes out through grille into room (convectional currents)

flow

return

flow of cold air through the panel heated by contact with hot water pipes

Figure 5.64 *Thermal panel*

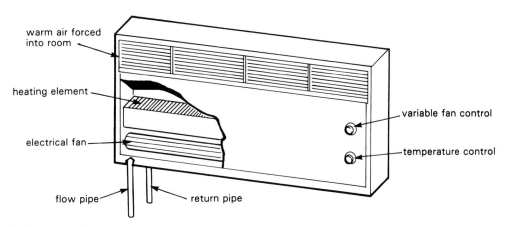

warm air forced into room

heating element

electrical fan

variable fan control

temperature control

flow pipe

return pipe

Figure 5.65 *Fan assisted heater*

forced through the heating element by a fan incorporated in the unit and out into the room (Figure 5.65).

These heat emitters are comparatively small and are therefore extremely useful where wall space is limited or where a larger unit would affect the decoration of the room. They are very attractive and neat in appearance and are also very effective: because they are fitted with a variable fan they are able to boost the heat of the room quite quickly. They are manufactured in a number of sizes, some capable of heating the largest room in a house, or in offices or shops. The one disadvantage is that when in operation there is the continual hum of the motor driving the fan, particularly when on high speed. However, the fact that the air in the room is subjected to a slight raise in pressure tends to prevent cold draughts entering that room. A further advantage over the ordinary radiator (emitter) is that the surface temperature is lower, no one can be burned, and fabrics adjacent to the heater are unaffected.

In the summer the unit can be usefully employed as a cooling fan.

Insulation

For a number of years the insulation of systems was carried out to prevent damage and possible waste of water caused by frost action. Today with the vast majority of homes having hot water and heating systems, and bearing in mind the high cost of fuel, the term insulation takes on a new meaning – that of 'saving' by preventing unnecessary heat loss from the hot water and heating system. Insulation is therefore of paramount importance in the home of today to assist in:

1 Prevention of loss of heat
2 Prevention of frost damage
3 Prevention of condensation.

Function of an insulator

It is well to realize that all insulators are of a spongy nature, containing minute air pockets, and by surrounding the appliance with this type of material, and due to the fact that still air is a bad conductor, the passage of heat through the material is very small. Broadly speaking, all insulators function in the same way, that is, to encircle the pipe or appliances with *still air*.

Efficiency

A good insulator will have an efficiency of approximately 80 per cent. It must be appreciated that this will be governed to a large extent by its thickness: the greater the thickness the better the insulation. To maintain its effectiveness the insulator must be kept dry; therefore should there be a possibility of its becoming wet it must be protected by a water-resisting material.

Colour

Colour is also an important factor with regard to heat emission: bright colours emit less heat than dull ones; therefore insulators are usually painted bright, shiny colours, e.g. red, whereas emitters (radiators) should be painted dull matt colours.

Properties of a good insulator

1 Should be durable under diverse influences.
2 Should be rot-proof.
3 Should resist mould growth.
4 Should be able to withstand the temperature range to which it will be subjected.
5 Should be able to withstand vibration and other rough treatment to which it may be subjected.
6 Should be water-repellent or rendered so.
7 Should not have a corrosive effect on the materials being covered.
8 Should be incombustible or at least fire-resistant.
9 Should not give off offensive smells at working temperature.
10 Should, if possible, be clean to handle and apply.
11 Should be vermin-proof.
12 Should have a surface finish easy to clean.
13 Should be pleasing to the eye.

Insulation of cylinders (hot storage vessels)

Modern cylinders are supplied pre-insulated with Polyurethane foam.

Thermal insulators

The insulating materials used today have improved tremendously over, the past few years. There is now a wide selection of insulators available ranging from modern spun glass to foams and plastics. Insulators come under two basic headings:

1 Cellular
2 Fibrous.

They are supplied for use in three main ways:

1 Pre-formed sections
2 Loose in-fill
3 Plastics powder.

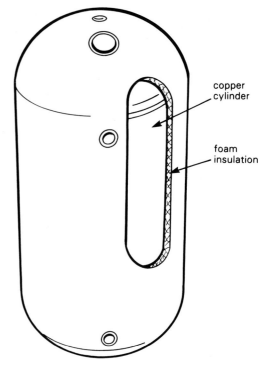

Figure 5.66 *Cylinder with foam insulation cover*

Safety

Great care must be taken when handling and working with some of the materials. Face masks should be worn when working with powders, and no skin exposed to contact with some of the fibrous materials, e.g. glass fibre.

Pipe insulation

Pipe insulation is made in various wall thicknesses for each pipe size and in both flexible and rigid sections. It is manufactured to fit standard copper and steel pipe sizes.

Flexible insulation

Foamed polyurethane provides one such flexible insulator which is both simple and easy to fit. It is available in lengths up to 3 m for pipes up to 76 mm in diameter.

Several types of flexible polyurethane insulators are marketed. Figure 5.67 shows one type which splits horizontally for ease of fitting; it is then slipped on to the pipe and sealed with waterproof adhesive tape. Another type has a nib along its full length to lock the insulation in place and is finished with a protective PVC sheeting. Flexible sections of glass fibre and mineral wool covered with felt or plastics and galvanized wire netting are recommended when the pipe is fixed externally, and when it may be subjected to adverse conditions. The insulating material is protected from the weather by felt which is secured by wire; it is then reinforced with wire netting. The whole is then painted with a bitumen solution as illustrated in Figure 5.68.

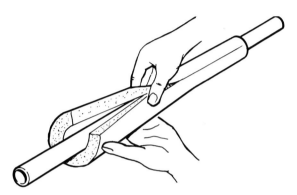

Figure 5.67 *Flexible insulation*

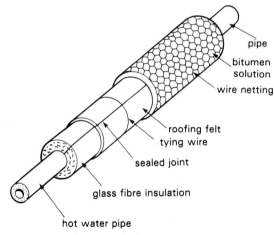

Figure 5.68 *Built-up insulation*

Rigid insulation

There are several types of rigid insulation available such as calcium silicate, cellular glass, glass fibre, cork, mineral wool or expanded plastics. They are usually supplied in half sections in 900 mm

lengths. Some of the insulation is jointed and simply snaps on and locks around the pipe. The usual method is to secure the sections with fixing bands every 450 mm, with additional bands at elbows and tee pieces; alternatively the sections may be stapled or fabric covers may be used.

The mitres are cut with a fine-toothed saw (hacksaw); joints may be sealed with adhesive tape.

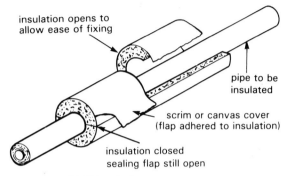

Figure 5.69 *Rigid insulation*

Loose in-fill insulation

As already mentioned there are several very good loose in-fill insulators such as cork, vermiculite, plastics, mineral and glass wool. These are very useful for insulating pipes running between joists as indicated in Figure 5.71.

1 The space between the joists housing the pipes is simply filled by the chosen loose in-fill insulating material. It is advisable to seal the space with a suitable cover. Gently tamp the loose in-fill material but under no circumstances should the material be rammed in hard or compressed. See Figure 5.71(A).

2 An alternative method would be to insulate the space between the joists in the normal way, then insulate the pipes individually with one of the approved pipe insulators and lay them on top of the roof insulation as shown in Figure 5.71(B).

3 Another alternative method is to lay, the pipes on the ceiling and to cover them with the normal roof insulation as in Figure 5.72.

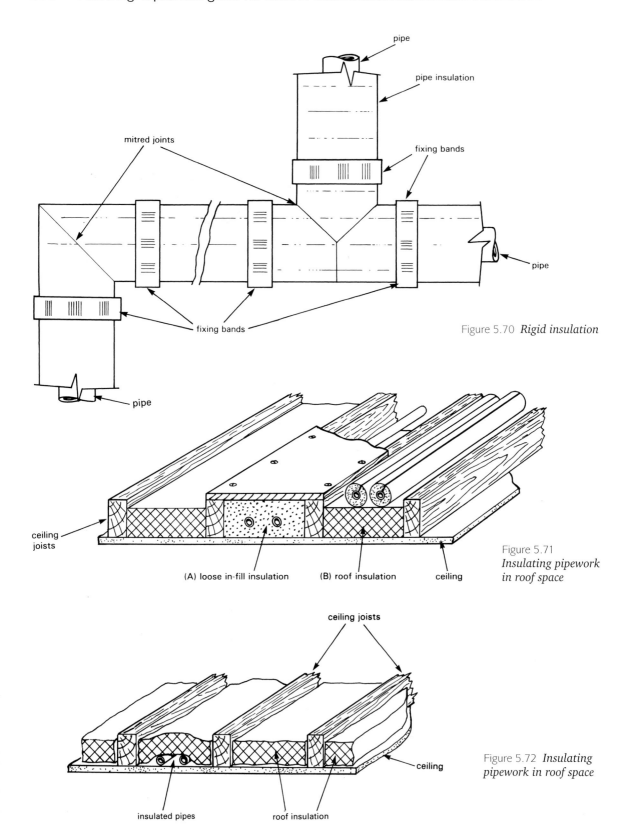

Figure 5.70 *Rigid insulation*

Figure 5.71
*Insulating pipework
in roof space*

Figure 5.72 *Insulating
pipework in roof space*

Self-assessment questions

1 The diameter of the cold feed in a domestic hot water system should be:
   (a) the same size as the supply to the feed cistern
   (b) the same size as the bath supply
   (c) not less than the diameter of the service pipe
   (d) not less than the diameter of the hot water draw off.

2 The total hardness content of water relates to the amount of:
   (a) temporary and permanent hardness together
   (b) permanent hardness only
   (c) temporary hardness and scale deposit together
   (d) carbon dioxide gas present in the water.

3 In a hot water installation the primary circulation is between:
   (a) towel rail and cylinder
   (b) boiler and cylinder
   (c) cylinder and taps
   (d) feed cistern and cylinder.

4 It is advisable to use an indirect system of hot water supply when the water supply:
   (a) is at a low pressure
   (b) has temporary hardness
   (c) has permanent hardness
   (d) is intermittent.

5 In a cylinder system of hot water supply the draw-off must come from:
   (a) open vent pipe
   (b) primary flow
   (c) cold feed
   (d) primary return.

6 An indirect system of hot water may be installed to prevent:
   (a) long runs of pipe
   (b) using too much water
   (c) the furring up of pipes
   (d) electrolytic corrosion.

7 A warning pipe is fitted to a:
   (a) feed cistern
   (b) hot water tank
   (c) hot water boiler
   (d) hot water cylinder.

8 The connection of the secondary return to a hot water cylinder should be:
   (a) adjacent to the secondary flow
   (b) level with the primary flow
   (c) opposite the cold feed pipe
   (d) not more than a third from the top of the cylinder.

9 Gravity circulation in a domestic hot water system depends upon:
   (a) conduction currents being set up
   (b) the pressure due to the head of water in system
   (c) residents drawing of water at the taps
   (d) the difference in density of the flow and return columns of water.

10 Heat is transferred through a metal by:
   (a) convection
   (b) radiation
   (c) emission
   (d) conduction.

11 Subjected to the same temperature increase which one of the following materials would expand the greatest overall length?
   (a) 10 metres of steel tube
   (b) 10 metres of copper tube
   (c) 10 metres of PVC gutter
   (d) 10 metres of stainless steel tube.

12 Condensing boilers are specifically designed to:
   (a) provide hot water directly
   (b) extract more heat from the flue gases
   (c) provide hot water at low temperatures
   (d) fit into a small space.

13 The boiler control interlock:
   (a) fixes the boiler securely to the wall or floor
   (b) ensures the boiler will stop when all TRVs are closed, and start again when open
   (c) ensures that the boiler does not fire when there is no demand for heat
   (d) enables more than one boiler to be connected together.

14 Which one of the following has a direction arrow → to show the position in which it MUST be fitted?
   (a) Non-return valve
   (b) Fullway valve
   (c) Drain valve
   (d) Float operated valve.

15 A combi boiler differs from a regular boiler because a combi boiler:
   (a) can be used on any size of heating system
   (b) must have a feed and expansion cistern
   (c) provides instant hot water at mains pressure
   (d) produces more hot water than a regular boiler.

16 Temporary hard water can be softened by:
   (a) adding fluoride
   (b) adding alum
   (c) filtration
   (d) boiling.

17 The mixing of the waters in a single feed indirect cylinder may be caused by:
   (a) primary water boiling due to thermostat failure
   (b) secondary water boiling due to thermostat failure
   (c) insufficient head pressure for the system
   (d) too much head pressure for the system.

18 The reason why the hot feed pipe from the top of a hot water storage cylinder must be positioned horizontally with a minimum of vertical rise before it connects to the vent pipe, is that this arrangement:
   (a) makes it easier to remove the cylinder for repairs
   (b) uses less pipe than other arrangements
   (c) allows the pipe to expand and contract better than other arrangements
   (d) reduces the loss of heat by convection from the top of the cylinder.

19 Storage systems differ from instantaneous hot water systems because they:
   (a) can provide lower flow rates which last until the store is exhausted
   (b) can provide higher flow rates which last until the store is exhausted
   (c) use a plate heat exchanger and a cylinder
   (d) cannot provide balanced pressures for hot and cold water services.

20 A strap-on cylinder thermostat should be fitted on a cylinder:
   (a) one-third the height of the cylinder down from the top
   (b) one-half the height of the cylinder
   (c) one-third the height of the cylinder up from the base
   (d) two-thirds the diameter of the cylinder down from the top.

6 Sanitation systems

After reading this chapter you should be able to:

1 Name common materials used for the manufacture of sanitary appliances.

2 Recognize different types of domestic sanitary appliances.

3 Select particular sanitary appliances for stated locations.

4 Understand the basic design criteria applied to sanitaryware.

5 Describe the working principles of domestic sanitary appliances.

6 State the recommended fixing heights of sanitary appliances.

7 Demonstrate Knowledge of the main provisions of the Building regulations, British and European Standards relating to sanitary pipework systems.

8 Identify different types of traps, understand and state the reasons for loss of seal in traps.

9 Describe preventive measures to eliminate loss of water seals in traps.

10 Recognize and identify systems of above ground discharge pipework and sanitation for dwellings, small industrial, commercial and public buildings.

11 Name common materials used for pipework systems and understand jointing techniques, including connections between above ground and below ground discharge systems.

12 Illustrate by diagrams the layout of above ground discharge and sanitation systems, appliances and components.

13 Recognize and identify different types of urinal.

14 State the purpose of domestic waste disposal units and describe the operating cycle of these units.

15 State the purpose and functional requirements of above ground discharge pipework and sanitary systems.

16 Describe and apply the correct procedures for testing and commissioning sanitary pipework systems.

Introduction

Sanitary fittings are usually divided into two groups:

1 Those which are intended to receive waste products from the human body.

2 Those required for dirty, soapy or greasy water.

The first group, which can be referred to as 'soil fitments', includes water closets and urinals, while the second group, often referred to as 'ablutionary' or 'waste fitments', includes baths, wash basins, showers, sinks and bidets.

The general design of sanitary fitments has reached a high standard. For domestic work a great diversity now exists in style and pattern, from costly coloured suites to the more orthodox fitments complying with the basic requirements of the appropriate British and European Standards.

A good sanitary fitment should be of the simplest possible design, constructed so as to be self-cleansing, and, as far as possible, free from any moving working parts. It should be capable of being easily connected to the appropriate waste pipe or drain, be completely accessible and free from all insanitary casings. The outlet should

Table 6.1 *Materials used for sanitary appliances*

Plumbers need to be able to identify and name materials used for the manufacture of sanitary appliances. Clients often ask about the selection of materials available and advantages and disadvantages of each.

Material	Use	Advantages	Disadvantages
Vitreous china	WC pans WC cisterns Bidets Bowl urinals Wash basins Drinking fountains	Easily available Medium cost Medium weight Stain free Non-corrosive	Will chip and crack
Fireclay	Belfast sinks Cleaner's sinks Shower trays Stall and slab Urinals	Strong Hard wearing Stain free Non-corrosive	Heavy Expensive Glaze will chip
Cast-iron – Porcelain enamelled	Baths	Strong Stain free	Heavy Glaze will chip
Pressed steel – Porcelain enamelled	Baths Sink unit tops Vanity wash basins	Lightweight Stain free	Easy to scratch and chip
Stainless Steel	Sink unit tops Bowl and trough Urinals WC pans Wash basins	Lightweight Very strong Non-corrosive	Expensive Noisy in use
Plastics – Acrylic, Resin Bonded Fibreglass	Baths Sink unit tops Wash basins Shower trays WC cisterns	Lightweight Medium cost Non-corrosive	Easily damaged Will burn
Armastone – Armacast, Polymer concrete	Baths Shower trays	Strong Non-corrosive	Heavy

be simple and capable of permitting the rapid emptying of the fitment. The overflow, where provided, should be efficient and accessible for cleansing purposes.

Sanitary fitments should be constructed of hard, smooth, non-absorbent and incorrodible materials. For domestic use sanitary appliances are usually made from one of the following materials (Table 6.1):

Vitreous china

Earthenware

Cast iron

Pressed steel

Plastics (perspex, glass-reinforced polyester, polypropylene)

Stainless steel

Ceramic fireclay

Synthetic compounds.

Sinks

Sinks for domestic purposes are available in a range of sizes and are manufactured from the following materials:

Ceramic glazed fireclay or earthenware

Vitreous enamelled steel

Stainless steel

Fibreglass

Plastics

Synthetic compounds.

Ceramic sinks are available in two patterns.

1 London pattern

2 Belfast pattern.

The London pattern sink is different to the Belfast pattern in that it does not have an overflow. The Belfast sink (see Figure 6.1) has emerged as the most popular of the ceramic sinks and is available with depths of up to 255 mm. These sinks are fitted with an integral weir type over-flow (see Figure 6.2). Ceramic sinks should be supported on cast iron cantilever type brackets. Fixing heights for sinks may vary according to particular requirements, but are generally fitted

900 mm from floor level to the top front edge of the sink.

The water supply to ceramic sinks is usually via bib valves fixed to the wall above the sink, but if the sink has been incorporated into a unit or worktop, the supply may be from pillar standard valves fitted to the worktop or unit. Valves fitted above the sink should be at a height to enable a bucket to be filled.

The outlets to ceramic sinks may be rebated or bevelled. The waste fitting will need to be slotted if the sink has an overflow. The fitting itself is math watertight with the use of silicon or mastic compound and nylon, rubber or plastics washers.

Figure 6.1 *Belfast sink*

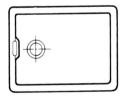

Figure 6.2 *Belfast sink*

Vitreous enamelled steel and stainless steel are popular materials for sinks. Patterns with single and double drainers and single or double sinks can be obtained. Metal sinks can be supported on cast iron cantilever brackets, a wooden base unit or inset into a worktop. Most metal sinks are provided with valve holes for 12 mm diameter pillar standard valves or a mixer valve with a swivel arm. Additional fittings such as flexible hand sprays to assist with washing-up are available.

Kitchen sinks are supplied with cold water direct from the service pipe and where mixer valve are fitted they must be of the biflow pattern which does not allow the cold and hot water to mix within the valve. Mixing takes place after the water has left the swivel arm or spout outlet.

Most metal sinks have an overflow combined with waste outlet.

Wastes for domestic sinks should not be less than 38 mm diameter. Control of the outflow is by plug and chain, or a pop-up plug assembly.

Showers

A shower is considered to be more hygienic than a bath because the used water is continually, washed away. They are particularly well suited for cleaning a number of people simultaneously or in quick succession. A shower occupies less floor space than a bath, and consumes less water. Shower trays (Figures 6.3, 6.4 and 6.5) are manufactured from:

Acrylic

Glass-reinforced polyester

Ceramic glazed fireclay.

Shower trays are available in a wide range of sizes.

When the appliance is used only as a shower tray, it will not require a plug and chain and the waste fitting can be flush topped (see Figure 6.6), but when the tray is also used as a foot bath the waste fitting must be recessed to receive a plug (see Figure 6.7). The tray must also be fitted with an integral overflow.

Prefabricated shower cabinets are available complete with all fittings, and when assembled

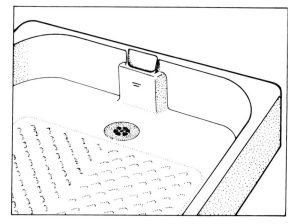

Figure 6.4 *Ceramic shower tray with integral overflow*

Figure 6.5 *Acrylic shower tray*

Figure 6.3 *Ceramic shower tray*

Figure 6.6 *Flush top waste fitting*

Figure 6.7
Recessed waste fitting

Figure 6.8 *Vitreous china wall-hung basin*

Figure 6.9 *Vitreous china pedestal basin with integral back overflow*

Figure 6.10 *Vitreous china inset basin with integral front overflow, concealed mixer valve and pop-up waste*

only require connection to water supply and waste services to be ready for use.

Wash basins

Wash basins are made in a wide variety of shapes and sizes, ranging from small cloakroom basins to large bathroom fitments. Each type offers different refinements to the basic functional requirements of a wash basin.

For identification purposes wash basins may be classified into three groups:

1 Wall basins (Figure 6.8).
2 Pedestal basins (Figure 6.9).
3 Vanity or inset basins (Figure 6.10).

Wall basins

These are available with various means of support The fixing may be concealed, employing the use of secret brackets, or cantilevered from the wall by means of a corbel which is part of the basin. This corbel is built into the wall with cement/ sand mortar. Traditional bracket fixing involves the use of cast iron support brackets which may be built in (Figure 6.11) or screwed to the supporting wall (Figure 6.12).

The main advantage with wall fixed basins is that the floor beneath is left clear.

Wall basins are manufactured to fix to a flat wall surface or to be contained within a 90° corner. This type of basin is called an angle basin (Figure 6.13).

Figure 6.11 *Built-in cantilever bracket*

Figure 6.12 *Screw to wall bracket*

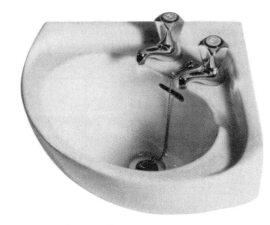

Figure 6.13 *Vitreous china wall-hung angle basin*

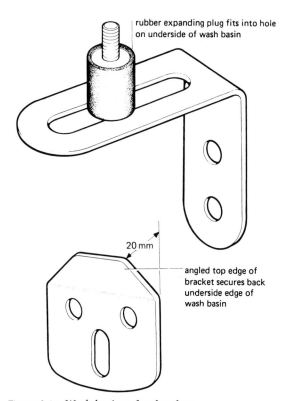

rubber expanding plug fits into hole on underside of wash basin

20 mm

angled top edge of bracket secures back underside edge of wash basin

Figure 6.14 *Wash basin safety brackets*

Pedestal basins

The pedestal provides a means of support for the wash basin and should be large enough to conceal the waste services and supplies. The basin should also be secured to the back wall. This is usually done by screw fixing through holes provided under the back edge of the basin or by using safety brackets (Figure 6.14). Many manufacturers provide a bracket to secure the basin to the pedestal. The pedestal itself should also be secured to the floor.

Vanity basins

These are manufactured to be inset into a cabinet, worktop or shelf unit. Some types of basin have a metal trim to make the seal between basin and surrounding surface (Figure 6.15), others are self-rimming (Figure 6.16).

These basins can be used singly or in range form, often saving valuable space. They may also be used to provide supplementary facilities in bedrooms, dressing rooms and cloakrooms.

Wall basins and pedestal basins are usually manufactured from vitreous china. Vanity basins may also be made from this material and pressed steel, stainless steel and plastics.

The control of the water supply into a wash basin is usually by 12 mm pillar valves which are

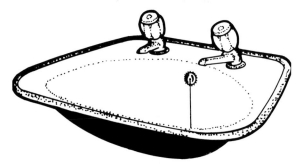

Figure 6.15 *Vitreous china inset basin with integral back overflow and recessed slotted waste fitting*

Figure 6.16 *Pressed steel inset basin with stainless steel trim, combined waste and overflow*

connected to the hot and cold service pipes. Several manufacturers supply spray valves and mixer valves suitable for wash basins.

The valves discharging water to a wash basin must have their outlets above the flood level of the appliance to obviate the risk of siphonage or contamination.

The waste outlet from a washbasin should never be less than 32 mm in diameter and should incorporate a slot to accommodate the overflow integral to the basin. Control of the outlet is usually by plug and chain, although pop-up waste units are available. Self draining recesses to hold soap are formed within the horizontal top edge of the basin.

Baths

Baths designed for domestic use vary little in basic shape whether made from cast iron, pressed steel or acrylic materials (see Figure 6.17). However, refinements in design are available beyond the standard rectangular shape: handgrips, recesses for soap, non-slip bottoms and drop fronts to make access easier and safer are features now incorporated by most manufacturers. Whirlpool and Spa pattern baths are also becoming more popular in domestic locations.

Also available are corner baths which use up more floor space and are generally more expensive than a rectangular bath, and 'sitz' baths which are short and deep and incorporate a seat. This type of bath may be used where floor space is limited and is also suitable for elderly people, since the user maintains a normal sitting position.

The following materials are used for the manufacture of baths: Vitreous enamelled cast iron Vitreous enamelled pressed steel, Acrylic plastics (perspex) and glass-reinforced polyester.

The control of the water supply into a bath is usually by 19 mm pillar valves or via a mixer valve which may also incorporate a diverter to direct the flow of water to a shower rose.

The waste outlet from a bath should never he less than 38 mm in diameter. The waste fitting may incorporate a slot to receive any water which leaves the bath via the overflow. Alternatively the overflow may connect into the trap itself (if

Figure 6.17 *Typical domestic bath complete with waste overflow, handgrips, side and end panels*

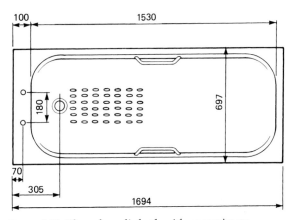

Figure 6.18 *Plan of acrylic bath with approximate dimensions in millimetres*

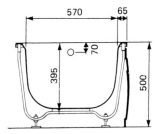

Figure 6.19 *Section through acrylic bath showing approximate dimensions and typical support detail*

permitted by the water authority), or discharge through the external wall.

Rectangular baths are manufactured in standard sizes ranging from 1.5 m to 1.85 m in length and 0.7 m to 0.85 m in width (see Figure 6.18). Most baths have four or five feet which are adjustable to gain the required height and level position (see Figure 6.19). Baths are fitted level as the necessary fall is provided in the bottom by the manufacturer. The height at which a bath is fitted may be governed by several factors which may include:

1 Structural limitations

2 Height of bath panel (if moulded)

3 Customer requirements

4 Sufficient space to allow waste pipe connection to be made.

Generally baths are fitted as low as is possible for ease of access, safety and good appearance. When fixing it must be remembered that using a bath exerts considerable force on its feet and if the feet are resting on a timber floor settlement may take place. This can be eliminated by fixing steel plates underneath the feet to distribute the load and building in the side and end sections of the bath to their adjoining walls. Some manufacturers provide brackets for securing the bath to the wall.

There are several types of leg, cradle, or support available for baths and when fixing any bath it is most important to adhere to the manufacturer's fitting instructions.

Water closets

A water closet consists of a pan containing a quantity of water and a vessel or device capable of providing a flush of water to remove excreta, wash any soiled surfaces clean hand re-seal the trap in the pan.

Water closet pans are manufactured from:

Vitreous china

Glazed earthenware

Stainless steel.

Flushing cisterns are manufactured from:

Vitreous china

Glazed earthenware

Stainless steel

Pressed steel

Cast iron

Plastics material.

Water closet pans may be divided into *two* categories: *washdown* (Figure 6.20) and *siphonic* (Figure 6.21).

Within these categories, there are several types of each, determined by details such as whether it is floor supported (pedestal) or wall supported (corbel), and the type and position of outlet: Most modern WC pans are manufactured with a horizontal outlet, there are still many of the 'P' and 'S' type pans still in use.

Washdown pans are the most common form of water closet, and the modern pedestal washdown pan has been developed from earlier short and long hopper types.

Corbelled washdown closets are used in situations where hygiene is particularly important or where a clear floor area is required. Two types are available, one requiring a firm solid back wall into which the corbel extension of the pan is built, and the other which rests on a metal frame fixed to the floor and back wall.

Siphonic pans

Siphonic water closets are more positive and silent in operation than are washdown closets, and rely on siphonic action to remove the contents from the pan. Siphonic closets are either single trap or double trap (as in Figure 6.21).

Single trap

Single trap siphonic pans are designed so that the bowl of the pan and its outlet form the short and long leg of a siphon. The outlet or long leg is shaped so as to restrict the outlet flow of water. The operation is as follows. The cistern is flushed and water passes via the open flushing rim into the pan and through to the long leg of the siphon. Due to the partial restriction in the outlet, this momentarily fills with water which, as it runs away, causes siphomc action which removes most of the water and contents of the trap. The trap is resealed with water from a special 'retention' or 'after flush' chamber built into the pan itself.

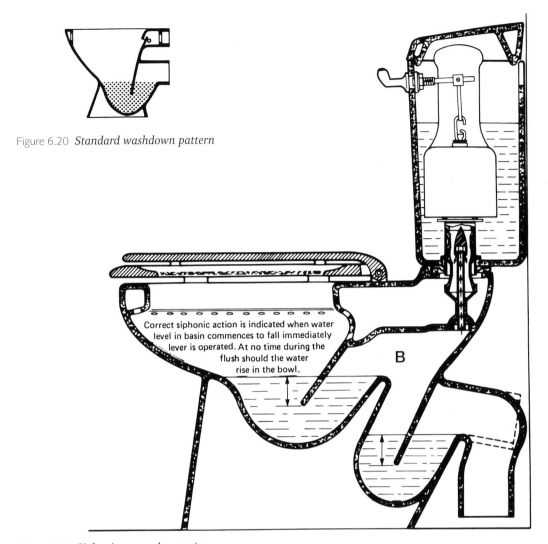

Figure 6.20 *Standard washdown pattern*

Correct siphonic action is indicated when water
level in basin commences to fall immediately
lever is operated. At no time during the
flush should the water
rise in the bowl.

B

Figure 6.21 *Siphonic water closet suite*

Double trap

Some double trap siphonic pans (see Figure 6.21) use a pressure reducing valve which is located between the outlet leg of the cistern siphon and the air chamber between the two traps. The operation of this type of suite is as follows.

The cistern is flushed, and water flows past the pressure reducing valve. Air is drawn through the valve from chamber B so creating a drop in air pressure in chamber B. This causes the water level in the pan to drop and siphonic action commences. The water from the cistern enters the pan via the perforated flushing rim, assisting with the siphonic action, and washing clean the pan surfaces. As the siphon ceases, this water reseals the traps in the pan.

Pedestal type pans

Pedestal type pans are manufactured with two or four holes through their base to provide screw fixing to the floor beneath. Ideally, a pan should be fixed so that it can be disconnected at a later date if necessary. The pan must be securely fixed to the floor, eliminating the possibility of movement.

The method of jointing the pan outlet to the drain/discharge pipe will depend upon factors such as:

1 Whether the supporting floor is solid (concrete) or suspended (timber).

2 The material from which the drain/discharge pipe is manufactured.

3 Whether the pipe connection is spigot or socket/collar.

When a WC pan is supported by and fixed to a solid floor, the method of jointing may be rigid or flexible. If the floor supporting the WC pan is suspended and movement or settlement may occur, the jointing method should provide flexibility to allow for any subsequent structure movement (see Figure 6.22). A wide range of multifit WC outlet connectors are available to suit most situations encountered by the plumber.

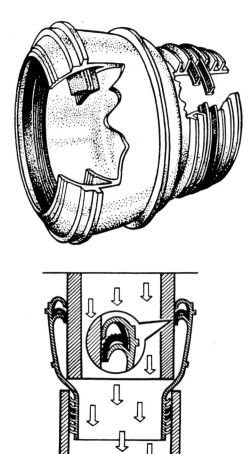

Figure 6.22 *Flexible jointing for WC outlet*

Flushing cisterns

Modern types of flushing cisterns are also known as water waste preventers (WWPs) because, under normal conditions, water cannot leave the cistern other than by siphonic action. There are two main types of WC flushing cistern:

1 Plunger or piston (Figure 6.23)

2 Bell (Burlington) (Figure 6.24).

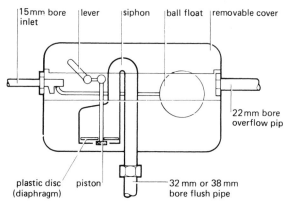

Figure 6.23 *Plunger or piston flushing cistern*

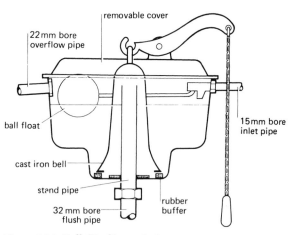

Figure 6.24 *Bell (Burlington) cistern*

Both types employ siphonic action to remove the water from the cistern to the flush pipe/WC pan. The capacity of the cistern is governed by the Water Regulations which stipulate a 6 litres flush of water.

The cisterns operate as follows.

Plunger or piston cisterns

When the lever handle, push button or chain pull is operated, the diaphragm plunger is lifted. As it rises, it displaces water over the crown of the siphon. This water falls down the long leg of the siphon taking some air with it which creates a reduction in air pressure in the siphon. The greater air pressure acting upon the surface of water in the cistern forces water past the diaphragm and through the siphon until the water level is low enough in the cistern to permit air to enter the siphon. This breaks the siphonic action and causes the flush to cease. The cistern refills via the valve to its working water level.

Some manufacturers produce a dual flush cistern. These incorporate an air relief tube (Figure 6.25) in the diaphragm chamber which breaks the siphonic action when the cistern is half emptied so reducing water consumption. The dual flush operates as follows.

When the lever handle is depressed and immediately released, the siphonic action commences and empties the cistern until air is drawn in through the air relief tube. This breaks the siphonic action at this level, so that a smaller quantity of water is used.

When a full flush is required, the lever handle is depressed and held in this position. This causes the diaphragm to seal the air relief tube and a normal flush takes place, the handle being released when the siphonic action ceases.

Bell flushing cistern

A pull action on the chain causes the lever arm to move on the fulcrum and raise the bell inside the cistern. When the chain is released, the bell drops and displaces water which flows down the funnelled standpipe into the flush pipe. As this water drops, it takes air with it creating a reduction of air pressure inside the bell. Atmospheric pressure acting upon the surface of the water in the cistern forces this water under the lower edge of the bell through the bell and down the standpipe. This siphonic action ceases when the water level in the cistern is low enough to permit air to enter the bell and equalize pressure.

Fixing of cisterns

Cisterns may be fixed at either high level, low level, or close coupled to the WC pan. The high level position is most efficient for washdown WC pans and should be fixed 1.8 m above floor level to the underside of the cistern to ensure a good flush. Plunger or bell type cisterns may be used for high level suites.

Low level and close coupled cisterns should be fitted according to manufacturers' instructions, and are of the plunger type. Bell type cisterns are only suitable for high level installation Modern practice favours low level or close coupled positions because of their neater appearance and ease of accessibility for maintenance and cleaning. Inlet and overflow connections may be either side or bottom according to site circumstances.

Flush pipes

Because of the greater head available above the WC pan, high level cisterns require a flush pipe of only 32 mm in diameter, whereas low level cisterns require a flush pipe diameter of 38 mm to compensate for loss of head. Close coupled suites do not have a flush pipe. All flush pipes should be as straight as possible as bends impair the efficiency of the flush.

Flush pipes, have been made from various materials including:

Lead

Copper

Plastics

Galvanized mild steel

Stainless steel

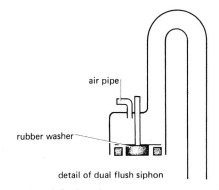

Figure 6.25 *Dual-flush siphon incorporating air relief tube*

air pipe

rubber washer

detail of dual flush siphon

Enamelled pressed steel.

(Modern flush pipes are usually plastics.)

They are identified by the number of sections, i.e. one-, two- or three-piece, and the location where the cistern is fitted in relation to the WC pan, i.e. low level, high level, backwall fixing, side wall fixing.

The flush pipe is usually jointed to the cistern outlet in one of two ways. The first method is by a 'cap and lining' type connection, where the end of the flush pipe has a formed shoulder which is sealed on to the cistern outlet via a washer. The second method uses a compression type joint where the flush pipe passes up inside the cistern outlet and the seal is made via a neoprene 'O' ring which may be captive in the siphon outlet, or loose to be compressed via the tightening of a nut.

The flush pipe may be jointed to the WC pan by: an internal rubber or plastics cone (Figure 6.26); an external rubber or plastics cone (Figure 6.27) or a Suttons collar (Figure 6.28) Suttons collar may only be found on old installations.

With all flush pipe connections, it is essential that the pipe enters the inlet collar centrally and that no jointing compound is allowed to enter into the flushing rim. Otherwise a faulty, ineffi-

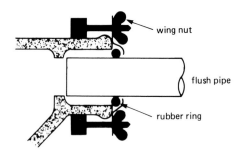

Figure 6.28 *Clamping ring (Suttons collar)*

cient flush will result, and one side of the pan will receive more water than the other.

Bidets

A bidet is a sanitary fitting used for washing the lower parts of the body and in appearance resembles a pedestal WC – but is shaped differently to suit its special purpose. The bidet provides a convenient means of cleaning the excretory organs and is useful to women during periods of menstruation. A secondary but nevertheless important use of the bidet is a footbath. For identification purposes, bidets are usually classified in two distinct types:

1 submerged inlet (Figures 6.29 and 6.30).

2 over-rim-supply (Figure 6.31).

Most bidets are designed in such a way that the waste can be plugged and the bowl filled with water and the lower parts of the body washed much more conveniently than by use of a conventional bath. There is also not such a waste of water because the contents can be quickly run off and the ascending spray used as a swilling douche.

The valve controls are conveniently placed to enable the bather to adjust the flow rate and temperature of the water.

With through-rim-supply bidets (a type of submerged inlet bidet), the rim seat is warmed with a supply of hot water which then fills the bowl for washing. Alternatively, the water supply can be diverted through the douche, which is available in either spray or jet nozzle, whichever is preferred. Water Regulations have special requirements for bidets with douche attachment (submerged inlet).

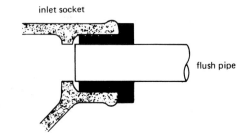

Figure 6.26 *Internal connector*

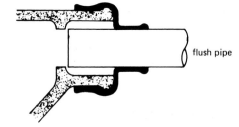

Figure 6.27 *External connector*

Figure 6.29 *Vitreous china bidet with concealed mixer valve, rim supply, centre spray douche, pop-up waste and integral overflow*

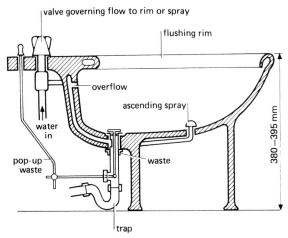

Figure 6.30 *Section through bidet showing flushing rim, ascending spray and pop-up waste*

Figure 6.31 *Vitreous china rimless bidet with over-rim-supply and integral overflow*

Bidets with over-rim-supply are simpler. They have no douche, just the bowl for washing purposes, and are supplied with water in the same manner as a wash basin.

Bidets for domestic use are made from vitreous china.

Definitions

To enable the reader to understand sanitary pipework systems, some knowledge of the terms used is necessary.

The Building Regulations, British and European Standards use the following definitions relevant to the design and installation of soil and waste discharge systems:

1 *Access cover.* A removable cover on pipes and fittings providing access to the interior of pipework for the purposes of inspection, testing and cleansing (see Figure 6.32).

2 *Branch discharge pipe.* A discharge pipe connecting sanitary appliances to a discharge stack (see Figure 6.32).

3 *Crown of trap.* The topmost point of the inside of a trap outlet (see Figure 6.32).

4 *Discharge pipe*. A pipe which conveys the discharges from sanitary appliances (see Figure 6.32).

5 *Soil appliances*. Fitting which receives the waste products of the human body, includes water closets, urinals, and slop sinks.

6 *Soil pipe*. Means a pipe (not being a drain) which conveys soil water either alone or together only with waste water, or rainwater or both (this term is not generally used now, the pipe being designated a discharge pipe).

7 *Stack*. A main vertical discharge or ventilating pipe (see Figure 6.32).

8 *Trap*. A fitting or part of an appliance to retain water or fluids so as to prevent the passage of foul air (see Figure 6.32).

9 *Ventilation pipe*. Means a pipe (not being a drain) open to the external air at its highest point (see Figure 6.32), which ventilates a drainage system, either by connection to a drain, or to a soil pipe or waste pipe, and does not convey any soil water, waste water or rainwater.

10 *Waste appliance*. A fitting which receives waste water, including wash basins, bath, shower tray, bidet, drinking fountain, sink.

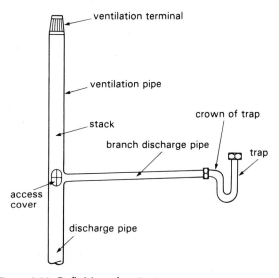

Figure 6.32 *Definition of sanitation terms*

Regulations relevant to sanitary discharge pipe systems

Regulations are necessary in the construction and building industry to ensure that a structure and its component parts are suitable and safe for the purpose for which they are designed. Building Regulations cover the main legislation relating to building drainage for above- and below-ground systems of discharge pipework. The following is a summary related to above-ground discharge pipework and sanitation systems.

1 All discharge pipes must be of adequate size for their purpose and must not be smaller in diameter than that of the largest trap or branch discharge pipe connected to and discharging into it.

2 Provision must be made, where necessary, to prevent the loss or destruction of trap seals.

3 All pipes and fittings used for the discharge of soil or waste and the ventilation of above-ground discharge systems must be made of suitable materials, these having the required strength and durability for this purpose.

4 All joints must be made in such a way as to avoid obstructions, leaks or corrosion.

5 Bends must have an easy radius and should not have any change of cross-sectional area throughout their length.

6 Pipes must be adequately secured to the building structure or fabric without restricting their movement due to thermal expansion or contraction.

7 The discharge system must be capable of withstanding an air test when subjected to a minimum pressure equivalent to 38 mm head of water for 3 minutes minimum.

8 Pipework and components must be accessible for repair and maintenance, and means of access must be provided for clearing blockages in the system.

9 Every sanitary appliance must be fitted with a suitable trap as close as possible to its outlet. Each trap must have an adequate water seal and access for cleaning. This does not apply to appliances which have an integral trap as part of the appliance, e.g.

water closets, or where appliances such as wash basins, sinks or baths are connected to form a range and each appliance discharges into an open half-round channel which discharges to its own trap – usually a gully – or where the range connects to a common waste pipe which is itself trapped.

10 No discharge pipes on the exterior of a building may discharge into a hopper head or above the grating of an open drain inlet or gully.

11 Discharge pipes carrying soil or waste water from a sanitary appliance must not be fitted on the exterior of a building except (a) on low-rise buildings of up to three storeys in height; or (b) where the building was erected before the 1976 Building Regulations came into force, and the discharge system is being extended or altered; or (c) on the lower three storeys of a high-rise building.

Above ground discharge pipework and sanitation systems

General principles

Soil and waste pipe discharge systems should comprise the minimum of pipework necessary to carry away the foul and waste water from the building quickly and quietly and with freedom from nuisance or risk of injury to health. The system should satisfy the following requirements:

1 Efficient and speedy removal of excremental matter and urine plus other liquids and solids without leakage.

2 The prevention of ingress of foul air to any building whilst providing for their escape from the pipework into a 'safe' position.

3 The adequate and easy access to the interior of the pipe for the clearance of obstructions.

4 Adequate protection against extremes of temperature.

5 Adequate protection against external or internal corrosion attack.

6 Correct design and installation to limit siphonage (if any) to an acceptable standard, and to avoid deposition of solids.

7 Correct design and installation procedure to prevent damage from obstructions or blockages.

8 In areas where the combined system of drainage is permitted, it may be advantageous to connect rainwater outlets directly to discharge pipes, providing it is practicable and economical to do so; ventilation must be able to take place even if the rainwater outlet is obstructed or blocked.

9 Economy and good design are essentials: both are aided by compact grouping of sanitary appliances in both horizontal and vertical positions.

Domestic sanitation systems (historical development)

The function of a well-designed soil and waste discharge system in a building is to take away efficiently all waste from the sanitary fitments to the main drains, without allowing foul air to enter the building via the system of sanitation pipework.

Three basic systems have evolved over the years to fulfil these requirements. The order in which the systems were developed was:

1 The two-pipe system (see Figure 6.33)

2 The one-pipe system (see Figure 6.34)

3 The single-stack system (see Figure 6.35).

Each system is different, and consequently has its own respective merits and limitations, which must be observed to ensure correct functioning.

The two-pipe system

From the beginning of the nineteenth century until the late 1930s most of the buildings above one storey in height utilized the two-pipe system of sanitation. In this system, the discharges from the ablutionary fitments, such as baths, bidets, wash basins, showers, sinks, etc., were kept separate from the discharges from soil appliances such as water closets, urinals, and slop sinks.

1 The waste stack received the discharge from the ablutionary fitments and conveyed this to ground level where it was delivered above the water seal in a trapped gully connected to the drainage system.

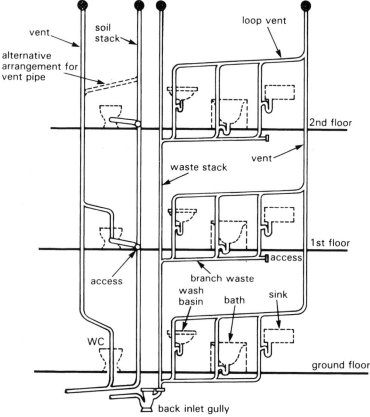

Figure 6.33 *Two-pipe or dual-pipe system*

2 The soil stack received the discharge from soil appliances and delivered it direct to the underground drainage system.

The waste and soil waters did not combine until they reached the below-ground drainage system. All pipework was fully vented to avoid the unsealing of traps and the risk of foul gases entering the building. Because the waste system was fully ventilated, shallow seal traps were adequate on baths, sinks, wash basins, etc. The two-pipe system functioned efficiently, but due to the duplication of pipework and the excessive labour involved in installation, it was expensive and was gradually replaced by the one-pipe system.

The one-pipe system

Since the late 1930s this system of sanitation has grown in popularity, particularly in multi-storey buildings. Originally used for domestic installation, it had virtually become standard practice in the non-domestic sector (hospitals, schools, offices, etc.). The name of the system is misleading, as two stack pipes are required, one for the combination of waste and soil, and the other as the main ventilating stack.

In the early stages of development double seal traps were often used on soil fitments. It was found, however, that with the aid of ventilating (relief) pipes standard seal traps could be used, as the ventilating system of pipework safeguarded the seals against variation in pressure. In this system all soil and waste water discharged into one common pipe and all branch ventilating pipes into one main ventilating pipe. This system largely replaced the two-pipe system and lent itself very well to use in multi-storey developments. It is far more economical than the two-pipe system. Rainwater will only connect into the same drain at the bottom of the building if the local authority permits a combined system of drainage. It is essential that all traps from waste fitments shall be capable of maintaining a water seal of 75 mm.

The one-pipe system fulfilled the requirements of the 1965 Building Regulations, but still proved more costly than the modified one-pipe system, also used in multi-storey buildings.

The modified one-pipe system

This system differed from the one-pipe system only in the provision of branch ventilating pipes. The modified system is extensively used in multi-storey buildings, and, where advantage can be taken of the design considerations of the single-stack system, enables certain branch ventilating pipes to be omitted.

The single-stack system

The single-stack system is a one-pipe system from which, subject to the observance of certain rules, all or most of the trap ventilating pipes are omitted.

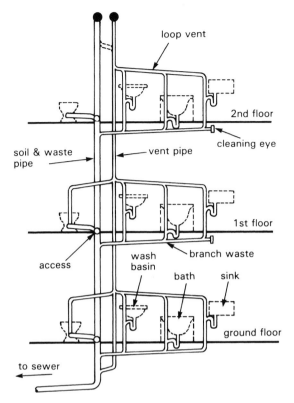

Figure 6.34 *The one-pipe system*

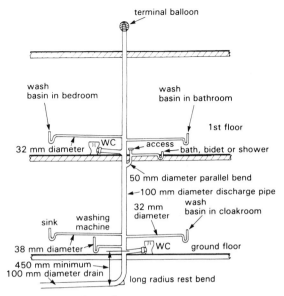

Figure 6.35 *Single stack system*

dwellings. Distinct savings are claimed over the one-pipe system.

Modern above-ground drainage systems

In 1994 BSI published the final version of BS 5572 'Code of practice for Sanitary Pipework', sixteen years after the previous edition. Since then there has been a great deal of work throughout Europe trying to produce Standards that can be applied to all countries but do not conflict with sound local traditions. Although the laws of physics, fluid dynamics and bodily functions are generally universal, within Europe there was great diversity of practice regarding pipe sizing loading and ventilation.

The European Standards committees, who had the task of formulating the new standards, tried to build harmonized standards upon common ground agreed by consensus. This required that every country's rules and guidance had to be examined against all the other countries' rules and the laws of physics. The result was that, in most situations, existing rules were simplified, as there was not sufficient justification for imposing on existing rules or traditions.

The single-stack system was developed by the Building Research Establishment and was formerly identified as the simplified system. The research which commenced in the mid-1940s soon proved that the unsealing of traps did not occur as readily as had been presumed previously and that, under certain circumstances, trap ventilating pipes could be omitted from the one-pipe system. It also proved that when properly designed it becomes basically a one-pipe system without vent pipes, hence the name *single stack*.

Originally this system was developed for low-cost housing, but, having proved successful, it was further developed for multi-storey buildings. The design of the pipework is very important and success relies upon close grouping of single appliances (each with a separate branch) around the stack. Providing the design is correct, vent pipes can be omitted except venting via the main stack, which is continued upwards to a safe position. The single-stack system is a simple system particularly suitable for housing, and for multi-storey

However, the diversity of practices, prejudices, perceptions and pipe sizes has meant that these first European Standards are frequently an amalgamation of established National Standards. In the future, when harmonization of practices, infrastructure, qualifications and products has become more noticeable, the production of harmonized codes of practice will be easier and even more necessary, but until then we will all have to come to terms with the current Standards.

For internal plumbing in the UK the principles and rules of BS 5572 have been established over many years. The designation of specific discharge units to an appliance, the use of swept branches and tees, 75 mm deep seal traps and full-bore flow branch pipes were all accepted features of the British Standard.

BS EN 12056

This European Standard is titled 'Gravity drainage systems inside buildings', however, its scope includes the outside of the building and any pipes running beneath the building. It generally covers all the pipes that would lie within a projection down from an overhanging roof. The parts of BS EN 12056 are:

- General and performance requirements
- Sanitary pipework, layout and calculation
- Roof drainage, layout and calculation
- Wastewater lifting plants, layout and calculation
- Installation and testing, instructions for operation, maintenance and use.

In the new European Standard, four basic above ground drainage systems are described. There are parallels with the common systems in the UK.

Within BS EN 12056-2, System III – Single discharge stack system with full bore branch discharge pipes – is basically the traditional UK system.

Although the minimum trap seal depth is specified as 50 mm, the tradition in the UK has been for 75 mm traps and a 100 per cent filling degree (full-bore flow).

So, in many ways the new BS EN 12056 Part 2 is an improvement on BS 5572 It provides

- A simple unified, and much more straightforward, approach to sanitary pipework system design
- The option to understand and use alternative pipework systems that have been proven in other countries
- All the background information to system III design (that was previously in BS 5572) in the National Annex
- It is one part of the five part European Standard 'Gravity drainage systems inside buildings'.

This Standard, along with BS EN 752 and others, forms the building block for current and new sanitary pipework systems. It is called up in the air admittance valve European Standard and in the latest Building Regulations of Scotland, England and Wales.

The Standard also defines different categories of water.

- Waste Water – water that is contaminated by use and all water discharging into a drainage system, e.g. domestic and trade effluent, condensate water and also rain water when discharged into a waste water drainage system.
- Domestic Waste Water – water that is contaminated by use and normally discharged from domestic sanitary appliances, e.g. WC, sink, bath, washbasin, shower or floor gully.
- Grey Water – waste water not containing faecal matter or urine.
- Black Water – waste water containing faecal matter or urine.
- Rainwater – water resulting from natural precipitation that has not been deliberately contaminated.

System types

The systems may be divided into four types, although there are variations in detail within each system (hence the need to refer to the national and local regulations and practice).

System I   Single discharge stack system with partly filled branch discharge pipes

Sanitary appliances are connected to partly filled branch discharge pipes. The partly filled branch discharge pipes are designed with a filling degree of 0.5 (50 per cent) and are connected to a single discharge stack.

System II   Single discharge stack system with small bore discharge branch pipes

Sanitary appliances are connected to small bore branch discharge pipes. The small bore branch discharge pipes are designed with a filling degree of 0.7 (70 per cent) and are connected to a single discharge stack.

System III   Single discharge stack system with full bore branch discharge pipes

Sanitary appliances are connected to full bore branch discharge pipes. The full bore branch discharge pipes are designed with a filling degree of 1.0 (100 per cent) and each branch discharge pipe is separately connected to a single discharge stack.

System IV   Separate discharge stack system

Drainage systems type I, II and III may also be divided into a black water stack serving WCs and urinals and a grey water stack serving all other appliances.

Configurations

Each system may be configured in a number of ways, governed by the need to control pressure in the pipework in order to prevent foul air from the waste water system entering the building. The principal configurations are described below but combinations and variations are often required.

Primary ventilated system configurations

Control of pressure in the discharge stack is achieved by air flow in the discharge stack and the stack vent (see Figure 6.36). Alternatively, air admittance valves may be used.

Secondary ventilated system configurations

Control of pressure in the discharge stack is achieved by use of separate ventilating stacks and/or secondary branch ventilating pipes in connection with stack vents (see Figure 6.37). Alternatively, air admittance valves may be used.

Unventilated discharge branch configuration

Control of pressure in the discharge branch is achieved by air flow in the discharge branch (see Figure 6.38).

Ventilated discharge branch configurations

Control of pressure in the discharge branch is achieved by ventilation of the discharge branch (see Figure 6.39). Alternatively, air admittance valves may be used.

Loss of water seal in traps

The loss of a trap seal will usually result in foul and objectionable gases and odours entering a building from the system of discharge pipework. These gases will at least be a nuisance and in extreme cases create a health hazard to the occupants of the building. It is for these reasons that much importance is placed upon retention of the water seal in all traps.

Trap seals may be lost in several ways. Some reasons may be attributed to bad system design or installation techniques, others are more natural or the result of particular circumstances. It is important that a plumber understands the particular circumstances which may cause the loss of water seal, the common circumstances are:

1 Self-siphonage
2 Induced siphonage
3 Compression or back pressure
4 Capillary action
5 Evaporation

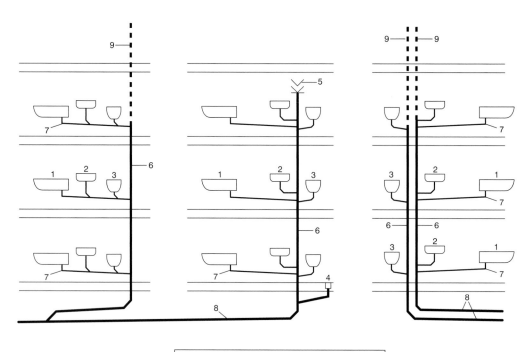

Figure 6.36 *Primary ventilated
system configurations*

Key

1	Bath	6	Stack
2	Wash basin	7	Branch discharge pipe
3	WC	8	Drain
4	Floor gully	9	Stack vent
5	Air admittance valve		

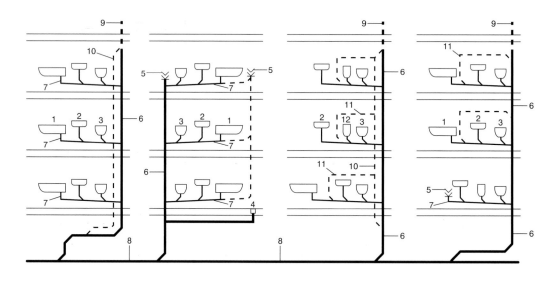

Figure 6.37 *Secondary ventilated
system configurations*

Key

1	Bath	7	Branch discharge pipe
2	Wash basin	8	Drain
3	WC	9	Stack vent
4	Floor gully	10	Ventilating stack
5	Air admittance valve	11	Branch ventilating pipe
6	Stack	12	Urinal

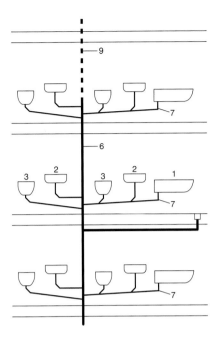

Figure 6.38 *Unventilated discharge branch configuration*

Key
1 Bath
2 Wash basin
3 WC
4 Floor gully
5 Stack
7 Branch discharge pipe
9 Stack vent

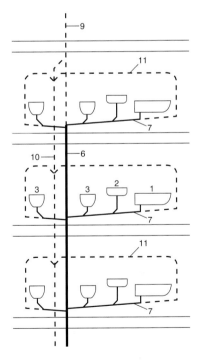
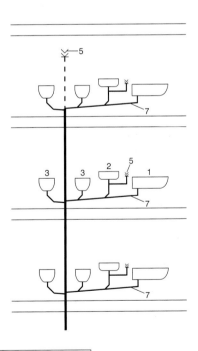

Figure 6.39 *Ventilated discharge branch configurations*

Key

1 Bath	7	Branch discharge pipe
2 Wash basin	9	Stack vent
3 WC	10	Ventilating stack
5 Air admittance valve	11	Branch ventilating pipe
6 Stack		

6 Leakage

7 Momentum

8 Wavering out.

Self-siphonage

This is caused by a moving plug or charge of water running out of a steep-sided sanitary appliance such as a wash basin. The plug of water moves through the trap, pushing the air on the outlet side of the trap in front of it, thereby creating a partial vacuum in the branch discharge pipe which causes siphonage to occur and loss of the trap seal. The technical term for the partial vacuum or negative pressure zone is 'hydraulic jump' as illustrated in Figure 6.40. Self-siphonage of a trap seal is usually indicated by an excessive amount of noise as the sanitary appliance discharges its final quantity of waste water. Self-siphonage is usually prevented by one or more of the following:

1 By fitting a P trap to the sanitary appliance (thus avoiding vertical branch discharge pipes)

2 Ensuring that the branch discharge pipe length and slope do not exceed those recommended

3 By fitting a ventilating or anti-siphon pipe adjacent to the trap outlet

4 By fitting a larger diameter branch discharge pipe to the trap outlet

5 By correct design of sanitary appliance

6 By fitting a resealing or anti-siphon trap to the sanitary appliance.

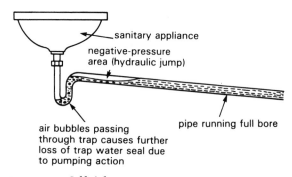

Figure 6.40 *Self-siphonage*

Induced siphonage

This is caused by the discharge of waste water from one sanitary appliance, pulling or siphoning the seal of a trap of another appliance connected to the same branch discharge pipe as shown in Figure 6.41. This form of siphonage is most common in buildings where ranges of appliances are fitted or where it is necessary to connect several appliances to a common branch discharge pipe. The main causes of induced siphonage are poor system design, inadequate pipe sizes or bad installation techniques.

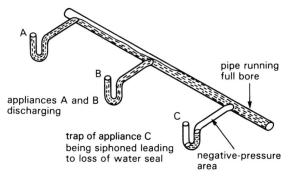

Figure 6.41 *Induced siphonage*

Compression or back pressure

Compression of air at or near the base of a discharge pipe may occur as shown in Figure 6.42. As water flows down a vertical stack it takes some air with it. As this water changes direction at the base of the stack (from the outside of the vertical pipe to the invert of the near-horizontal drain pipe) the pipe bore becomes momentarily full of water, so preventing the free flow of air up and down the discharge stack. Under these conditions a hydraulic wave is formed and air at the base of the stack is compressed by the water falling from above, thus creating air pressure in excess of atmospheric pressure. This pressure will attempt to escape through any branch discharge pipe connected in the compression zone, thus blowing the seal of the trap and allowing foul air from inside the discharge pipework to enter the building. Often the displaced water will run back into the trap to form a seal, but continued blowing of the seal will cause a noise nuisance and allow foul gases into the room where the appliance is fitted. Compression and back pressure are prevented by not fitting small or sharp radius bends or having branch discharge pipes connected near to the base of a discharge stack. Correct practice is to fit large radius bends and to ensure inlet connections are not made close to the base of the stack.

Capillary action

Loss of seal by capillarity occurs when a piece of porous material such as threads or string from a mop or dishcloth are deposited into the water seal and over the outlet invert of a trap, as shown in Figure 6.43. This will most commonly happen to appliances such as kitchen or cleaners' sinks fitted with S pattern traps. Capillarity is prevented by regular cleaning of the inside of the trap and branch discharge pipe, or by laying a loose mesh strainer into the waste fitting of the appliance to catch and retain loose porous strands which may be contained in certain types of waste water.

Evaporation

This occurs usually in traps connected to sanitary appliances which are not used regularly or where the ambient temperatures are relatively high. Evaporation is more usual in the summer months when temperatures are higher or when buildings are left empty or unattended due to holiday periods.

Leakage

Traps occasionally lose their seals due to a leak on the fitting below water-seal level. The leak may be due to a loose or badly jointed access bowl or cleaning eye, or, in the case of soft materials, caused by impact resulting in fracture and damage to the trap body.

Momentum

The most usual cause of loss of trap seal by momentum is when a quantity of water is quickly discharged into a gully or washdown WC pan and the discharging water carries away the water forming the trap seal.

Wavering out

The effect of a high-velocity fluctuating wind passing over the top of an exposed discharge stack ventilating pipe (see Figure 6.44) will create varying air pressures and draughts within the pipework system which may cause trap seals to fluctuate or waver, resulting in loss of water from the seal. Fluctuating pressures caused by wind are best prevented by locating the vent terminal position away from exposed locations and ensuring

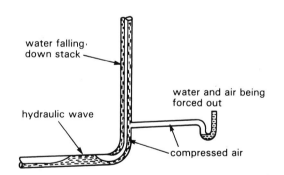

Figure 6.42 *Compression or back-pressure*

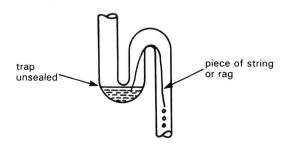

Figure 6.43 *Capillary attraction*

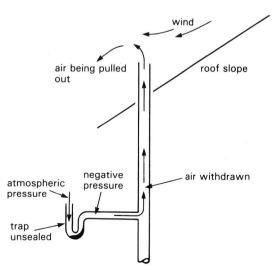

Figure 6.44 *Wavering out*

that a terminal grating is fitted to all ventilating pipework.

Inspection and testing

Inspections and testing should be carried out during the installation of a discharge system to ensure that the pipework is properly secured and clear of obstructing debris, and that all work which is to be concealed is sound and free from defects before it is finally concealed. On completion, the discharge system installation should be visually inspected for damage and then tested for soundness and performance.

As mentioned earlier in this chapter, the Building Regulations require that a completed discharge system should be capable of withstanding an air (pneumatic) test for a minimum period of 3 minutes at a pressure equivalent to a head of not less than 38 mm of water.

Air test (Figure 6.45)

Normally this test should be carried out as one operation, although on large installations it may have to be done in stages or sections or as the work proceeds. The water seals of the traps of all connected appliances should be fully charged and test plugs or inflatable bags inserted into the open ends of the pipework to be tested. To ensure that there is a complete air seal at plugs or bags at the base of the stack, a small quantity of water can be emptied into the system. Plugs located inside vertical ventilating pipework can be sealed by a small quantity of water poured around the outside of the plug. A flexible tube should be fitted to one of the test plugs as shown and be complete

with a valved T piece, manometer and air pump. An air test is very thorough and searching and will indicate any leak in the system. It is therefore necessary to ensure that all the testing equipment is well sealed and sound before testing of the pipework commences. Air is then pumped into the system until the required test pressure is recorded on the manometer. The valve adjacent to the air pump should then be closed to isolate the air pump while the test is in progress; the pressure within the installation should remain constant for a period of not less than 3 minutes. If it is not convenient to connect the flexible tube to one of the test plugs, the tube can be passed through the water seal of a sanitary appliance and the test procedure applied from this position.

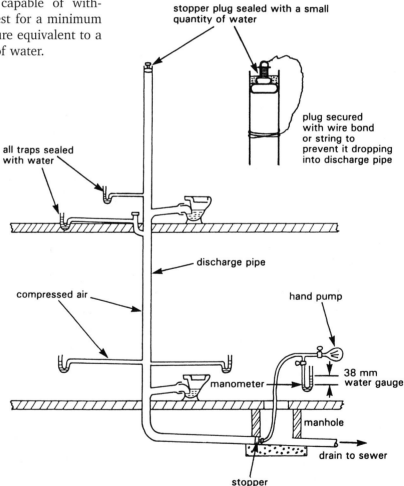

Figure 6.45 *Air test on soil and waste system*

Leak detection

Any leak in the system will be indicated by a drop in pressure reading at the manometer. Detection is best carried out with the aid of a soap solution which is brushed on to the joints and pipes, the leak position being indicated by the formation of air bubbles in the soap solution liquid. Smoke may be used as a trace to detect a leak. Either a smoke-producing machine or a smoke-generating cartridge can be used. The leak is detected visually as the smoke escapes from the system.

Smoke cartridges should be used with caution, and care must be taken to ensure that the ignited cartridge is not in contact with the installation pipework, and that the effects of combustion do not have a harmful effect upon the material used for the discharge pipe system. Testing or tracing of plastics pipework using smoke is not recommended and should be avoided due to naphtha having a detrimental effect, particularly on ABS, UPVC and MUPVC. Rubber jointing components can also be adversely affected.

Water test

There is no justification for a water test to be applied to a complete above-ground discharge system as the part of the system mainly at risk is that below the lowest sanitary appliance connection. Some authorities may request that this lower part of the system is water-tested and this is carried out by inserting a test plug or sealing bag into the lowest point of the pipework system and filling the pipe with water up to the flood level of the lowest sanitary appliance, provided that the static head imposed does not exceed 6 m. The water level is observed during the test period and any leak will be indicated by a drop or lowering of the water level; the leak itself is detected by a visual inspection of the installation.

Performance test

The main purpose of these tests is to check the stability and retention of trap seals under peak working conditions and that all appliances, whether discharged singly or in groups, drain speedily, quietly and completely. Immediately after each test a minimum of 25 mm of water seal should be retained in every trap. Each test should be repeated at least three times, with the trap or traps being fully recharged before each subsequent test. The maximum loss of water seal in any one test, measured by a small diameter transparent tube, or dry dip-stick, should be taken as the significant result.

Tests for self-siphonage and induced siphonage in branch discharge pipes

To test for the effect of self-siphonage the sanitary appliance should be filled to overflow level and then discharged by removing the waste outlet plug; WC pans should be flushed. The seal remaining in the trap should be measured and recorded when the discharge of water has finished. Ranges or multiple appliances, connected to a common discharge pipe, should also be tested for induced siphonage in a similar way, and the seals remaining in all the traps measured at the end of the discharge period.

Waste disposal units

These mechanical units are designed to macerate and dispose of organic food such as vegetable or fruit peelings. The units must not be used for the disposal of wooden objects, plastics, string, rag, glass or metal. The disposal unit connects to a sink outlet and is housed beneath the sink. Most units require a larger diameter sink waste outlet hole than the standard opening, and it is normal practice to connect waste disposal units to sinks specially manufactured to accommodate these units.

The operating cycle of the unit is as follows:

1 Cold water is run into the sink and the unit is switched on.

2 Waste food is pushed through the rubber splash guard and is washed down by the flowing water on to the cutter rotor.

3 The waste food is thrown by centrifugal force (the tendency of a rotating body to throw liquids and solids away from the force centre) on to the cutting blades, thereby cutting and shredding the waste substances into very small particles.

4 The small particles are washed by the flowing water through the cutting ring into the sloping

discharge chamber and pass out of the unit and into the waste pipe via the tubular trap connected to the outlet connection.

The outlet invert of the connected trap should be lower than the inlet so that the water or liquefied waste is not trapped inside the discharge chamber. Tubular traps should always be used for waste disposal units. Bottle traps are not recommended as they tend to clog and block, due to the nature of the waste discharged from the unit. The slope of the discharge pipe must be enough to ensure good velocity of waste flow, thereby preventing deposition of solid particles in the pipe. The discharge pipe should be as short and straight as possible and any bends or changes of direction should be large-radius with cleaning access available. Disposal units connected to ground-floor sinks should discharge to the drainage system via a back inlet gully as shown in Figure 6.46. Units located on upper floors in a building will connect to the discharge system. Figure 6.47 shows a typical domestic waste disposal unit. Modern units are provided with a thermal overload cut-out switch in case of jamming, and a de-jamming tool is provided should the unit lock. The unit is supplied with electricity via a fused control switch, usually incorporating a pilot light which illuminates when power is on and the motor is running.

Urinals

Urinals are sanitary appliances for the reception and flushing away of urine.

Bowl urinal (Figure 6.48)

This is a wall-hung receptacle shaped like a bowl, frequently with an extended lip. It should be fitted with an individual spreader for flushing water which should wash the whole internal surface of the bowl. These appliances can be fitted on their own or in a range (see Figure 6.49).

Pedestal urinal

A pedestal urinal is a bowl-type urinal supported on a pedestal.

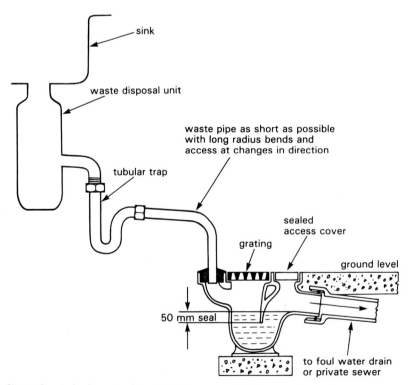

Figure 6.46 *Waste disposal unit discharging direct to below ground discharge pipe system*

Trough urinal (Figure 6.50)

These should have a back slab or edge extending to at least 450 mm above the level of the front lip of the trough and preferably integral with the trough. Flushing is by sparge pipe or spreaders, and discharge is to a trapped outlet.

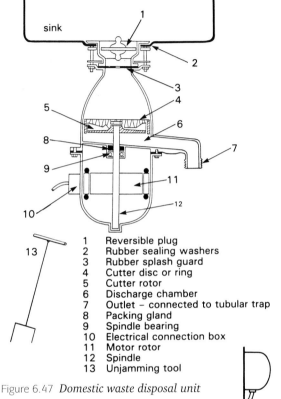

	1	Reversible plug
	2	Rubber sealing washers
	3	Rubber splash guard
	4	Cutter disc or ring
	5	Cutter rotor
	6	Discharge chamber
	7	Outlet – connected to tubular trap
	8	Packing gland
	9	Spindle bearing
	10	Electrical connection box
	11	Motor rotor
	12	Spindle
	13	Unjamming tool

Figure 6.47 *Domestic waste disposal unit*

Slab urinal (Figure 6.51)

This consists of an impervious slab fixed against a wall. It may have division slabs of the same material, usually for only part of its height. The slab discharges to a channel which has an outlet provided with a trap. Flushing is by sparge pipe or individual spreaders.

Stall urinal (Figure 6.52)

This consists of a back slab curved in plan with integral side divisions. These divisions are generally supplemented by separate extension rolls or wings to cover the joints when stalls are fixed in ranges. Stalls are complete with integral channels. A spreader should be fitted to each stall for flushing. The discharge is via a trapped outlet.

Bowl, trough and slab urinals are made from glazed ceramic materials, stainless steel or suitable plastics. Stall urinals are made from glazed fireclay or stainless steel. Outlet grids should be made of gun-metal, brass, glazed ceramic or suitable plastics, all preferably domed, and either hinged or capable of being fixed (Figure 6.53). Sparge pipes and spreaders should be of corrosion-resisting metal or suitable plastics.

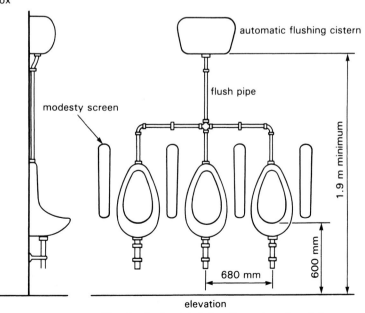

Figure 6.48 *Bowl urinal*

Figure 6.49 *A range of bowl urinals with fixing dimensions for adult use*

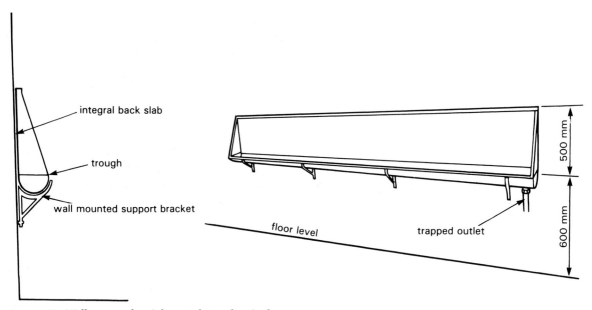

Figure 6.50 *Wall-mounted stainless steel trough urinal*

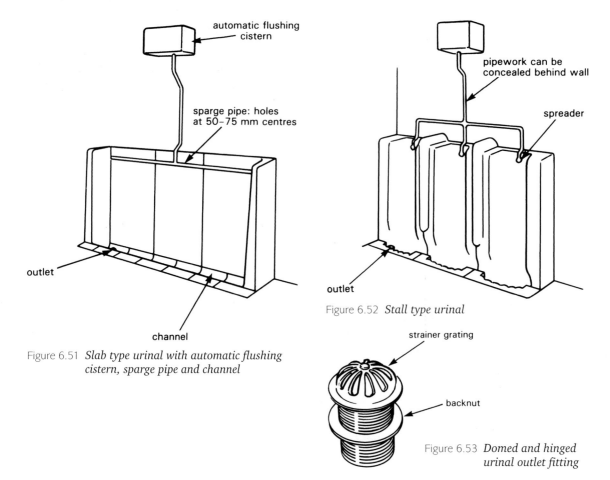

Figure 6.51 *Slab type urinal with automatic flushing cistern, sparge pipe and channel*

Figure 6.52 *Stall type urinal*

Figure 6.53 *Domed and hinged urinal outlet fitting*

Selection

The slab type is simple and usually less difficult to clean than the others.

The stall type is heavy, takes up more space and is more difficult to clean. It affords greater privacy than other types and is better able to-withstand rough usage.

Bowl urinals are less restrictive to planning and are more suitable for use where floor/wall movement may occur.

In hard water districts, the waste pipes are more liable to blockage.

The combination of hard water and urine results in rapid deposition of encrustation which should be removed by frequent and thorough cleaning. Acid-based cleaning powders and fluids should be used with caution to avoid damage to the appliance or injury to the cleaning operatives.

Urinal flushing

All urinals should be fitted with a means of flushing. Where a single unit is installed a hand-operated flushing cistern may be used.

Where more than one urinal is fitted, flushing is achieved by the use of an automatic flushing cistern. This is designed to discharge its contents, by siphonage, at intervals determined by the rate at which water is fed into the cistern (see Figure 6.54).

The cistern should hold at least 4.5 litres of water per bowl, stall or 0.6 m length of slab. All flushing pipework and spreaders should be manufactured from non-ferrous materials. The cistern should fill at such a rate as will produce a flush of water at 20 minute intervals or depending on frequency of use. The supply of water into the cistern should be controlled by a valve approved by the Regional Water Authority.

Urinals and waste of water

Since 1981 all supplies to urinals in new buildings have required some form of electrically or hydraulically operated control valve to shut off the supply of cold water to urinal flushing cisterns where the sanitary accommodation or the building is not in use.

The usual way of achieving this control is:

1 An electrically operated motorized valve controlled by a time clock or programmer (the clock closing the valve when the building is vacant).

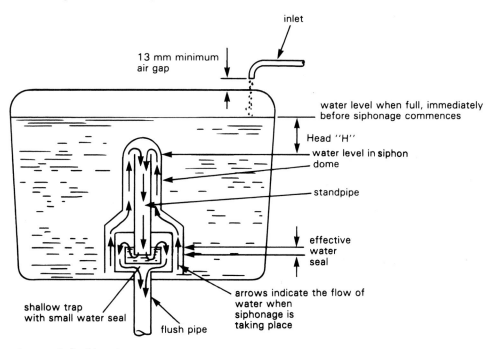

inlet

13 mm minimum air gap

water level when full, immediately before siphonage commences

Head "H"

water level in siphon dome

standpipe

effective water seal

arrows indicate the flow of water when siphonage is taking place

shallow trap with small water seal

flush pipe

Figure 6.54 *Automatic flushing cistern*

2 A valve or flushing cycle operated and actuated by the breaking of a beam or ray of light (photo-electric).

3 A hydraulically operated flush control valve.

Urinal outlet details

Urinal compartments should have an impervious floor, draining into the urinal channel (if fitted) or into a separate gully let into the floor. This is to facilitate cleaning. All compartments should be well ventilated.

With stall and slab ranges one outlet (75 mm) should not serve more than seven stalls or 4.3 m length of slabbing and should be at or near the middle of the range if there are more than three stalls or 1.8 m length, of slab.

Table 6.2 indicates the recommended outlet sizes.

Bowl or trough urinals should be provided with an outlet fitting of not less than 32 mm diameter. Outlet fittings for all types of urinal should incorporate either an inbuilt strainer or a hinged dome (Figure 6.55) to prevent possible blockage of the outlet, trap or waste pipe.

Table 6.2 *Recommended outlet sizes*

1 or 2 stall	0.6–1.2 m slab	50 mm outlet
3 or 4 stall	1.8–2.4 m slab	65 mm outlet
5 to 7 stall	3.0–4.3 m slab	75 mm outlet

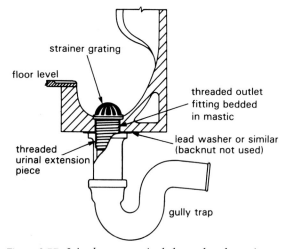

Figure 6.55 *Joint between urinal channel and cast-iron drain*

Tubular traps are best for all types of urinal as they offer less restriction to flow of water than other types of trap.

Traps

A trap is a fitting or part of an appliance or pipe designed and manufactured to retain a quantity of water which forms a seal to prevent the flow of gases or foul air from the discharge pipe into the building. The 'water seal' is that depth and quantity of water which would have to be removed from a fully charged trap before gases or air could pass through.

For identification purposes traps are usually classified in three distinct types:

1 Common traps (see Figure 6.56)

2 Resealing traps (see Figure 6.57)

3 Anti-siphon traps (see Figure 6.58).

A trap should be fitted to every sanitary appliance and omitted only in special cases.

Principles of operation

Figure 6.58(a) shows the trap under normal operating conditions with full water seal. Figure 6.58(b) when subjected to severe siphonage conditions the automatic hydraulic action allows air through the by-pass tube without any major loss of water. Figure 6.58(c) when normal conditions return the remaining water falls back to reseal the trap.

Some sanitary appliances are manufactured with the trap as part of the appliance, i.e. WC pan. This arrangement is called an *integral trap*. The majority of sanitary appliances do not have integral traps and these require a trap to be fitted immediately below the fitting, i.e. bath, wash basin, sink, shower, bidet.

An efficient trap should:

1 Be self-cleansing

2 Have a smooth internal surface

3 Be made from an incorrodible material

4 Allow access for cleaning

5 Have a water seal

6 Have a uniform diameter or bore.

Traps are made from several materials:

1 Copper
2 Plastics
3 Brass
4 Galvanized low carbon steel
5 Cast iron
6 Aluminium
7 Glass.

The type of trap required for a particular location depends on several factors:

1 The type of sanitary appliance
2 The position of the appliance in relation to the discharge pipe or gully
3 The type of discharge pipe system.

Traps may be obtained with a water seal of varying depths. For the single stack or one pipe system, traps must have a minimum water seal of

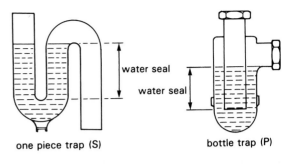

one piece trap (S) bottle trap (P)

two-piece swivel trap (P) three-piece running trap

Figure 6.56 *Common traps*

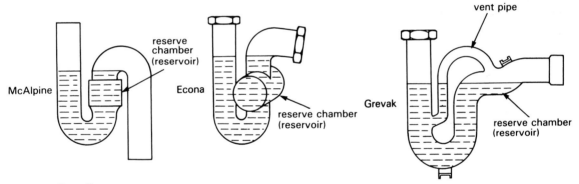

McAlpine — reserve chamber (reservoir)

Econa — reserve chamber (reservoir)

Grevak — vent pipe — reserve chamber (reservoir)

Figure 6.57 *Resealing traps*

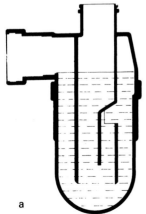
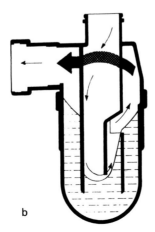
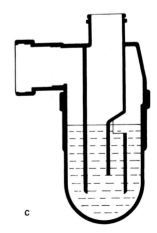

a b c

Figure 6.58 *Anti-siphon bottle trap*

75 mm for pipework up to 50 mm in diameter. These traps are generally called 'deep seal'.

Traps of 50 mm diameter and above must have a minimum water seal of 50 mm for any system of sanitary pipework.

Trap outlets are identified by their position relative to the trap inlet and may be specified in degrees, i.e. 92°, 135°, or 180° or by the corresponding letters P, Q or S (see Figure 6.59).

Tubular traps such as the one-, two- and three-piece common traps shown in Figure 6.56 are essential in locations where the least resistance to flow is required, i.e. sinks, washing machines, waste disposal units. Bottle traps are most useful in locations where space is limited or where the trap is visible – because of its neater appearance. Resealing traps (Figure 6.57) are used to maintain the seal of the trap in location, where loss of seal may occur, and they do away with the need for trap ventilating or anti-siphon pipes.

With most types of anti-siphon or resealing traps, the method of maintaining the water seal is accomplished by means of a reservoir or reserve chamber integral to the trap. Water is retained in this chamber during siphonage conditions and reseals the trap when inlet and outlet pressures are in equilibrium. The term anti-siphon is misleading as this type of trap does not prevent siphonic action from taking place, but preserves the seal of the trap when siphonic conditions occur. Other types of anti-siphon trap incorporate an air valve (see Figure 6.60) which opens to allow air into the trap should a reduction of air pressure occur, thus equalizing pressures and preventing siphon-age of water.

Materials used for soil and waste systems

Cast iron

This material has the advantage of resisting mechanical damage better than most other materials. The pipes are heavy and require sound support, but they do not expand or contract as much as the other materials commonly used. The pipes are protected from corrosive attack by a coating of pitch. An extensive range of fittings is available.

Jointing, the traditional method, is by caulked lead (see Figure 6.61). Tarred gaskin is first placed in the joint and hammered firmly in with a yarning chisel. The gaskin helps to centralize the spigot in the socket and prevents molten lead from entering into the bore of the pipe. Molten lead is then poured in to fill the remainder of the joint (approx. $2/3$ lead to $1/3$ gaskin). When the lead has cooled, it must be caulked with special chisels (see Figure 6.62) to compress the poured lead. A semicircular groove inside the collar of the pipe or fitting provides a key for the lead jointing, preventing it from slipping out over a period of time due to expansion and contraction. *Note:* An alternative type of caulked joint may be made with a cold caulking compound. This joint consists basically of cement and fibre which is caulked into the joint with a damp caulking chisel – the moisture causing, a chemical action with the cement, so making the joint material 'set'. (The

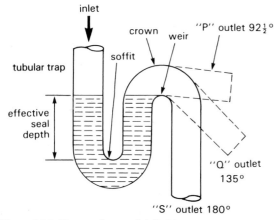

Figure 6.59 *Trap outlets and definitions*

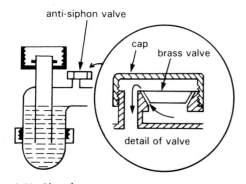

Figure 6.60 *Air valve*

manufacturers' instructions should be adhered to when making this type of joint.)

Flexible joints may also be used: these are quicker to make and permit easier thermal movement than caulked joints. The joint is also easier to disconnect should this be necessary.

Figure 6.63 shows a 'Rollring' type joint.

Figure 6.64 shows a 'Timesaver' joint, used for jointing spigot ends.

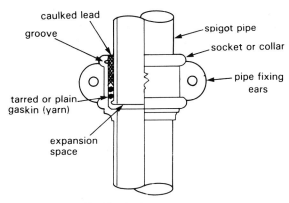

Figure 6.61 *Caulked lead joint*

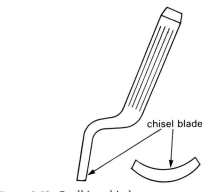

Figure 6.62 *Caulking chisel*

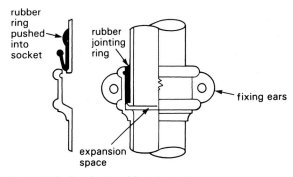

Figure 6.63 *Synthetic rubber ring joint*

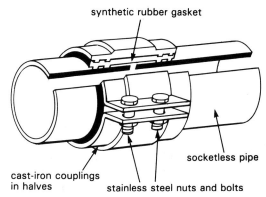

Figure 6.64 *Flexible (Timesaver) joint for cast-iron pipe*

Lead

The use of this material for soil and waste systems has decreased over the years and is now generally only used for historic and listed type buildings and installations.

Copper

The main advantage of this material is that the pipes can be obtained in long lengths, so reducing the number of joints required. Copper pipes have a smooth internal and external surface and are strong. They are also resistant to corrosive attack from most building materials. The thermal expansion of copper is higher than that of cast iron and expansion joints may be required on long runs of pipework.

Jointing

Copper pipes may be jointed by several techniques, including hard and soft-soldered, capillary and non-manipulative fittings. Figure 6.65 shows a selection of these methods of jointing.

Synthetic plastics

The advantage of these materials is their light weight and the simple methods used in the jointing process. The term 'plastics' can be applied to a very wide range of materials. It is important for a plumber to be able to identify the type of plastic which is being used as only a few of these materials are suitable for sanitary pipework systems.

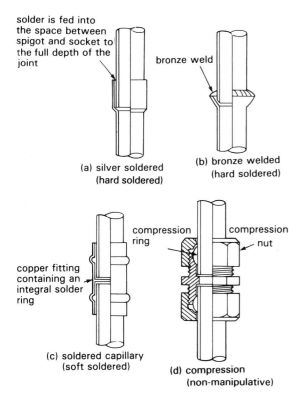

solder is fed into the space between spigot and socket to the full depth of the joint

bronze weld

(a) silver soldered (hard soldered)

(b) bronze welded (hard soldered)

compression ring

compression nut

copper fitting containing an integral solder ring

(c) soldered capillary (soft soldered)

(d) compression (non-manipulative)

Figure 6.65 *Joints on copper pipes*

The most popular plastics materials used for sanitary pipework systems and components are:

1 Polyvinyl chloride (PVC)

2 Acrylonitrile butadiene styrene (ABS)

3 Polythene

4 Polypropylene.

Polyvinyl chloride

This is used for discharge system pipework and fittings, and is capable of receiving discharge water at high temperatures without softening. Unlike other plastics materials, it will not burn easily and, if set alight; usually self-extinguishes.

Acrylinitrite butadiene styrene (ABS)

This is another material suitable for discharge system pipework and is usually employed for small diameter branch discharge and waste pipes. It has the advantage over PVC in that it can withstand higher water temperatures for a longer period of time than PVC. Some manufacturers do

produce complete ABS waste systems, but as it is a more expensive material than PVC its use is restricted to smaller diameter pipework.

The usual method of jointing ABS is by solvent welding as shown in Figure 6.66, although 'O' ring joints are satisfactory. The adhesive used for ABS solvent welding is not generally suitable for PVC solvent welding, and joints made using incorrect adhesive may fail after a short period of time. ABS is often manufactured in the same colour as PVC, but its surface finish is not as shiny or polished as PVC and appears usually to have a dull or matt surface. If ABS is ignited it will burn with a bright white flame and will give off small particles of carbon soot. PVC is usually jointed by 'O' ring push-fit joints (see Figure 6.67). The advantage of this method of jointing is the easy connection or disconnection of pipes and fittings plus the expansion

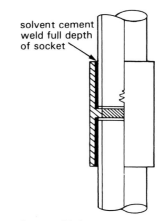

solvent cement weld full depth of socket

Figure 6.66 *Solvent welded joint*

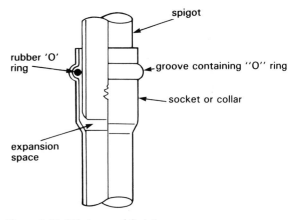

spigot

rubber 'O' ring

groove containing "O" ring

socket or collar

expansion space

Figure 6.67 *'O'-ring pushfit joint*

gap which can be accommodated in every fitting during assembly of the joint. This expansion space or gap is very important and must be incorporated when using materials which have a high rate of linear expansion such as PVC and ABS.

Polythene

These pipes are not so rigid as other thermoplastics used for sanitation discharge systems, and their use is usually limited to specialist work such as laboratory wastes because of their high resistance to chemical attack.

High-density polythene is more rigid and can withstand slightly higher temperatures than the low-density material. Pipes are usually jointed by 'O' ring pushfit-type joints or non-manipulative-type nut and rubber cone fittings.

Polypropylene

This is a tough, rigid lightweight plastic in the same family group as polythene. Its main characteristics are its surface hardness and its ability to withstand high temperatures. Components manufactured from polypropylene are able to withstand boiling water for short periods of time without damage or deformation, which makes the material superior to PVC, ABS or polythene when high temperatures are a consideration. Manufacturers produce waste fittings and traps from this material, the methods of jointing being similar to those used for polythene. Solvent welded jointing is not suitable.

Self-assessment questions

1 The water seal of a washdown WC pan should have a minimum depth of:
   (a)   5 mm
   (b)   50 mm
   (c)   100 mm
   (d)   1000 mm.

2 After a discharge pipe system has been installed in a partly completed building it is advisable to:
   (a)   leave all open ends and vents clear
   (b)   seal all inlets with cement mortar
   (c)   insert expanding drain plugs on the vents only
   (d)   fit temporary plugs or seals to all open ends.

3 It is recommended that lever valves should be fitted to wash basins in:
   (a)   laboratories
   (b)   medical rooms
   (c)   factories
   (d)   offices.

4 The advantage of using a corbel WC is that:
   (a)   it is silent in action
   (b)   it has a siphonic action
   (c)   the outlet is 65 mm in diameter
   (d)   the floor below the pan can be cleaned easily.

5 Water discharged from a sanitary appliance and passing a branch discharge pipe lower down the stack may cause loss of the trap seal by:
   (a)   induced siphonage
   (b)   self-siphonage
   (c)   evaporation
   (d)   capillarity.

6 A special steel stand would be used to support a corbel WC pan where the:
   (a)   WC outlet connects to a plastics junction
   (b)   back wall provides inadequate support
   (c)   supporting floor is of timber
   (d)   flush pipe has a vertical entry.

7 The test applied to above-ground discharge pipework to ensure that adequate water seals are retained during working conditions is the:
   (a)   air test
   (b)   hydraulic test
   (c)   soundness test
   (d)   performance test.

8 Which type of valve should be fitted to a wash basin?
(a) globe
(b) pillar
(c) bib
(d) lockshield.

9 The purpose of the after-flush in a syphonic water closet pan is to:
(a) wash the sides of the appliance
(b) make up the trap water seal
(c) prevent induced syphonage
(d) assist the momentum.

10 Water regulations require that the outlet of a valve is to be above the spillover level of the sanitary appliance it serves. The reason for this is:
(a) makes the appliance easier to use
(b) prevents waste of water
(c) prevents the possibility of back syphonage
(d) avoid splashing.

11 The most popular pattern of ceramic sink is called:
(a) Chelsea
(b) Glasgow
(c) Belfast
(d) Croydon.

12 The internal diameter of a waste pipe from a domestic sink should not be less than:
(a) 25 mm
(b) 32 mm
(c) 38 mm
(d) 50 mm.

13 An important consideration when fixing plastics discharge pipes is:
(a) to provide the required allowance for thermal movement
(b) not to use metal fixings which will corrode the discharge pipe
(c) to have as few joints as possible
(d) not to damage the protective coating.

14 A suitable material for the manufacture of WC pans is:
(a) vitreous china
(b) pressed steel
(c) vitrified clay
(d) galvanized mild steel.

15 Ventilating pipes are connected to traps to prevent:
(a) loss of water seal
(b) capillarity
(c) evaporation
(d) corrosion.

16 An overflow from a flushing cistern should not have a bore less than:
(a) 8 mm
(b) 12 mm
(c) 16 mm
(d) 19 mm.

17 An approved method of jointing a plastics flush pipe to a WC pan is to use:
(a) sand and cement joint
(b) solvent weld joint
(c) rubber or plastics cone
(d) putty joint.

18 Where movement or settlement may occur the method of jointing a WC pan to its drain or discharge pipe should be:
(a) rigid
(b) flexible
(c) permanent
(d) cement mortar joint.

19 A pipe used for the conveyance of water from a flushing cistern to a WC pan is called:
(a) feed pipe
(b) service pipe
(c) overflow
(d) flush pipe.

20 The recommended capacity for a WC flushing cistern is:
(a) 4.5 litres
(b) 6 litres
(c) 9 litres
(d) 12.5 litres.

7 Sheet lead weatherings

After reading this chapter you should be able to:

1 Recognize types of roofing materials used in the industry, i.e. lead, copper and aluminium.

2 Appreciate thickness grade and size of sheet for differing purposes.

3 Recognize different tools and understand their use.

4 Describe methods of fixing and jointing.

5 Appreciate the preparation of building surfaces to receive sheet weathering.

6 Understand the principles of chimney weatherings.

7 Appreciate the use of sheet weathering as a damp proof course.

8 Understand the principles of fixing domestic eaves gutters and rainwater pipes.

9 Understand the safety requirements in the setting up and use of high pressure welding equipment.

10 Have knowledge of the safety requirements when performing lead welding operations.

11 Have knowledge of the working principles of regulators and blowpipes, pressure adjustment and control.

12 Recognize and name the component parts of high pressure equipment.

13 Have knowledge of the jointing process by:
   (a)   Soft soldering
   (b)   Hard soldering
   (c)   Fusion welding.

14 Demonstrate knowledge of the preparation and jointing of sheet lead.

Introduction

The introduction of the new *Plumbing Qualification* has brought about changes some of which affect this subject. The new qualification requires knowledge only on sheet lead work, it was felt that to enable the student to fully comprehend and appreciate the complex roof workers' skills part of this chapter should include some information on other roofing materials such as Copper and Aluminium. The fixing of cast iron and plastics gutter and rainwater pipes should also be included.

The covering of complete roofs and the weathering of component parts of buildings with the use of sheet weathering material forms an important part of the work of the present-day plumber.

Although the covering of large roofs and intricate weatherings has over the past few years tended to become the work of the roofing specialist, that specialist has generally progressed from the plumbing craft. In any case, the plumber must of necessity understand the principles of roofing and be able to perform to a satisfactory level in the manipulation and fixing of all the different types of material used for this purpose. There are many different types of roofing material in use today, the most common of which are:

1 Lead
2 Copper
3 Aluminium
4 Zinc.

Lead

Most of the lead sheet used in building is manufactured on rolling mills and is best described as rolled lead sheet. Slabs of refined lead about 125 mm thick are first rolled out to a thickness of 25 mm then cut into suitable sizes and passed backwards and forwards through the mill until reduced to the required thickness.

Lead sheet is manufactured to BS/EN 12588 in Sheet and Strip form for Building Purposes. BS lays down requirements that control the quality of the material. These include stipulations that the material be free from defects such as inclusions and laminations, that tolerance on thickness be not more than ±5 per cent and that the chemical composition be not less than 99.9 per cent lead. The preference, as traditionally, is that the composition of lead sheet should not contain any alloying elements that would significantly affect the characteristic softness and malleability of lead.

Sizes

In the past the thickness of lead sheet was defined by the weight per square foot (3 lb lead, 4 lb lead, 5 lb lead, and so on). However, measurement in metric units was introduced following the publication of a metric version of BS 1178 in 1969, which provides for a range of six sizes of lead sheet defined by thickness in millimetres. For easy

Table 7.1 *Range of thickness of lead*

BS No.	Thickness mm	Weight Kg/m	Colour
3	1.32	14.97	green
4	1.80	20.41	blue
5	2.24	25.40	red
6	2.65	30.05	black
7	3.15	35.72	white
8	3.55	40.26	orange

identification and because the range of metric sizes corresponds closely to the traditional range expressed in lb/sq ft, they have code numbers – 3, 4, 5, 6, 7 and 8. The substance of lead sheet is therefore specified by its BS code, e.g. no. 4 lead sheet (thickness 1.80 mm).

The standard range of thickness is given in Table 7.1 together with colour code markings for easy recognition in store or on site.

Rolled lead sheet is supplied by the manufacturer cut to dimensions as required or as large sheets 2.40 m wide and up to 12 m in length. Lead strip is defined as material ready cut in widths from 75 mm to 600 mm. Supplied in coils, this is a very convenient form of lead sheet for many flashing and weathering applications.

Cast lead sheet

Cast lead sheet is still made as a craft operation by the traditional method of running molten lead over a bed of prepared sand. A comparatively small amount is produced by specialist lead working firms, largely for their own use, in particular for replacing old cast lead roofs and for ornamental leadwork. There is no British Standard for this material. The available sizes of sheets made in recent years have varied from 2.75 m × 1 m to 5.5 m × 2 m. Skilled casters can cast sheets to an accuracy of 0.02 mm thickness on average in a range of thicknesses corresponding to the standard sizes for milled lead sheet code numbers 6, 7 and 8. Apart from the surface texture and lack of constancy in thickness, cast lead sheet, for all practical purposes, is not significantly different from milled lead sheet.

Machine-cast lead sheet

Thin lead sheet for building applications is made by a continuous casting process as well as by milling. A rotating water cooled drum is partly immersed in a bath of molten lead and picks up a layer of solid metal which is removed over a knife edge that scrapes the drum as it rotates. The thickness of the sheet being cast is controlled by varying the speed and temperature of the drum. Lead sheet made by this process is generally limited to thicknesses between 0.4 and 1.2 mm with a maximum width of about 1.2 m.

Properties of lead

Malleability

Lead is the softest of the common metals and in a refined form is very malleable. It is capable of being shaped with ease at ambient temperatures without the need of periodic softening or annealing, since it does not appreciably work harden. Lead sheet can, therefore, be readily manipulated with hand tools without risk of fracture and it is in taking advantage of this property that skill in working and fixing lead sheets has largely evolved. By the technique of bossing, lead sheet can be worked into the most complicated of shapes. Lead flashings can be readily dressed *in situ* to get a close fit to the structure even when the surface is deeply contoured as is the case with some forms of single-lap roof tiling.

Thermal movement

Most of the uses of lead sheet in building are those where it is fixed externally and is thus subjected to conditions of changing temperature. The coefficient of linear expansion for lead is 0.0000297 for 1°C and it would not be rare in the summer. where the lead is continuously in the sun, for it to vary in temperature daily by 40°C. A 2 m length of lead sheet could therefore, increase in length by 2 m × 40 × 0.0000297, i.e. about 2.3 mm. If the expansion and resultant contraction on cooling of the lead cannot take place freely, there will be a risk of it distorting and of a subsequent concentration of alternating (fatigue) stress which, over a long period, can cause cracking of the lead. It is, therefore, of first importance with lead sheet fixed externally, as with other metals, to limit the size of each piece so that the amount of thermal movement is not excessive and also to ensure that there are no undue restrictions on this movement. Long experience has shown that it is quite practical to make provision for the thermal movement of lead and take full advantage of its other outstanding properties to get external leadwork that will last a long time. Recommendations for limiting the size of pieces of lead sheet are given with the descriptions in this book of the various uses. Nowadays, somewhat thinner lead sheet is used than was the common practice in the past and as a general rule the thinner the lead the smaller each piece should be.

Fatigue and creep resistance

A factor that affects the strength of lead and its dilute alloys is the size of and uniformity of the grain structure of the metal. Lead, like other metals, is crystalline in nature and grain size describes the size of the crystals. In lead, the grain is readily visible after a small magnification when the surface is treated by cleaning and etching. Basically, the purer the lead the coarser the grain and variation in its size.

The presence of very small amounts of some other metals can modify the grain structure, in particular make it smaller and more uniform so that the fatigue resistance of the metal is improved and it is then better able to cope with stresses arising from thermal movement. The cracking of the lead as a result of excessive fatigue stressing is intercrystalline.

'Creep' is the tendency of metals to stretch slowly in the course of time under sustained loading, and is a factor of significance in external leadwork. The term creep should not be applied to the slipping of lead down a pitched roof when the fixings have failed to give adequate long-term support.

Although the composition of the lead for making lead sheet for building purposes as laid down in BS/EN does not stipulate the inclusion of grain refining agents, the usual practice of manufacturers is to use such compositions. Usually they use a copper-bearing lead which can include up to 0.06 per cent copper in the lead of basically 99.9 per cent purity required by BS WC. For all practical purposes this modification in composition does not affect the malleability of lead sheet.

Fatigue can occur, causing premature failure of lead sheet, because of faulty design or method of fixing. For instance, the use of oversized sheets or allowing too little freedom of movement for the sheet to expand and contract due to changing temperature, can both cause fatigue. Where failure occurs in lead sheet on an existing building due to bad design or installation it is most likely to be a fatigue failure since this can be caused by moderate oversizing of sheets or by the restriction of free movement due to incorrect fixing. Creep failure can only result from considerable oversizing of sheets. It would therefore, be preceded by a very premature fatigue failure, i.e. the sheet would fail by fatigue before the creep problem had time to become serious.

Patination of external leadwork

Lead is extremely resistant to corrosion by the atmosphere whether in town, country or coastal areas. In time, lead develops a strongly adhering and highly insoluble patina, the natural colour of which is silver-grey. Because of the insolubility of the patina, rainwater running off from weathered lead takes nothing into solution to stain or harm adjoining materials such as stonework. The patina of old leadwork often appears darker than the natural silver-grey when there is a coating of grime. Noticeably, parts of the surface of the old leadwork that are less exposed than others to the scouring action of wind and rain will appear darker and, of course, this is a condition more likely to be seen in towns than in the country or coastal areas. However, since grime forming emissions from the burning of fuel have been greatly reduced, external leadwork, like buildings in general, can be expected to present a more natural weathered appearance in the future.

While lead weathers so well it is nevertheless important to bear in mind its behaviour in the first stages of exposure. When freshly cut and exposed to the air lead forms a surface film of one of its oxides which imparts a dark grey appearance. Generally it will then slowly develop an even coloured and adherent patina by reaction with carbon dioxide and more importantly with sulphur dioxide in the atmosphere. Recent investigations have established that the permanent patina of external leadwork, even on leadwork many years old, is largely lead sulphate, irrespective of the exposure, whereas in the past it was thought to be predominantly a lead carbonate, even in town environments. However, the initial patina may begin to form somewhat patchily, particularly when the weather is showery shortly after the lead has been fixed. Such patches and streaks of light grey patina develop quite quickly. This initial patina is a lead carbonate which, while soluble in atmospheric moisture, is only loosely adherent to the lead and can wash off. Eventually the permanent patina will develop and in many forms of leadwork incidence of streaky initial patination will be of little consequence.

Patination oils

These oils have been specifically developed to prevent unsightly staining by new lead weatherings of roof surfaces, and should be applied to all new lead sheet flashings and components. Patination oils should be applied as per the manufacturers' instructions.

Compatability with other metals

The general experience is that lead can normally be used in close contact with another metal – such as copper, zinc, iron and aluminium – without corrosion by electrolysis. For example, no corrosion problems arise in the traditional use of copper nails and clips as fixings for lead and there is wide experience of the satisfactory use for lead flashings with patent glazing formed from aluminium bars. In marine and some industrial atmospheres it may be advisable to avoid direct contact between lead and aluminium because of the danger of accelerating corrosion of the aluminium and the guidance of the makers of the aluminium components should be sought.

Corrosion from timbers

Dilute solutions of organic acid leached from hardwoods can cause lead to be slowly corroded, Furthermore the corrosive effect of continuous condensation on the inner face of roofing and cladding (see below) can be exacerbated when it takes up organic acid from hardwood members of the substructure. Investigation has shown that impregnation of softwoods with preservative and fire retardant solutions does not, in itself, increase the risk of attack.

There are organic acids in cedar roofing shingles which can be taken up by light rain and dew to form a dilute acid solution which will slowly corrode lead flashings on to which it runs. In this situation it is advisable to protect the lead with a coating of bitumen paint for a few years during which natural weathering of the shingles will leach out the free acid.

Condensation

Condensation can exist in well heated buildings in which warm moist air will filter through the external walls and roof structure and, unless prevented, condense on the inner face of an impermeable cladding or roof covering. In the long term, condensation can cause significant corrosion of lead by slowly converting the metal mainly to lead carbonate. The importance of incorporating a vapour barrier in the external fabric where there is a risk of such condensation arising is well understood as is also the need to ventilate the space behind metal roof coverings and claddings where the substance is of frame construction. Past experience is that the incidence of significant corrosion of lead roof coverings and cladding through condensation is rare but the need to take these precautions against condensation should, nevertheless, be borne in mind.

Lichen on roofs

Slow corrosion of non-ferrous metals by dilute organic acids also arises with gutter linings of old buildings in country areas when lichen or similar mosses are growing on tiled or slated roofs. The attack on the metal gutter lining takes the form of narrow clean-cut grooves. What happens is that heavy dew or light rain dripping slowly off the roof picks up organic acid from the vegetable growth and where the solution falls on to the gutter lining it dissolves the normal protective patina on the metal. Repetitive dissolving and reforming of the patina results in grooves being cut into the metal, although in the case of lead it may be many years before the gutter lining is penetrated. There are modern solutions that can be applied to the roof covering to kill the lichen and prevent it growing again for a very long time. However, if the lichen has to be retained for appearance a periodic coating of bitumen paint on the affected area will arrest the attack or alternatively a sacrificial lead flashing can be fitted.

Effects of Portland cement concrete

Concretes and mortars made from Portland cement contain some free lime that can initiate a slow corrosive attack on lead in the presence of moisture. Direct contact between lead and new concrete or mortar should, therefore, be avoided in situations where drying out and carbonation of the free lime by reaction with atmospheric carbon dioxide is likely to be slow. Lead sheet built into brickwork or concrete as a damp proof course or impermeable membrane should be protected with a thick coat of bitumen paint. Flashings tucked into brick joints do not need any protection, since here carbonation of free lime is rapid and no risk of alkali attack then exists. When claddings, roof coverings and weatherings are applied to concrete surfaces, a sealing coat of a hard drying bitumen paint together with an underlay gives adequate protection during the drying out period.

Thermal insulation

Thermal conductivity of lead is 34.7 W/m °C and for practical purposes the effect of lead cladding or roofing can be ignored when calculating thermal resistance.

Table 7.2 summarizes the physical properties of lead.

Bossing

Bossing is the term applied to the general shaping of malleable metals. In particular it is the term used to describe the shaping of lead sheet with hardwood hand tools in its application as a building material. The hardwood tools are still in current use, although quite recently similar tools

Table 7.2 *Physical properties of lead*

Atomic weight	207.2
Atomic number	82
Density	11.34 g/cm
Coefficient of linear expansion	0.0000293 per °C
Thermal conductivity	34.76 W/m °C
Melting point	327 °C

manufactured from a tough plastics have become available and found to be an acceptable alternative.

Lead at ordinary ambient temperatures is only 300°C below its melting point and for this reason it behaves in many ways at ambient temperatures similarly to harder metals at higher temperatures.

Notably, the malleability of lead without the application of heat is exceptional, and it is the easiest of the base metals to shape by bossing with hand tools.

The malleability of lead is outstanding because it has the following properties. It is the softest of the common metals; it is very ductile in that it will stretch substantially before fracture under comparatively low tensile forces; and most importantly, it does not harden significantly when worked, being relatively quickly self-annealing at ambient temperatures – the warmer the day, the more quickly it will revert to its normal softness. Thus the lead can be readily shaped by bossing without the application of heat, and yet without it cracking or becoming hard and brittle.

Malleability and, therefore, ease of bossing can vary according to the chemical composition of the metal and thus the grain structure. The purer the lead, the more malleable it will be. The lead-worker should, however, find no significant difference for all practical purposes in the ease with which they can boss any lead sheet conforming in composition to the BS/EN 12588.

Leadworkers skilled in bossing can work lead sheet into the most complicated of shapes but in its practical application the basic aim is to achieve the required shape without undue thinning or thickening the substance of the lead sheet being used. In bossing there can be a surplus of lead to be bossed out and cut away or a shortage that has to be provided for when setting out the work so that the risk of thinning can be avoided.

The range of tools used for bossing lead sheet components to shape and dressing them into position are basically as follows, and they are made in different sizes to suit the thickness of lead sheet to be worked.

Bossing tools

The use of the correct hardwood (or plastics) tools for the working of sheet metals is essential.

Even when using the proper tools work must be carried out using great skill and care to prevent damage or bruising to the sheet. Many different types of hardwood have been used but the most common are hornbeam and beech.

The wooden tools should have all the sharp edges and corners removed and to preserve them they should not be allowed to dry out excessively or become soaked with water. They should be given an initial soaking in linseed oil, followed by periodical attention. Wooden tools should not be transported together with steel tools. They should never be struck with metal tools or used in any way that may damage or score the working surface as this will be transferred to the sheet metal and detract from the finished appearance of the work.

The introduction of plastics has brought to the craft a range of tools for the working of sheet metal. These appear to be proving very satisfactory and have several advantages over the traditional wooden tools. They are not affected by atmosphere; no soaking in oil is required and they are less easily damaged.

The following are the tools most commonly used to boss lead and are illustrated in the tools chapter:

Dresser	Step turner
Bossing stick	Chalk line
Bending dresser	Metric square
Bossing mallet	Straight edge rule
Tinmans mallet	Tinmans snips
Setting-in stick	Hammer
Chase wedge	Steel chisels
Drip plate lead	Compasses
Lead dummy	Lead welding equipment.

Bossing operation (external corner)

The technique of bossing sheet lead is a skill that must be taught and learned with great patience as there is no short cut. Once it is appreciated that by proper manipulation of the tools the lead can be made to move from one point to another, a tremendous satisfaction can be achieved.

A practical demonstration by a skilled crafts-man is by far the best way to learn this skill. The following notes should be helpful in the setting out and the procedure to follow.

1 Using chalk, chalk line or felt pen and square, set out the corner (Figure 7.1).

2 The two upstands form the sides of the corner. It will be evident that if the sides were turned up, the square in the corner is not required. It is therefore necessary to boss this square of lead out of the corner. To reduce the amount of work part of the corner is cut off (shaded portion in Figure 7.1).

3 Using a piece of timber as a support place it along the fold line of the upstand and carefully raise each side (upstand) leaving the corner open. Always keep the corners rounded – no sharp angles. To assist in the lead bending at the correct line, the fold line can

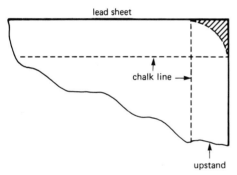

Figure 7.1 *Setting out an external corner for bossing*

be set in using a setting-in stick and mallet. This must be done gently and with care.

4 A helpful step although not absolutely essential is that of raising part of the base to form what is known as a stiffener. This helps to hold the corner in place and is done by striking the lead with the dresser.

5 We are now ready to begin the bossing operation which can be carried out using any combination of the lead working tools, for example bossing mallets, bossing stick, bending dresser and lead dummy. Again as a matter of preference, lead can be placed in its normal position with the bossing operation in an upwards direction, or alternatively the lead can be turned upside down and the bossing done in a downwards direction (see Figures 7.2 and 7.3). The bossing is performed using one tool to strike the lead and the other to act as a support. The bossing must start at the base working the surplus lead upwards (note direction of arrows). The shape of the lead must be carefully checked to make sure that it is still within defined limits of that required. It is of paramount importance that creases and thickening of the lead are not allowed to take place. Should the bossing be carried out correctly this becomes obvious to the skilled worker as the corner keeps on growing as the surplus lead is worked out of the corner.

6 The corner is checked for square in all directions (Figure 7.4). Note the corners

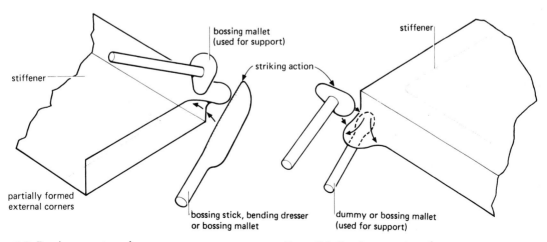

Figure 7.2 *Bossing an external corner*

Figure 7.3 *Bossing an external corner*

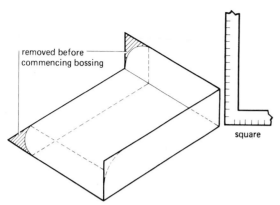

Figure 7.4 *One complete bossed corner: the operation is repeated until the box is complete*

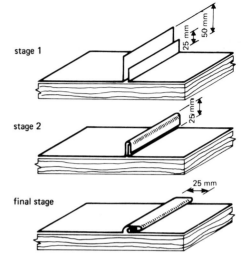

Figure 7.6 *Single lock welt*

should not have square angles but have a small radius. Finally, it should be trimmed to size.

Jointing of sheet lead

Single welt

The single welt or fold is used to weather and stiffen the edge of the sheet, for example the edge of a valley gutter, top edge of a back gutter, edge of dormer cheek, etc. (see Figure 7.5).

Single lock welt

This is in fact two single welts locked together. It is suitable for situations where the welt will not be submerged by water because it is unlikely to remain watertight over a period of time. The method of fixing is by means of 50 mm strips of sheet copper screwed to the roof surface at 500 mm intervals. Figure 7.6 shows the method of making the single lock welt, while Figure 7.7 shows the method of fixing.

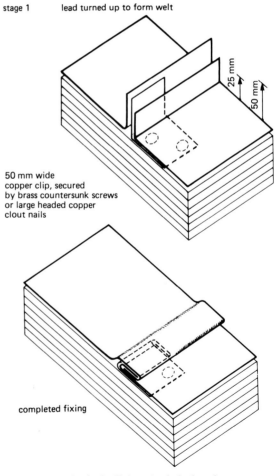

Figure 7.7 *Method of affixing single lock welt*

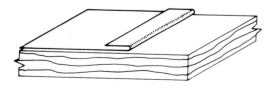

Figure 7.5 *Single welt*

Double locked welt

The double locked welt with its extra fold of metal is a water-tight joint. The method of fixing is by means of 50 mm strips of copper as detailed with the single lock welt. Figure 7.8 shows the method of making this welt.

Standing seam

Although this joint is not usually associated with sheet lead, it can have an application. It should however only be used in cases where it would not be walked on and never in any situation where the seam may be subjected to distortion. It is made and fixed similarly to the welt (see Figure 7.9).

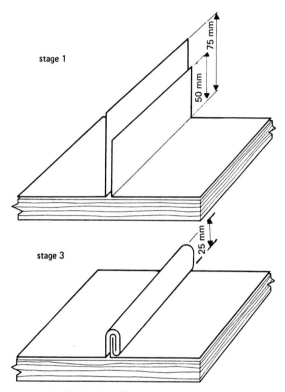

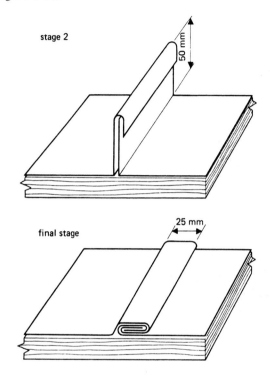

Figure 7.8 *Double lock welt*

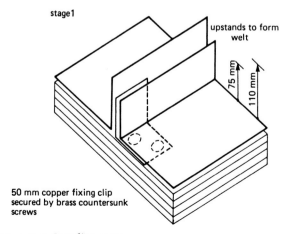

Figure 7.9 *Standing seam*

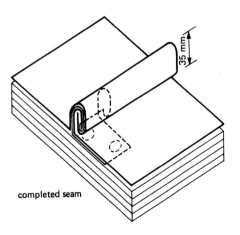

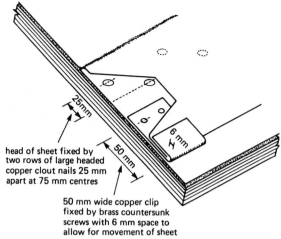

head of sheet fixed by
two rows of large headed
copper clout nails 25 mm
apart at 75 mm centres

50 mm wide copper clip
fixed by brass countersunk
screws with 6 mm space to
allow for movement of sheet

Figure 7.10 *Method of fixing a lap*

Lap

This type of joint is simply the laying of the sheet to form an overlap.

The method of fixing is shown in Figure 7.10. Where two metals are in close contact, water can be drawn up between these two surfaces by capillary attraction. To overcome this problem the sheets must overlap sufficiently to give a vertical lap of 75 mm. It is shown in Figures 7.11 and 7.12 that the slower the pitch the greater the length of the lap necessary to fulfil the requirements.

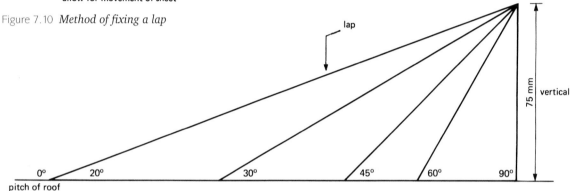

Figure 7.11 *The slower the pitch of the roof, the greater the length of the lap*

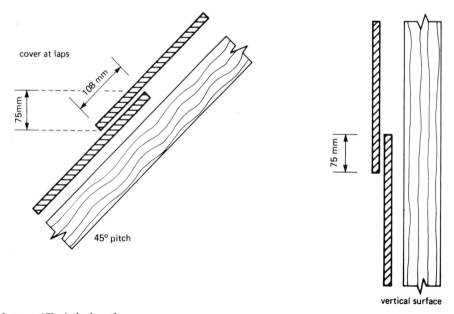

Figure 7.12 *Lap on 45° pitched roof*

Drips

These are used to form joints across the fall of flat and very low pitched roofs (see Figure 7.13). The drip should be not less than 50 mm deep or anti-capillary grooves will have to be incorporated in the joint as shown in Figure 7.14. By dressing the undercloak into the groove a space is formed between undercloak and overcloak. This prevents capillary attraction taking place. The top edge of the undercloak is dressed into a prepared rebate and copper nailed at 50 mm centres.

Figure 7.15 shows a sloping drip joint with fixing and jointing method.

Note: The 75 mm vertical height requirement.

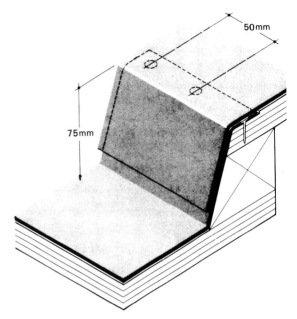

Figure 7.15 *Sloping drip joint with fixing and jointing method*

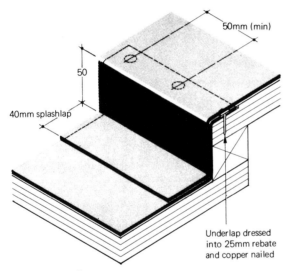

Figure 7.13 *Drip*

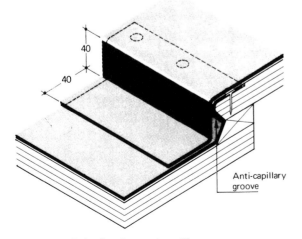

Figure 7.14 *Drip showing anti-capillary grooves*

Wood-cored roll

This joint is formed by dressing the edges of the adjoining panels over a shaped wooden core. The wood roll should not be too square in shape since this will prevent a good turn-in for the overcloak. For flat roofs the dimensions given in Figure 7.16 can be seen as standard but bigger rolls, giving a more bold appearance, are sometimes preferred for pitched roofs: the wood-cored roll is the accepted joint for flat and low pitched roofing, since it will stand up well to foot traffic. The undercloak is turned well over the roll and nailed with copper clout nails about 150 mm apart for a distance about one-third to a half the length of the panel, starting from the head. The overcloak is dressed fully over the roll, and with flat roofs it is extended as a splash lap which serves to stiffen the free edge and keep it in position. While the overcloak should be dressed to fit the shape of the roll, it should not be forced in tightly along the inside of the roll. The wood-cored roll formed in this way is most suitable for roofs up to 30° pitch, but wood-cored rolls above this pitch are preferably formed without the splash lap and with the free edge secured with copper clips.

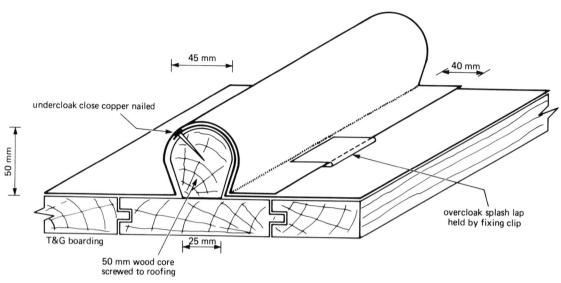

Figure 7.16 *Solid wood core roll*

Hollow roll

This joint is suitable for pitched roof coverings that have few abutments or complicated joint intersections. The hollow roll is made by forming a tall welted seam and turning it into a roll (see Figure 7.17). The undercloak is turned up 100 mm at right angles and 50 mm wide copper clips fixed in the required positions as previously described. The overcloak is turned up 125 mm and the edge welted, not too tightly over the edge of the undercloak, then the ends of the copper

clips are welted over at the same time and pinched to the lead. The prepared and clipped upstand is then turned to form a hollow roll.

Conclusion of jointing

All the above joints enable the material to move when heated. They are therefore known as expansion joints. To enable this to take place only one edge of the sheet is fixed, i.e. *undercloak*. The *overcloak* is left free to move. The overcloak may be held in place by means of clips but must never be firmly secured.

Fixings

The traditional method of fixing to a timber substrate at the head of panels of lead sheet that make up a pitched roof covering (above 15° pitch) or wall cladding is with copper clout nails 75 mm apart in two rows and staggered (see Figure 7.18). A single row will be adequate when the panel does not exceed 400 mm in height. The large heads of the nails pinch on the lead and the nailing is covered by overlapping lead sheet, and this fixing secures and supports the lead well for a very long life.

In the past, heavy lead sheet was sometimes supported at the head by dressing it over a wrought iron bar, round or rectangular in section,

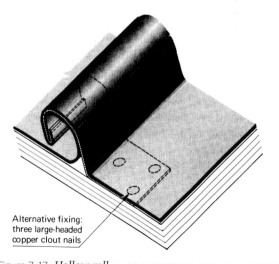

Alternative fixing:
three large-headed
copper clout nails

Figure 7.17 *Hollow roll*

which was tightly secured to the substrate so that the lead was firmly pinched across the full width of the panel. However, the simpler copper nailing method is used in modern leadwork where the lead is being fixed to a substrate of timber boards, plywood or other materials into which nails can be driven and will hold firmly.

When the substrate is concrete softwood timber battens impregnated with preservative can be inset in the concrete to take the copper nailing (see Figure 7.19). However, in some urban areas the authority may require the lead to be fixed direct to the concrete to give maximum fire resistance. In this case brass or stainless steel screws with washers to give a good pinch on the lead are inserted into plugs in the concrete at the same spacing as for copper nails (see Figure 7.20).

The plugs should be of a kind unaffected by moisture and preformed plastics plugs are commonly preferred. The heads of these screw fixings should be dished below the surface of the lead or the overlapping lead will tend to take up the shape of the projecting heads and show small bumps.

Copper clips that are incorporated in joints as fixings at the side of panels are cut from 0.6 mm thick soft temper copper sheet. They should be 50 mm wide and fixed to the substrate with three copper clout nails or two brass or stainless steel screws to each clip. In forming the joint, the lead sheet and copper clip should be lightly pinched together so that although the fold will tend to open with thermal movement of the lead, the clip will continue to give some support to the panel as well as securing it to the substrate.

Retaining clips to secure free edges against wind lift should be 50 mm wide and cut either from lead sheet or from copper sheet. Copper clips are best in situations of severe exposure, and where clips are visible (see Figure 7.21) they

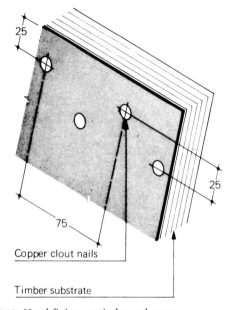

Figure 7.18 *Head fixing on timber substrate*

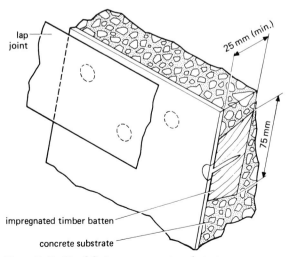

Figure 7.19 *Head fixing on concrete substrate*

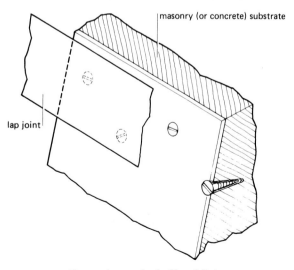

Figure 7.20 *Alternative method of head fixing on concrete substrate*

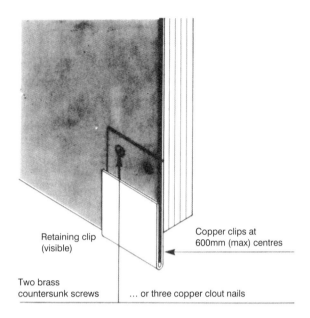

Retaining clip (visible)

Copper clips at 600mm (max) centres

Two brass countersunk screws

... or three copper clout nails

Figure 7.21 *Retaining clips*

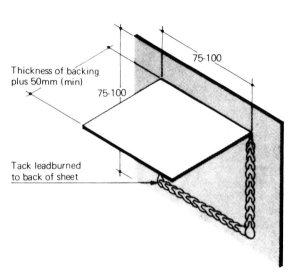

Thickness of backing plus 50mm (min)

75-100

75-100

Tack leadburned to back of sheet

Figure 7.22 *Lead tack method: first stage*

should be hot-dip coated with a high lead content solder. The retaining clip shown in Figure 7.21 should be fixed with some freedom between the clip and the bottom of the panel to allow for thermal movement. At the bottom of an area of wall cladding a continuous lead or copper clip can be used to provide drip edge finish.

For the lead tack method, a piece of lead 100 mm wide and about 200 mm long is lead welded to the back of the panel (see Figure 7.22). This tack is passed through a slot formed in the backing material, turned down on the inside and secured with brass screws and washers (see Figure 7.23). This type of intermediate fixing is particularly suitable when making up preformed lead faced panels for cladding.

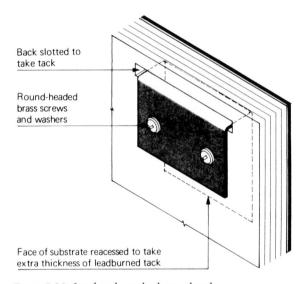

Back slotted to take tack

Round-headed brass screws and washers

Face of substrate reacessed to take extra thickness of leadburned tack

Figure 7.23 *Lead tack method completed*

Preparation of roof surfaces

Substructure and backing materials

Most structural materials make suitable substrates for lead coverings applied *in situ* provided they offer a continuous smooth surface that is strong enough to support the lead (and take superimposed loadings as required) and will hold the fixings firmly and permanently. See notes on

contact with other materials and similar aspects of corrosion resistance.

Timber

The traditional substrate for lead coverings fixed *in situ* is softwood boarding with an underlay. The boarding should be wrought, tongued and grooved, well-seasoned to give maximum resistance to warping and be fixed in the direction of the fall or diagonally (see Figure 7.24). Exterior

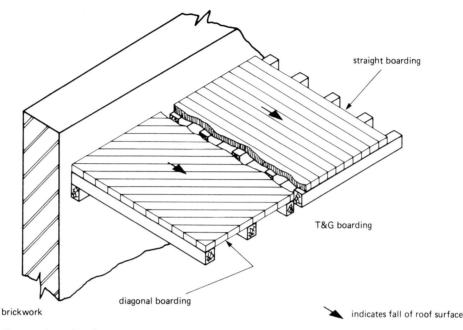

Figure 7.24 *Preparation of surfaces*

quality plywood or blockboard is also a satisfactory timber substrate providing it is rigidly fixed, and the use of this material has tended to supersede boarding. As plywood presents a smooth level surface, an underlay may be unnecessary.

The heads of nails used for fixing boarding and other timber members should be well pushed below the surface and similarly all screws should be counter-sunk. Sharp corners of external angles of timber substrates should be rounded off.

Concrete and masonry

The surface should have a smooth finish and an underlay should be provided both to isolate the lead from the substructure and, in the case of flat and low pitched roofs, to help the lead to move freely with temperature changes. Any materials for fixing the lead that are set in concrete or masonry should not be vulnerable to decay through the presence of retained moisture or condensation.

Other substrates

Where lead is laid on a thermal insulating material it is necessary to determine whether fixings made direct into such materials will be strong enough (equivalent at least to nail or screw fixings into timber). It is unlikely that fixings for the lead made directly into compressed cork or open texture woodwool slab will be strong enough, and timber battens will then have to be secured to the substructure in such a way that the fixings will give a long life.

With special proprietary thermal insulating substrates the question may also arise to whether the surface is firm enough to bear the dressing of the lead flat when it is being fixed.

Roof coverings

Underlays

The use of an underlay is to be recommended. This is usually a flame resistant polyester. Its purpose is as follows.

1 To act as an insulator thereby preventing heat loss in winter and also preventing heat entering the building in summer.

2 To act as a sound insulator absorbing much of the noise caused by heavy storms.

3 By acting as an insulator, to also prevent water vapour (condensation) from forming on the underside of the sheet. The condensation

could cause corrosion of the metal or rotting of the timber.

4 By acting as a separator between the sheet and (concrete) roof structures, to prevent the possible corrosion which may occur with new laid concrete.

5 To allow the free movement of the sheet when subjected to temperature changes.

Choice of thickness of lead sheet

Thicknesses of lead sheet normally appropriate for flat and pitched roof coverings are code nos. 5, 6, and 7. However code no. 4 may be acceptable in some cases, particularly for small roof areas. For wall claddings code nos. 4, 5, or 6 are appropriate.

Choice between these thicknesses should take into account the following factors:

- Quality of the building
- Necessary assurance of long life
- Design of the roof or wall cladding
- Size and shape of the panels required
- Exposure.

Sizes of lead sheet

The maximum sizes of the separate panels of lead sheet that are to form an area of roof covering, and hence the spacing of joints, will depend primarily on the thickness of lead to be used. Another factor, particularly when comparatively thin lead sheet is specified in line with modern practice, is the extent to which the covering will be exposed directly to the summer sun. The pitch of the surface is also taken into account, having in mind the need to support the lead adequately without excessive fixings.

Table 7.3 gives a general guide to the code of lead required for specific locations. For work in no. 4 and no. 5 lead sheet a range of maximum lengths is recommended according to the degree of exposure to the sun. The shortest length is suggested where there will be full day's long exposure to the summer sun.

The most important points to bear in mind when selecting the correct code of lead are those of the size of each piece fitted and the method of fixing.

Table 7.3 *Lead recommended for various jobs*

Job	Code
Soakers	No. 3
Step flashings	No. 4
Aprons	No. 5
Back gutter	No. 5 or no. 6
Small flat roofs	No. 5 or no. 6
Step and cover flashings	No. 5
Valley gutter	No. 5 or no. 6
Slate pieces	No. 4
Cladding	No. 4

Weathering to chimneys

For roofs covered with slates and plain tiles, the watertight connection is made by means of both soakers and flashings. Figure 7.25 illustrates their use in two different types of chimney.

In example A, the chimney passes through the slope of the roof. It requires:

Back gutter (back)

Apron (front)

Flashings (sides)

Soakers (sides).

In example B, the chimney passes through the ridge of the roof. It requires:

2 aprons (front)

Flashings (sides)

Soakers (sides).

Soakers

A soaker is a thin sheet of metal, i.e. zinc, copper or lead. Part of the soaker is fitted between the slates or tiles while the other side is turned up the side of chimney or brickwork abutment (see Figure 7.27). These are fitted by the roofer as the tiles are fixed. The 25 mm added to the length (Figure 7.26) is to allow for turning or nailing at the head.

The width of the soaker must be a minimum of 175 mm. This gives an upstand of 75 mm

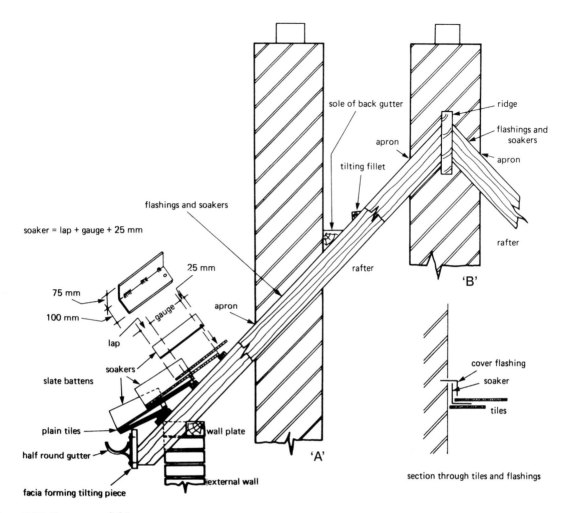

soaker = lap + gauge + 25 mm

75 mm

100 mm

25 mm

gauge

lap

apron

soakers

slate battens

plain tiles

half round gutter

facia forming tilting piece

flashings and soakers

sole of back gutter

tilting fillet

rafter

apron

wall plate

external wall

'A'

ridge

flashings and soakers

apron

rafter

'B'

cover flashing

soaker

tiles

section through tiles and flashings

Figure 7.25 *Two types of chimney*

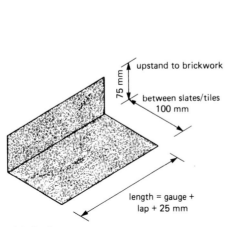

75 mm

upstand to brickwork

between slates/tiles
100 mm

length = gauge +
lap + 25 mm

Figure 7.26 *Soaker*

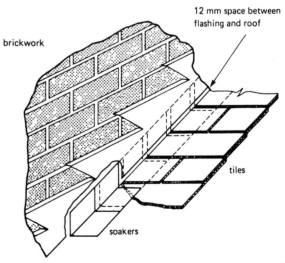

brickwork

12 mm space between
flashing and roof

tiles

soakers

Figure 7.27 *Step flashing*

against the brickwork and allows for 100 mm under the tiles. The length of the soaker can be obtained by calculation. First, it is necessary to find out the length of slate or tile and the lap.

Example
Calculate the length of soaker required for a slated roof. The size of slates are 510 mm × 225 mm, having a lap of 76 mm.

$$\text{length of soaker} = \frac{\text{length of slate} - \text{lap}}{2} + \text{lap} + 25$$

$$= \frac{510 - 76}{2} + 76 + 25$$

$$= \frac{434}{2} + 76 + 25$$

$$= 217 + 76 + 25$$

$$= 318$$

length of soaker = 318 mm

Slate fixing

Slate can be either head (Figure 7.28) or centre nailed (Figure 7.29). The lap allowed is 76 mm on roofs pitched up to 45°.

Slates must be laid to form a bond. Each slate must overlap another and the straight line vertical jointing should be staggered (see Figure 7.30).

The gauge is the centre of each batten to which the slate or tiles are fixed.

$$\text{Gauge of slate} = \frac{\text{length of slate} - \text{lap}}{2}$$

Example
Calculate the gauge of a roof to be covered with slates 510 mm × 255 mm with a lap of 76 mm.

$$\text{gauge} = \frac{510 - 76}{2}$$

$$\text{gauge} = \frac{434}{2}$$

gauge = 217 mm

Calculate the number of soakers required for a roof of 6510 mm covered by 510 mm × 225 mm slates laid to a gauge of 217 mm.

$$\text{number of soakers} = \frac{\text{length of roof}}{\text{gauge}}$$

$$\text{number of soakers} = \frac{6510}{217}$$

number of soakers = 30

It will be readily seen that the number of soakers is equal to the number of slate or tile courses.

Flashing

Figures 7.31 and 7.32 illustrate the method of setting out for continuous step flashing (sometimes known as running flashing). The recommended width of flashing is 150 mm with a length of not more than 1.5 m. Should a greater length be required, this is achieved by lapping the necessary pieces.

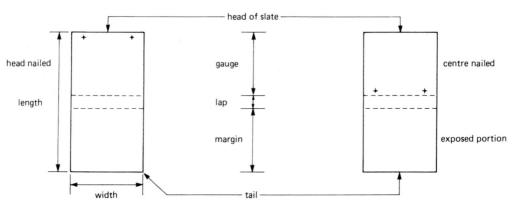

Figure 7.28 *Slate nailed at head* Figure 7.29 *Slate nailed in centre*

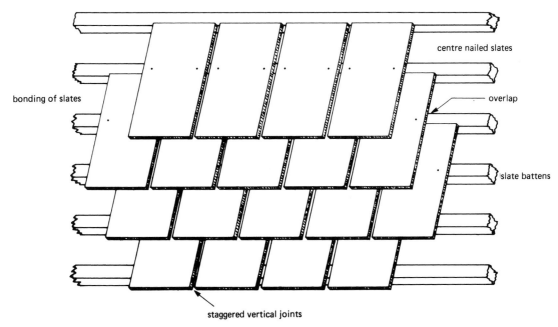

centre nailed slates

bonding of slates

overlap

slate battens

staggered vertical joints

Figure 7.30 *Bonding of slates*

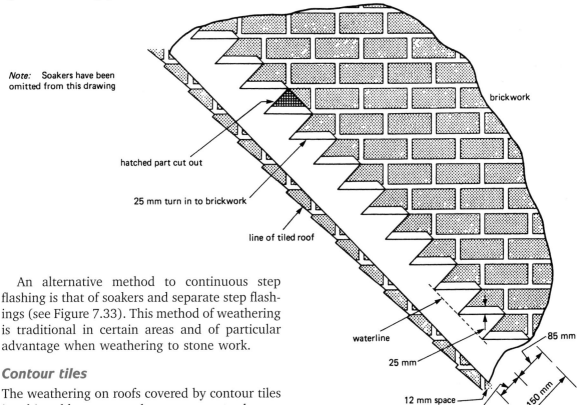

Note: Soakers have been omitted from this drawing

brickwork

hatched part cut out

25 mm turn in to brickwork

line of tiled roof

waterline

25 mm

12 mm space

65 mm

85 mm

150 mm

An alternative method to continuous step flashing is that of soakers and separate step flashings (see Figure 7.33). This method of weathering is traditional in certain areas and of particular advantage when weathering to stone work.

Contour tiles

The weathering on roofs covered by contour tiles is achieved by a process known as step and cover flashing. A single piece of metal takes the place of both the soaker and step flashing as described

Figure 7.31 *Continuous step or running flashings*

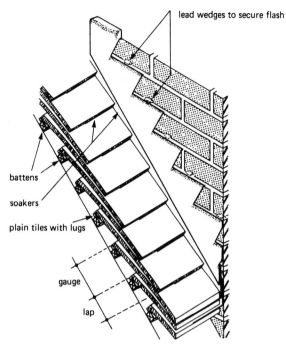

Figure 7.32 *Continuous step or running flashings*

for plain tiles (Figure 7.34). The weathering material should be dressed over a *minimum* of one contour. Two contours would be considered good practice.

Note

In all cases the weathering material must be dressed to fit snugly to the contour of the tiles.

Figure 7.35 shows a weathered chimney, complete with all its component parts. Figure 7.36 shows all these parts in detail.

Pipes passing through slated/tiled roofs

Where this occurs it is necessary to make a water tight connection between the pipe and the roof covering i.e. slates or tiles. This is achieved by making and fixing what is known as a lead slate piece.

A piece of lead large enough to give weathering to all sides of the opening is required. The size will be governed by the size of the pipe passing through the roof, the angle of the roof and the roof covering.

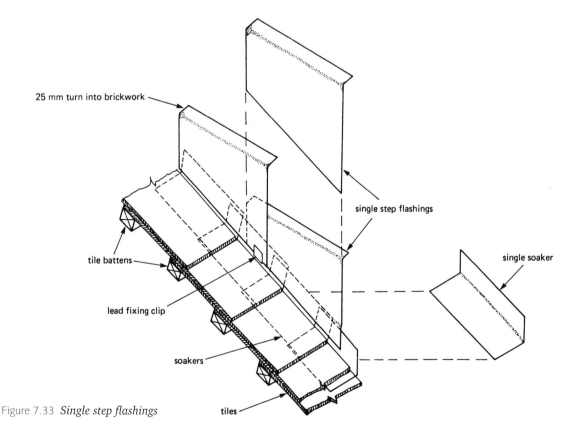

Figure 7.33 *Single step flashings*

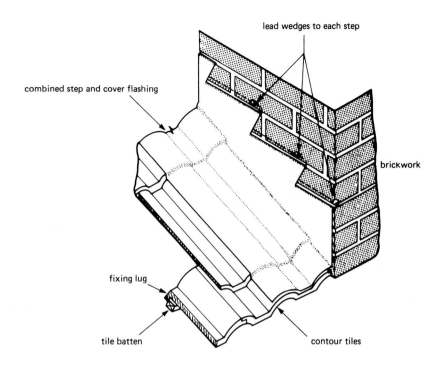

lead wedges to each step

combined step and cover flashing

brickwork

fixing lug

tile batten

contour tiles

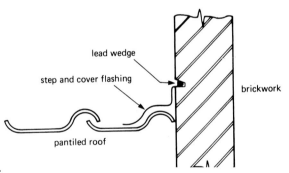

lead wedge

step and cover flashing

brickwork

pantiled roof

Figure 7.34 *Step and cover flashings*

It is possible to boss (work the lead) to form the slate in one piece (no joints) but this requires much time and skill. It is therefore common practice to fabricate the slate and join together by lead welding as shown in Figure 7.37.

The development of surface areas plays a big part in plumbing. It aids understanding of the shape so that calculations can be performed with accuracy. It is also necessary for setting out, cutting and forming to make items such as tanks or linings, cylinders, slate pieces and pipe intersections.

Development of cylindrical tank (Figure 7.38)

▪ Draw elevation (rectangle ABCD).

▪ Project plan (circle).

▪ Using 60°–30° set square divide plan circumference into twelve equal parts.

▪ Project AB to form line EF equal in length to circumference of circle. This is achieved by stepping off one of the equal distances on the plan twelve times along this line.

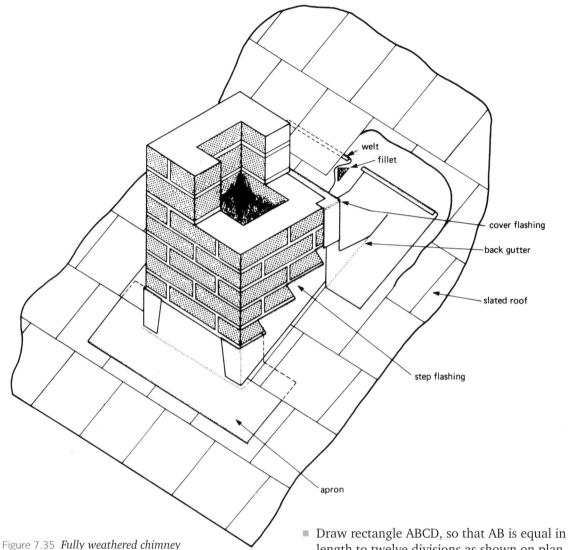

Figure 7.35 *Fully weathered chimney*

welt
fillet
cover flashing
back gutter
slated roof
step flashing
apron

- Draw perpendiculars at 0 and 12 to terminate line HG.
- Rectangle EFGH is the required development.

Development of slate piece (Figure 7.40)

- Draw the elevation of slate piece.
- Draw plan and divide circumference into twelve equal parts (using 60°–30° set square).
- Project points from plan up on to elevation to give intersection at roof line.

- Draw rectangle ABCD, so that AB is equal in length to twelve divisions as shown on plan, and AD is at least equal to height of slate piece. Project horizontal lines from intersections at roof line and vertical lines from divisions on AB to obtain intersections marked by crosses (follow arrows). The true shape of the development is obtained by joining the intersecting points by means of a freehand line as shown.

Development of true shape of hole through roof (Figure 7.39)

- Draw elevation of slate piece.
- Draw plan. Divide circumference into twelve equal parts as shown.

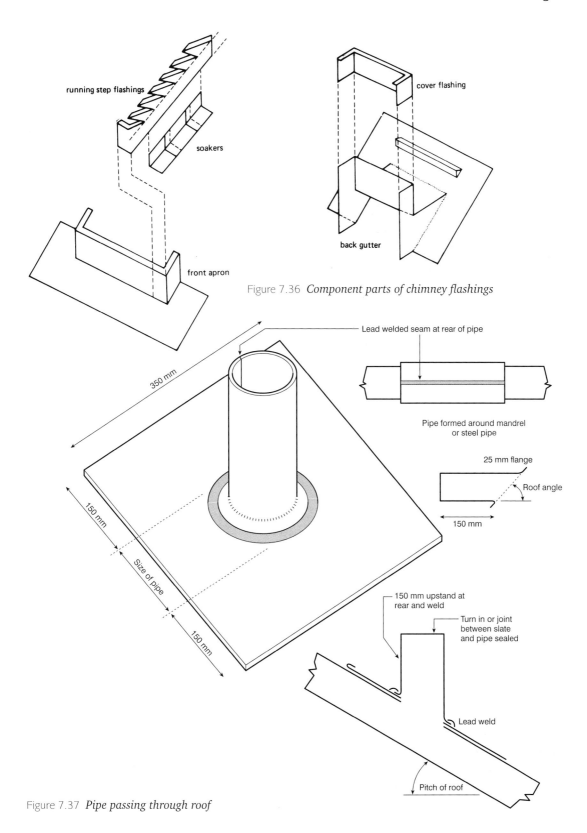

running step flashings

soakers

front apron

cover flashing

back gutter

Figure 7.36 *Component parts of chimney flashings*

Lead welded seam at rear of pipe

350 mm

150 mm

Size of pipe

150 mm

Pipe formed around mandrel or steel pipe

25 mm flange

Roof angle

150 mm

150 mm upstand at rear and weld

Turn in or joint between slate and pipe sealed

Lead weld

Pitch of roof

Figure 7.37 *Pipe passing through roof*

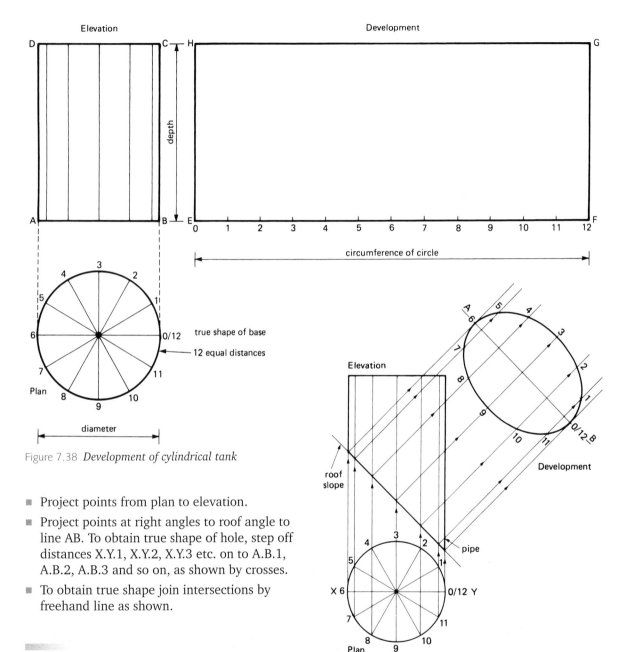

Figure 7.38 *Development of cylindrical tank*

■ Project points from plan to elevation.
■ Project points at right angles to roof angle to line AB. To obtain true shape of hole, step off distances X.Y.1, X.Y.2, X.Y.3 etc. on to A.B.1, A.B.2, A.B.3 and so on, as shown by crosses.
■ To obtain true shape join intersections by freehand line as shown.

Figure 7.39 *Development of true shape of hole through roof*

Copper

In designing a roof covering for a building, the architect will naturally look for a material that will give a long and reliable service. Such a material must be applied easily and quickly. It should add to the appearance of the finished job, possess a high degree of corrosion resistance, be economical and keep maintenance costs down to a minimum. Copper combines these qualities better than any other weathering material and is, therefore, an obvious choice.

There are many old buildings in existence possessing copper roofs, the age of which can be counted in hundreds of years – justifying the

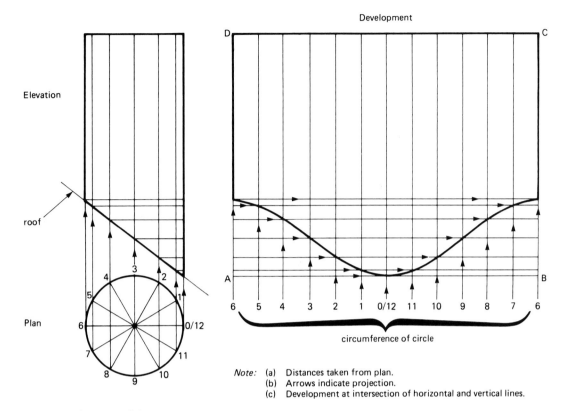

Note: (a) Distances taken from plan.
        (b) Arrows indicate projection.
        (c) Development at intersection of horizontal and vertical lines.

Figure 7.40 *Development of slate piece*

claim of durability for copper roofs. This long life is accounted for by the fact that copper develops, by natural processes, a surface film or patina, which forms a protection against corrosion by the effect of the sulphurous gases in the atmosphere. When exposed to the air, tarnishing takes place, resulting in a general darkening and blackening of the surface, due to the formation of copper salts. The main factor which determines the speed of this action is the quantity of sulphurous gases in the atmosphere.

When the black, tarnished layer has been formed, further changes, still caused by the sulphurous gases, gradually take place, forming an insoluble green layer on the surface. Thereafter, this film remains virtually unchanged, affording complete protection to the copper. It is not possible to forecast how long the development of the patina may take in any given district, but in London, copper laid six or seven years ago has already acquired a beautiful green film.

Properties of copper

Copper sheet and strip can be worked to conform to all the normal contours encountered in building, invariably enhancing the appearance of the building at the same time. The physical properties of copper are such that it will remain unaffected by changes in temperature and will not creep. This allied to its resistance to corrosion ensures that a copper roof will give a long and trouble free service. Another very important advantage of copper sheet and strip is its comparatively light weight per unit area and advantage may be taken of this in designing the substructure.

Where copper is laid over timber, it has been found that boring insects, such as the death-watch beetle, do not attack the timber. This is attributed to the fact that the metallic salts formed by condensation are lethal to such insects – it is well known that copper salts are used to a large extent in timber preserving compounds.

Copper can be used on oak or other timbers without any detrimental effect on the metal. However, it is essential that a roofing sheet should never be laid direct on to an understructure but always laid with an intermediate layer of a suitable felt.

The whole of the roof covering, including clips, flashings, saddle ends, etc., but excluding continuous fixing strips, should be made from fully annealed copper sheet or strip.

In exposed situations subjected to high winds, the width and length of the bays should be reduced and/or thicker copper used. This applies especially to gables and verges.

High winds may also retard the normal drainage of rainwater from the roof and this can result in welts and seams being temporarily submerged. Under these conditions it is recommended that the seams and welts be sealed with a non-hardening jointing compound (or mastic) before being folded.

Nails

Any nails used for the fixing of the copper and underlay must be made either of copper or a copper alloy such as brass. The nails should not be less than 25 mm long (measured under the head), not less than 2.6 mm thick, and weigh not less than 1.5 kg per 1000. The heads should be flat with a diameter of not less than 6 mm and the shanks barbed throughout their length.

Screws

Screws used in securing clips or other components should be made of brass. Where a batten roll joint is employed, the batten may be fastened by steel screws (or steel bolts and nut) provided that such fastenings are countersunk below the top surface of the roll and the exposed steel is suitably protected by painting with bitumen or covered with a felt ring or washer.

Clips

Copper clips of the same thickness as the roof sheeting should be used. They should be fastened to the understructure by two copper nails (or two brass screws) close to the turn up.

Clips for standing seams

For standing seam systems, clips should not be less than 38 mm wide, spaced at a maximum of 380 mm centres along the length of the standing seam.

Clips along a roll

Clips for rolls should not be less than 38 mm wide and spaced at not more than 460 mm centres. They should pass under the roll and be turned up on each side to retain the copper in position during fixing and service. Holes, in clips to receive screws, should be drilled or punched out with a parallel punch to the diameter of the shaft of the screw. Clips in standing seams and rolls should not be fixed closer than 75 mm from the junction with the cross welt.

Clips along ridge and hip rolls

Clips should be 50 mm wide and placed two per bay. They may pass under the ridge roll and turn up on each side, or be nailed on the side of the roll.

Clips in cross welts

Clips should be incorporated in all cross welts. Double lock welts require one 50 mm wide clip and single lock welts two 50 mm clips.

Clips at drips, eaves and verges

Clips should be 50 mm wide and placed two per bay in the drip edge and eaves welts, and at 300 mm centres in verge welts.

Clips against upstands

Upstand clips should not be less than 38 mm wide and fixed with two copper nails (or two brass screws) at not more than 460 mm centres.

Jointing

Many of the joints used for copper are very similar to those detailed in the section on sheet lead. The only main differences are the sizes recommended for the differing materials.

There are two main systems of traditional copper roofing, the standing seam system and the batten roll system. These terms relate to the methods used to join adjacent pieces of copper in the direction of the slope with cross welts or drips for the transverse joints.

Standing seams (see Figures 7.7–7.9) running from ridge to eaves may be used on all roofs where the pitch is 6 degrees or greater, while wooden rolls can be used on all pitches. On roofs of flat or low pitch, i.e. 5 degrees or under, the sheets must be jointed by means of wooden rolls (two types of wooden roll are illustrated in Figures 7.16–7.18). The reason for this is that if standing seams are used they may be trodden flat and the joints may then allow moisture through as a result of capillarity. Also, the standing seam can be vulnerable on flat or low pitched roofs located in exposed positions. In these circumstances wind can retain the rainwater on the roof causing some sections of the seams to be submerged.

Transverse joints may be either double or single lock welts or drips depending upon the pitch of the roof.

Sheet aluminium

Aluminium was first introduced as a roofing material at the end of the nineteenth century and by the middle of the twentieth century it had become a serious contender against the other metals for this type of work. The method of forming and jointing is similar in many respects to that of sheet copper using the same type of tools.

Contact with other materials

Metals

Direct contact with copper, and copper-rich alloys such as brass, must be avoided. In no circumstances should water be allowed to drain from a copper surface on to an aluminium one. The danger of electrolytic attack is less with lead or unprotected steel, but contact faces should, nevertheless, be painted. There is also a risk of attack, particularly in industrial or marine atmospheres, by water running from lead to aluminium. If this cannot be avoided by suitable design, the lead should be painted with bituminous paint. Zinc and aluminium may be used safely together. In order to avoid contamination, tools previously used with lead or copper should be well cleaned before use with aluminium.

Cement and lime mortars

Aluminium is subject to some attack when in contact with cement, lime mortars and concrete in the presence of moisture and it is, therefore, good practice to paint the metal before embedding in joints where continuous dampness is to be expected. Where the metal is used as a coping covering, steps must be taken to avoid moisture coming in contact with the underside of the metal. This can be achieved by painting the underside with bituminous paint or by inserting an impervious roofing felt with a covering of building paper.

Welding

Although the process of welding is a skill in its own right, and to become a qualified and certificated welder requires much skill and many hours of study and practice, the modern plumber must learn the basic requirements of this craft. Before welding was developed metals were joined together by riveting or by a blacksmith heating the metal to a very high temperature (but not to its melting point) then hammering or pressing it into unity. Today three methods of welding are in use in the plumbing craft:

1 Brazing
2 Bronze welding
3 Fusion welding (which includes lead welding).

Safety is of paramount importance when setting up the welding equipment and carrying out the process of welding. Since high pressure welding equipment together with the combination of oxygen and acetylene are in the most common use in both site and college, this is the method every plumber must be conversant with.

Equipment

The plumber must first be able to recognize all the components of high pressure gas welding equipment as detailed below and shown in Figures 7.41 and 7.42:

1 Gas cylinders:
  (a) Oxygen
  (b) Acetylene.
2 Regulators:
  (a) Oxygen
  (b) Acetylene.

3 Hoses:
  (a) Oxygen
  (b) Acetylene.

4 Welding blowpipes – size and type depends on the size and type of welding operation.

5 Set of spanners:
  (a) Regulator spanner
  (b) Outlet spanner
  (c) Cylinder spindle key.

6 Spark lighter.

7 Goggles.

8 Protective clothing:
  (a) Gloves – good quality chrome leather gloves should be available to give protection from cuts, burns and heat
  (b) Aprons
  (c) Trousers and jackets
  (d) Skull cap.

9 Wire brush.

10 Cylinder support.

In addition to the above list there are other items that are used in special welding situations and these will be described at the appropriate place.

In addition to the large welding equipment illustrated it is possible to obtain smaller portable outfits as shown in Figure 7.42 which shows the BOC Saffire Portapak, this portable equipment proves very useful being much easier to handle and move about the site. These portable outfits must be treated with the same respect as any other oxy-acetylene welding equipment with the same rules being observed.

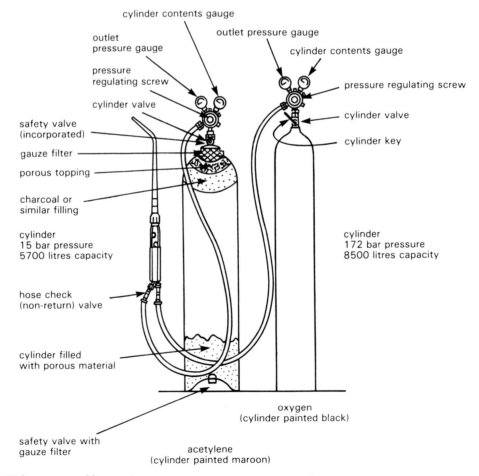

Figure 7.41 *High pressure welding equipment*

Flashback arrestor

This valve is fitted on the hose connection near the cylinder its function being to prevent the passage of the oxy-acetylene flame reaching the cylinder, the flame being arrested by the valve. The non-return valve will close and the explosion relief valve will vent the pressure and the products of combustion to the atmosphere.

The gas supply is automatically cut off and will start again only by re-activating the lever, this should only be done after the cause of the flashback has been established and rectified. Should the valve cut off at any time the equipment must be checked before being put back into use.

Maintenance

At regular intervals it should be tested with clean dry air or nitrogen against reverse flow and to ensure proper operation of the pressure activated cut-off mechanism.

Never use oil or grease and never test for leaks with a naked flame.

A reduction in flow could be due to a blocked inlet filter or flame trap.

A blocked filter in the inlet may be cleaned or replaced. Should the flow reduction be due to a build up of carbon due to repeated flashbacks, *on no account must this flame trap be removed*. The arrestor must be returned for Service Exchange.

Safety

Welding

1 Use protective clothing. Wear eye shield, goggles and gloves.

2 Make sure the cylinders are clear of falling sparks.

3 Have fire extinguisher nearby.

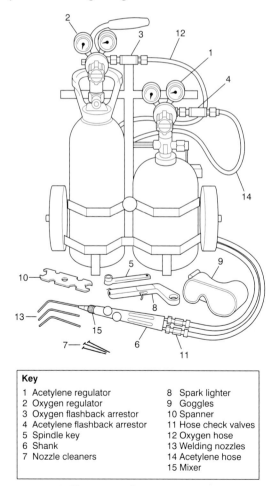

Figure 7.42 *Saffire Portapak*

Key

1 Acetylene regulator	8 Spark lighter
2 Oxygen regulator	9 Goggles
3 Oxygen flashback arrestor	10 Spanner
4 Acetylene flashback arrestor	11 Hose check valves
5 Spindle key	12 Oxygen hose
6 Shank	13 Welding nozzles
7 Nozzle cleaners	14 Acetylene hose
	15 Mixer

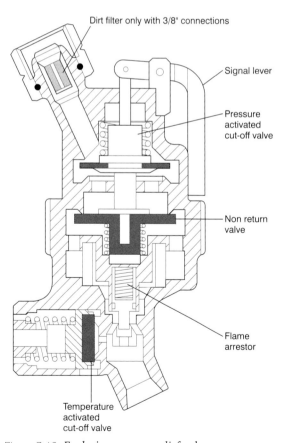

Figure 7.43 *Explosion pressure relief valve*

4 Use surrounding screens if other workers are near.

5 Electric equipment – see that the work is properly earthed.

Gas cylinders

1 Handle all cylinders carefully – never violently.

2 Store oxygen and acetylene separately in a cool place.

3 Acetylene cylinders must be stored and used vertically.

4 Do not stack oxygen cylinders more than four high, check for stability.

5 Keep grit, oil and dirt out of valves, otherwise it is impossible to prevent equipment leaking at the joints.

6 Never leave gas cylinders exposed to excessive heat or cold.

7 Never test for leaks with a naked light or flame. Use soapy water.

8 *Never* allow oil or grease to come into contact with cylinder valves or fittings because in the presence of oxygen under pressure oil and grease will ignite and may result in an explosion.

9 Shut off the valve before disconnecting the hose.

Safety precautions (HSW Form 1704)

1 Cylinders must not be roughly handled.

2 Grease and oil must not come in contact with valves or fittings.

3 No jointing materials or washers to be made from flammable materials.

4 Cylinders must be kept cool.

5 Cylinders must be kept away from naked flames.

6 Frozen equipment must not be thawed out by flames but by the use of hot water.

7 Ensure gas tight system:

(a) Leak of oxygen could endanger the operative by fire

(b) Leak of acetylene could cause an explosion or fire.

8 Oxygen must not be used in place of compressed air:

(a) To clear fumes from an enclosed area

(b) To test any system.

9 Oxygen must not be inhaled direct from the cylinder – it is injurious to health.

10 Adequate protection must be worn by operative:

(a) Eyes: tinted goggles

(b) Hands: leather gauntlets

(c) Head: leather skullcap or hat

(d) Body: leather apron or overalls.

11 Cylinders to be kept away from electric controls and switches to guard against possible arcing.

12 Excessive force must not be used when assembling welding equipment.

13 Precautions must be taken to prevent starting fires, i.e. remove or cover flammable materials, extinguishers to be available.

14 Ensure adequate ventilation when working in confined spaces.

The craftsperson must always bear in mind they are handling potentially dangerous equipment and treat it with great care and respect. Many accidents happen when the operator becomes complacent and thereby treats the equipment with contempt.

The setting up procedures are shown in Table 7.4.

Regulators

The regulator is a very delicate and important piece of welding equipment and great care must be taken when handling and fitting gas regulators to the cylinders or gas lines.

The function of the regulator is to reduce the high pressure of the cylinder content to a working pressure at the blowpipe. By means of the regulator the pressure is controlled to give a steady flow of gas with no fluctuation of the flame, even if the gauge does not register correctly. There are two types of regulators:

1 Single stage (seldom used)

2 Two stage.

Table 7.4 *Setting up procedures*

Procedure	Important points to note
Secure oxygen and acetylene cylinders in an upright position	Fasten to a trolley or cylinder stand or similar place
Open each cylinder valve momentarily then close again	This is called cracking the valve and is to ensure the removal of any foreign matter in the outlet
Fit the regulators in their respective cylinders and tighten the nut with the appropriate spanner	Do not use excessive force
Fit the respective hoses to the regulators and tighten, with appropriate spanners. Ensure hoses are connected correct way round, i.e. hose check valve connected to blowpipe. No gas will flow otherwise	As before do not use excessive force Oxygen equipment is coloured *blue* with right-hand threads. Acetylene equipment is coloured *maroon* with left-hand threads
Check that the regulator control is in the position that ensures it will not pass gas (the control will be rotated in an anti-clockwise direction). Slowly open the cylinder valve until gas is registered on the contents gauge. Turn the regulator control in a clockwise direction until a small amount of pressure shows on the line gauge. This will blow through the hose to ensure cleanliness. Turn the control anti-clockwise to stop the supply of gas. *Note:* It should not be necessary to close the cylinder valve at this point	1 Cylinder valves to be opened slowly and carefully otherwise damage to gauges will result 2 Do not stand in front of gauges; the faces may blowout 3 Ensure the gases released into the atmosphere will not prove a *hazard*, *i.e.* oxygen supports combustion; acetylene is a flammable gas
Now fit the welding blowpipe to the other ends of the hoses taking care as before and observing the same salient points	
Ensure acetylene blowpipe control valve is open, adjust regulator to give a pressure reading of 0.7 bar. Close valve, repeat for oxygen, check thoroughly using leak detection fluid that the equipment is gas tight	Bubbles will indicate leaks. Do not use excessive force should a leak be indicated

Two-stage regulators

These are the ones you will encounter in your normal situation on site and workshop. Figure 7.45 shows the working of the two stage regulator, the gas from the cylinder enters through the inlet and the pressure is measured on the first dial. It then passes through the first stage valve and diaphragm which is preset at approximately 20 bar (atmospheres). The gas then passes through the second stage valve, the pressure controlled by the second diaphragm and adjustment control being set at the working pressure. The two stage regulator gives a smoother and more constant flow of gas to the blowpipe and is therefore the one to be recommended.

Welding blowpipes

There are basically two types of blowpipes, high pressure and low pressure, but they are made in a variety of sizes and designs depending upon the size and type of welding to be performed and also the manufacturer. We will be dealing with the high pressure type only in this book as this is the one in common use.

High pressure blowpipes

This type of blowpipe is designed for use with oxygen and dissolved acetylene from cylinders (Figures 7.46 and 7.47). The blowpipe is designed to use equal volumes of the gases. The gases enter the mixing chamber where they are thoroughly mixed before issuing from the nozzle (see Figure 7.48). The blowpipe is fitted with control valves for each of the gases and interchangeable nozzles of varying sizes for welding different thicknesses of metal.

Note: High pressure equipment cannot be used on a *low pressure system.*

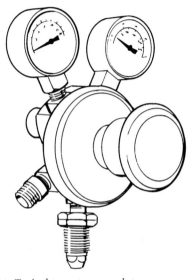

Figure 7.44 *Typical two-stage regulator*

Hoses

These are rubber-based with canvas reinforcement and are complete with brass or alloy connections for fixing to the regulators and blowpipes. The connectors are held in place by O-clips. The ends of the hoses connected to the blowpipes have 'hose check valves' fitted to them, which act as non-return valves in the case of a backfire taking place in the blowpipe. They are also known as 'hose protection valves'. Oxygen hose is *blue* in colour with right-hand connectors and acetylene hose is *maroon* in colour with left-hand connectors.

BOC hoses have 40 per cent rubber linings reinforced with cotton tape, covered with vulcanized abrasion-resistant rubber. They are kink and pinch resistant and have a bursting pressure of four times working pressure. The colour code is *maroon* for acetylene, *blue* for oxygen and *black* for non-combustible gases.

Note: Oxygen cylinders and old equipment are coloured black.

Safety

Flashback arrestor

The flashback arrestor can be fitted to the cylinder regulator (see Figure 7.50). It incorporates the following features:

1 Flame trap – a sintered stainless steel filter quenches the flame.

2 Non-return valve – a diaphragm is actuated under reverse pressure.

3 Safety valve – excessive pressures vented to atmosphere.

4 Flow cut-off valve – cuts off incoming gas supply.

5 Warning re-set lever – lever 'jumps out' indicating the fact that it has been triggered. Re-set after checking cause.

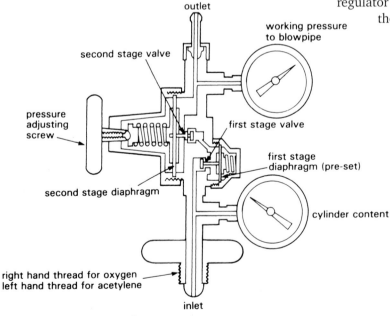

Figure 7.45 *Two-stage regulator*

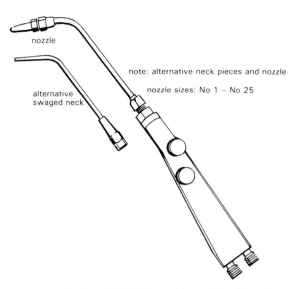

Figure 7.46 *Saffire (BOC) lightweight welding blowpipe*

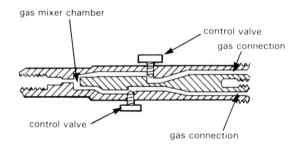

Figure 7.47 *Section through a high pressure blowpipe*

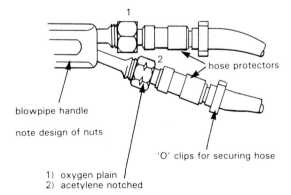

Figure 7.48 *Hose connection to blowpipe*

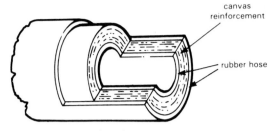

Figure 7.49 *Rubber-based hose*

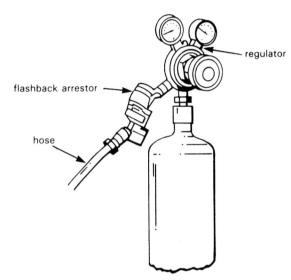

Figure 7.50 *Cylinder protection*

Backfires and flashbacks

Both these troubles are in many ways similar in their cause and also in the remedial methods required to correct them. They are caused by pre-ignition of the gases:

1 *Backfire*: the retrogression of the flame into the blowpipe neck or body with rapid extinction.

2 *Sustained backfire*: the gases ignite right back in the blowpipe and a loud squealing noise is emitted. The blowpipe itself will become red hot and if not turned off immediately will melt.

3 *Flashback*: the retrogression of the flame beyond the blowpipe body into the hose with the possibility of a subsequent explosion.

There are several ways in which these faults occur, mainly:

1 The nozzle overheating causes pre-ignition.

2 Welding in confined spaces (especially 'T' fillets).

3 Slag build-up on the end of the nozzle.

4 Nozzle too close to the weld.

5 Nozzle not correctly fitting into blowpipe.

6 Incorrectly set pressures.

Should either 'backfires' or 'flashbacks' occur:

1 Turn off both blowpipe valves immediately (oxygen first).

2 Check for causes 1–6 above and remedy.

3 Relight in accordance with correct procedure.

Note: If the nozzle has become overheated turn on oxygen valve *only*, plunge the nozzle and blowpipe head into a bucket of cold water. The oxygen prevents water entering the blowpipe.

Nozzles

Any skilled craftsperson has a natural pride in their work and knows that to achieve a good welding technique only good equipment should be used. The welding process depends a great deal for its efficiency and accuracy on the nozzle selected.

Nozzles should be carefully chosen for each particular job. Their design and manufacture are the result of years of practical experience and research.

Great care should be taken to ensure the nozzle does not become damaged; being made of copper and/or brass or sometimes copper alloys, this is easily done. Therefore, to avoid damage, the nozzle must not be handled roughly, dropped on the floor, or used to remove slag, etc. from welding jobs.

Should the nozzle become partially blocked this must be removed using only the proper nozzle cleaners manufactured for this job. Should the nozzle end become burred it should be cleaned and filed to return it to the proper shape and size. An imperfect nozzle will give a split or crooked flame, or an incorrectly burning flame, all of which will affect the welding process.

Lead welding

This skill is performed using oxy-acetylene welding equipment together with a model 'O' blowpipe with interchangeable nozzles depending upon the thickness of the lead and the type of job, i.e. sheet or pipe inside the building or outdoors.

The controlled melting of the lead and joining together with or without the addition of more lead (filler rod) has for many years been an established skill of the plumber and called lead burning.

This title has now been generally dropped and superseded by the term 'lead welding'.

Lead welding is in fact the fusion welding (autogenous) of lead, generally with the addition of a filler rod of the same composition as the parent metal. Fusion welding is carried out using a blowpipe which produces a fine concentrated flame of immense heat which melts the parent metal and the filler rod into a locally controlled pool.

In almost every way lead welding can be likened to fusion welding of steel and is performed using the *neutral flame*.

Table 7.5 covers the most commonly used joints, the ones omitted being the underhand and

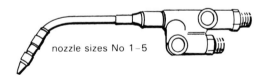

Figure 7.52 *Model 'O' blowpipe*

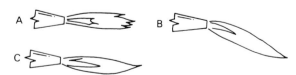

Figure 7.51 *Faulty flames caused by misuse or dirty nozzle*

Figure 7.53 *Lead welding nozzle*

Table 7.5 *Joints on lead sheet and recommended nozzle size*

Joint	Lead (code)	Nozzle size	Gas pressure (bars)
Lead to Brass	6–10	3–4	0.14
Flat butt	4–7	2–4	0.14
Flat lapped	4–7	2–4	0.14
Angle	4–7	2–3	0.14
Horizontal lap	4–7	2–3	0.14
Lap on inclined surface	4–7	2–4	0.14
Upright	4–7	1–3	0.14
Inclined on vertical face	4–7	1–3	0.14

the overhead, both of which are seldom used in normal plumbing work.

The nozzle sizes are given only as a guide, the skill of the operator and such factors as the position and climate (i.e. inside or outside of the building) will have a marked influence on the final choice. There is also a great range of flexibility in the setting of the flame for each nozzle size.

Preparation

The important points to be observed when preparing and welding lead sheet and pipe are:

1 *Clean surfaces*. All surfaces, including the edges, must be thoroughly shaved clean (shave hook recommended). The shaved area must not be touched. *No flux* is required, the work being carried out using a *neutral* flame. The flame as well as the metal must be clean (neither oxidizing nor carburizing). Keep preparation straight and neat using a straight edge.

2 *Correct penetration*. This is when the weld bead just penetrates the underside of the parent metal (flat butt joint) but in all cases the filler rod must be melted (fused) into a homogeneous mass with the parent metal. Insufficient penetration is due to too cold a flame caused by incorrect setting or too small

a nozzle size. Excessive penetration is due to too hot a flame, moving too slowly or unsupported work.

3 *Reinforcement bead*. Should be of herring bone pattern, even and uniform, extending to edge of shaved lead. Add approximately one-third thickness of lead as reinforcement.

4 *Undercutting*. This is the usual problem encountered by the beginner and is the melting away or thinning of the parent metal along the toe of the weld. It is generally caused by incorrect position or angle of blowpipe, too hot a flame or holding the flame in the same spot for too long.

Tacking

This is the holding together of the work prior to welding. It takes the form of individual tacks or a continuous tack (burning-in/fusing) performed with or without the addition of a filler rod (see Figure 7.55).

Filler rods

These are coils of lead wire of varying diameter (usually 3–6 mm) or strips of lead sheet 6–12 mm wide according to the thickness of the work and the type of joint being welded. The filler rod must be shaved clean immediately prior to use.

Technique

Wherever possible the lead should be fully supported. The lead is cleaned and tacked as shown in Figure 7.55. The blowpipe is directed on to the lead with the tip of the cone just clear of the surface of the metal. As soon as the molten pool is formed the filler rod is added, sufficient rod being melted off to form the joint with a reinforcement of approximately one-third the parent metal thickness. Sufficient heat is applied to extend the pool area to the width of preparation (sometimes by sideways movements of the blowpipe). The blowpipe should be directed down the centre of the seam with progress being made from right to left. There are two recognized methods:

1 Straight line progression

2 Side-to-side progression.

The latter method is generally done with a cooler flame and the finished bead has a more

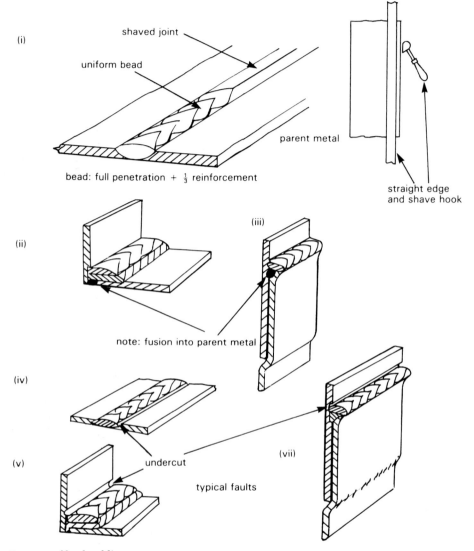

Figure 7.54 *Features of lead welding*

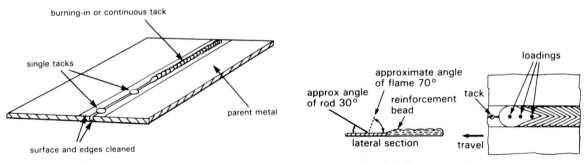

Figure 7.55 *Tacking*

Figure 7.56 *Straight line progression*

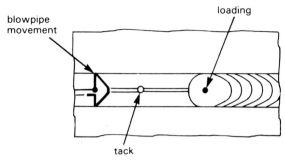

Figure 7.57 *Side to side progression*

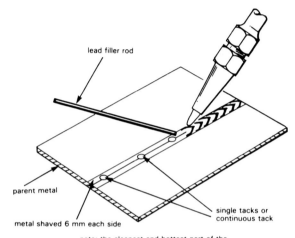

note: the cleanest and hottest part of the flame is immediately in front of cone

Figure 7.58 *Flat butt seam*

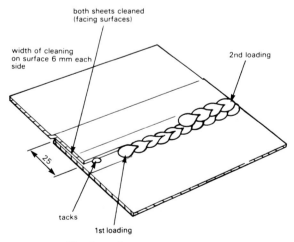

Figure 7.59 *Flat lapped seam*

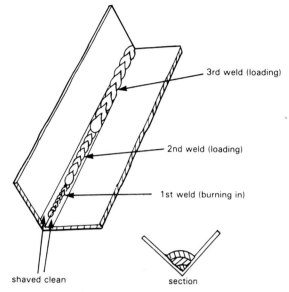

Figure 7.60 *Angle seam, both faces inclined*

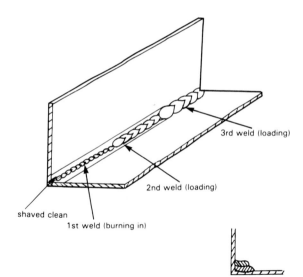

Figure 7.61 *Angle seam, one face vertical*

rounded appearance. The straight line progression method with the hotter flame has a more herring-bone shaped bead.

Sheet joints

Various sheet joints are illustrated in Figures 7.58–7.65.

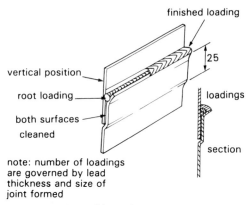

Figure 7.62 *Horizontal lapped seam*
Note: Number of loadings is governed by lead
thickness and size of joint formed.

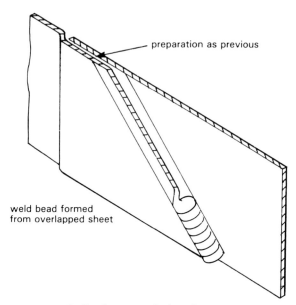

Figure 7.64 *Inclined on a vertical surface*

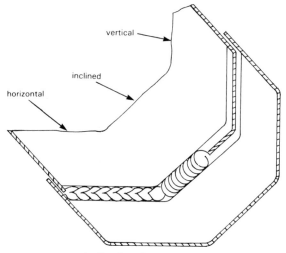

Figure 7.63 *Lap inclined surface*

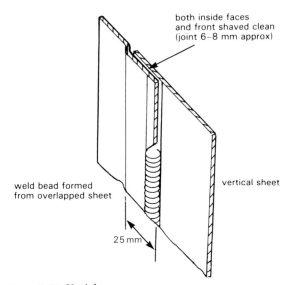

Figure 7.65 *Upright seam*

Safe working

1 Always ensure the hoses are purged (cleaned out). This is achieved by allowing a small amount of acetylene gas to blow into the atmosphere to expel any gas mixture which may cause an explosion. Acetylene gas is flammable, so ensure this is not done near any naked flame.

2 Always ensure use of correct nozzle size and gas pressure:

  (a) If gas velocity is low a backfire will occur. This could also happen if the nozzle is presented to a flame before the flow of acetylene is established.

  (b) If the gas velocity is too high the flame will blow off the end of the nozzle and be extinguished.

3 Do not use matches. These could be a hazard. Use some form of spark lighter or flint gun (Figures 7.66 and 7.67).

4 Only when you are sure the equipment is in a sound working condition can you proceed to

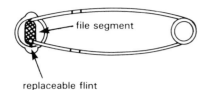

Figure 7.66 *Spark lighter*

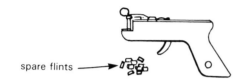

Figure 7.67 *Flint gun*

the pressure adjustment, lighting up and flame setting required.

The three flames

Before any kind of brazing or welding can be performed the operator must be aware of the *three* different types of flames and also know the type of flame required to carry out the task that has been set.

The three flames can be obtained by adjusting the controls on the blowpipe and so regulating the amounts of gas being emitted. They are:

1 *Carburizing* (used for hard facing of metals).
2 *Neutral* (used for fusion welding).
3 *Oxidizing* (used for bronze welding).

Oxygen and acetylene gases combine to produce a flame giving a temperature of 3200°C (Figure 7.68), which, by simple adjustment of the valves, can be used for a wide variety of purposes,

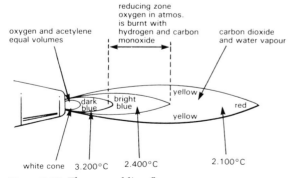

Figure 7.68 *The gas welding flame*

such as fusion welding, brazing, bronze welding, hard facing, flame cleaning, heating and cutting.

Carburizing flame

Assuming the gases have been previously turned on and adjusted to the correct pressure for the size of the nozzle fitted:

1 Open acetylene valve on blowpipe and purge the air.
2 Apply spark lighter to nozzle.
3 Adjust acetylene to get rid of the smoke and soot – but not enough to cause flame to lift off nozzle.
4 Open oxygen valve on blowpipe (slowly).

As the oxygen supply is increased a white diminishing cone with a feathery edge will be noted. This is caused by unburned particles of carbon. The degree of carburizing depends on the length of the acetylene feather.

Neutral flame

The procedure is the same as for the carburizing flame, up to the production of the acetylene cone. Continue slowly to open the oxygen control on the blowpipe and observe the flame. As the oxygen supply increases the acetylene feather cone decreases. This continues until the acetylene feather almost disappears inside the small white cone. At this point the flame will be neutral. The moment the acetylene feather disappears the increasing of the oxygen must be stopped. Some people recommend that a very slight feather should be seen on the small cone as a safeguard against excess oxygen.

Oxidizing flame

The process is the same as for the neutral flame. When this has been obtained the flame is changed into an oxidizing one by *reducing* the flow of acetylene. When the acetylene is reduced it will be noticed that the size of small white cone also decreases and it appears to he very sharply pointed. For normal bronze welding it is recommended that the cone be reduced by one-third its length. It is better practice to reduce the acetylene from the neutral position to obtain an oxidizing flame than to increase the flow of oxygen. The former gives a gentler and less noisy flame.

Flame characteristics

Carburizing flame

Examination of this type of flame makes the difference in structure between it and the other two very obvious. In the oxidizing and neutral flames there are only *two* really visible cones whereas the carburizing flame clearly shows *three*. This effect is brought about by an excess of acetylene, which gives the intermediate cone known as an acetylene feather (see Figure 7.69).

This type of flame should not be used for ordinary welding. It is a dirty flame, the carbon not burned away being deposited on the weld metal. This process shows as a boiling action and results in a hard brittle weld.

Carburizing flames are used for hard-facing of metals.

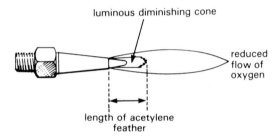

Figure 7.69 *Carburizing flame*

Neutral flame

The neutral or normal flame is the one most frequently used in welding. It is called neutral because it neither oxidizes nor carburizes the parent metal (i.e. it does not add or take anything away).

A neutral flame requires equal volumes of oxygen and acetylene to be burnt in complete combustion.

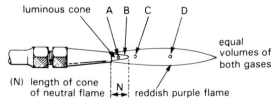

a) unburned mixture of oxygen and acetylene gases
b) primary combustion cone (slightly rounded)
c) hottest point of the flame
d) secondary combustion zone

Figure 7.70 *Neutral flame*

The neutral flame is a clean flame: the metal remains clean and flows easily. It is used for most welding jobs that require the metal to be melted and mixed together.

Oxidizing flame

At first view, particularly to the beginner, this flame may be mistaken for a neutral flame. However, the cone is short and more pointed, and it is more noisy and fierce. The greater the imbalance between the gases, the shorter and more pointed the cone will be. (Note N on the neutral and oxidizing flames.)

This type of flame should not be used for fusion welding steel, as it causes the metal to boil and spark. The additional oxygen in the flame causes the metal to burn resulting in a brittle weld. A slightly oxidizing flame is used for bronze welding.

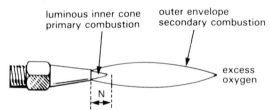

'N' length of cone on neutral flame

Figure 7.71 *Oxidizing flame*

Fusion welding

In fusion welding there is controlled melting of the two edges of the parent metals while a filler rod of the same or similar material is added, thereby producing a uniform mass along the length of the joint. The addition of the filler rod enables the welder to build up the weld face to give additional strength and the uniform finished bead.

Rainwater, gutters and downpipes

Gutters and R.W. downpipes were traditionally made of lead and examples of this skill can still be seen on many buildings such as churches, castles and public buildings. Over a period of time lead has been superseded by cast iron and steel, this in turn is now being replaced by plastics in this chapter we will indicate the differences but deal mainly with plastics. The sizing of gutters and rainwater systems are a very important task especially for large complex buildings.

There are many manufacturers of rainwater goods all supplying goods of a high quality and to a British and European Standard. The one selected to illustrate is the Geberit Terrain. The material which is known as PVC-u (Polyvinyl Chloride unplasticized) can be obtained in a range of colours, and manufactured in five shapes for domestic work, the most commonly used ones being the half round or square profile.

Gutter profiles

1 This is the half round gutter and is perhaps the one most commonly used on domestic work and small commercial buildings. It is capable of draining up to 122 square metres of roof a area to one downpipe with a flow rate of 2.54 litres/second per outlet.

2 Shows a square profile and again this shape is very popular for domestic work this shape enhances the view of the building, this gutter also has an improved draining and carrying capacity to that of the half-round type.

3 This is another rectangular shaped gutter with a very pleasing appearance it offers still greater draining ability being capable of draining a roof area of 294 square metres to one single downpipe with a flow rate of 6.11 litres/second per outlet.

4 Shows an elliptical profile, due to the depth its carrying capacity is greater than that of the half-round and is very similar to the rectangular gutter.

5 This is known as the Ogee pattern and again is very pleasing to the eye and often used on domestic dwellings where appearance is important. The drainage and flow rate is good and comes midway in the range.

Fixing

The usual tools required to cut and fix gutters are:

1 Bradawl
2 Screwdriver
3 Line (length of string)
4 Fine toothed saw
5 Ruler
6 Marker (felt pen or similar)
7 File.

The first job is to check that the eaves are level. Then fix a bracket at the highest point just below

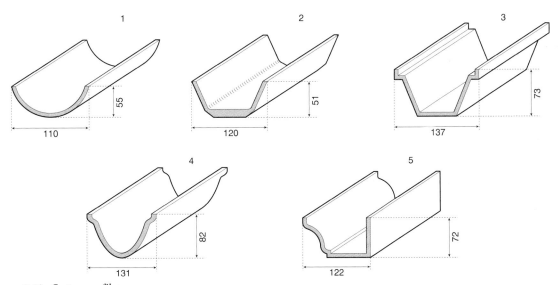

Figure 7.72 *Gutter profiles*

the tiles. Then go to the lowest point and fix another bracket, making sure there is a slight fall from the first bracket. The recommended fall is 1:350 approximately this would equate to 3 mm in 1 m. The string line is then fixed between the two brackets to enable you to fix the intermediate brackets to the correct fall. The fixing of the bracket will depend upon its type some are facia as shown in Figure 7.73 others are rafter brackets see Figure 7.73 there is also a drive in type of bracket, this kind is seldom used.

The brackets should be fixed every 0.8 m if the total length is longer than 4 m then an expansion joint should be fitted. Expansion is a very important factor and must be allowed for in every joint see Figure 7.73.

When all the brackets are fixed the gutter can then be fitted. Locate the back of the gutter under the clip of the bracket, pull the front of the gutter down until it snaps into position, make sure the gutter is located under the front clip.

The plastics rainwater goods should be cut using a fine toothed saw, any or all burrs should be removed using a file.

Each manufacturer produces all the necessary fittings and connections for their system. A sample is shown in Figure 7.73.

Jointing

This is achieved simply by the pressure applied by means of the clips holding the gutter on to the jointing rubber seal which is an integral part of the fitting, no jointing compound of any kind is needed or should be used.

Cast iron gutters

As stated previously this type of gutter was in common use and proved to be very successful. Its problem was unless it was regularly protected (painted) it rusted away, it was also, due to its weight more difficult to handle, fix and support. It has now been superseded by the plastics as described in this chapter.

To summarize the disadvantages of cast iron goods:

1 It is very heavy.

2 Needs greater support.

3 More difficult to work (i.e. cut and drill).

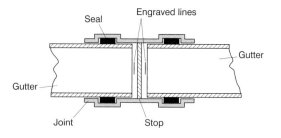

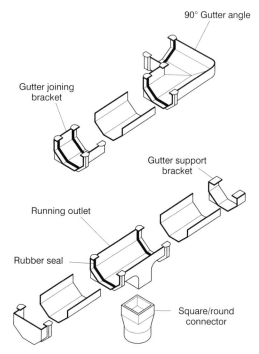

Figure 7.73 *Expansion joint*

4 Being cast it is fairly easy to break.

5 Rusts readily especially at the back where it is difficult to paint.

6 Needs regular maintenance (i.e. painting).

Fixing

The method of fixing is usually by means of rafter brackets, although facia brackets could be used, due to the weight of the gutter they are not recommended.

The jointing is by means of putty and paint or other jointing compound, one of the non-setting mastics is now favoured. The spigot (plain end) is placed on the bed of putty in the socket and the two are now secured by means of a 6 mm, bolt. Too much pressure should not be applied when tightening the bolt or the cast iron gutter can crack.

Downpipes

As described earlier for gutters the downpipes can be manufactured in both cast iron and plastics.

The pipes are made in either round or square in shape and in various colours. A range of fittings are manufactured such as: brackets, connections, branches, shoes, heads and bends. Offsets are not made. These are obtained by connecting two bends and inserting a piece of pipe should a greater length be required.

Fixing

Most domestic R.W. pipes are fixed to the building externally it is therefore essential that the pipe be fixed perpendicularly. This is achieved by the use of a chalk line. The line is fixed in line with the centre of the outlet. The plumb-bob on the bottom of the line ensures a perfectly vertical line. You can either work to the line or a chalk line on the wall.

For external work the joints are not sealed. In the case of internal work the joints are sealed by solvent cement.

As stated previously expansion of the material must be allowed for or damage will occur (see Figure 7.74).

Summary

1 Fully support all downpipes and fittings with clips positioned at the correct intervals.

2 Fittings fixed with sockets facing upstream.

3 Pipes to be cut square and deburred.

4 Position pipe and clip in moulded grove on fitting.

5 Check clip is correctly positioned by marking the expansion gap.

6 Use the sight hole to achieve the correct expansion gap.

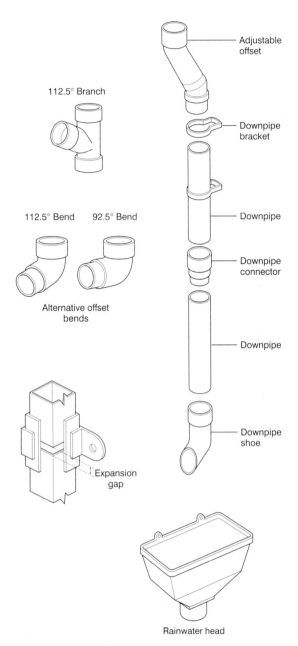

Figure 7.74 *Downpipes*

Self-assessment questions

1 Which of the following correctly describes the main purpose of a tilting fillet in roofwork?
   (a) it allows for thermal movement
   (b) it prevents capillary action
   (c) It prevents draughts under the tiles
   (d) It allows a circulation of air.

2 The metal least suitable for roofwork in a polluted atmosphere is:
   (a) copper
   (b) aluminium
   (c) lead
   (d) zinc.

3 When sheet copper is exposed to the atmosphere it forms a protective coating known as:
   (a) patina
   (b) verdigris
   (c) galena
   (d) polyurathene.

4 The purpose of a drip in a lead-lined gutter is to:
   (a) slow down the flow of water
   (b) speed up the flow of water
   (c) prevent the wind lifting the lead
   (d) allow for thermal movement.

5 The expansion of sheet metal is called
   (a) linear expansion
   (b) cubic expansion
   (c) volumetric expansion
   (d) superficial expansion.

6 Which of the following would be used to join the spigot ends of two lengths of gutter?
   (a) single socket
   (b) union clip
   (c) double stopend
   (d) double socket.

7 The term secret tack in sheet leadwork is used when:
   (a) concealing copper nails behind a welt

   (b) soldering over brass screws and washers
   (c) securing and supporting vertical sheets
   (d) securing a drip edge.

8 The correct type of underlay material for a metallic roof should be:
   (a) inlaid
   (b) spongy
   (c) flame retardant
   (d) indestructible.

9 Assuming the gauge has been established the length of a soaker on a plain tiled roof is determined by:
   (a) gauge + lap + 25 m
   (b) gauge − lap + 25 m
   (c) girth + lap + 25 m
   (d) margin + lap.

10 When fusion welding of lead is performed:
   (a) a powder flux is necessary
   (b) a flux is not required
   (c) a paste flux is necessary
   (d) oxygen is given off.

11 The correct jointing substance for plastics gutters is:
   (a) putty
   (b) rubber seal
   (c) mastic
   (d) paint.

12 When cutting plastics pipe by hand the saw blade should:
   (a) be fitted with teeth facing forward
   (b) have not less than two teeth per millimetre
   (c) be fitted with teeth facing the operator
   (d) be fitted with teeth facing either way.

13 Creep of sheet lead is caused by:
   (a) malleability of lead and friction with the roof
   (b) corrosive sub-surfaces and chemical resistance
   (c) weight of lead and temperature
   (d) absence of underlay and insufficient fall.

14 The thickness of sheet lead is recognized by a colour code. Number 5 lead is:
   (a) green
   (b) blue
   (c) red
   (d) black.

15 Which of the following physical properties is possessed by sheet lead?
   (a) ductility
   (b) malleability
   (c) tenacity
   (d) elasticity.

16 When laying sheet lead using lap joints, the vertical lap should be:
   (a) 75 mm
   (b) 50 mm
   (c) 85 mm
   (d) 100 mm.

17 The side weathering to a chimney passing through a slated roof is known as:
   (a) step and cover flashings
   (b) step flashings
   (c) soakers
   (d) soakers and step flashings.

18 The acetylene cylinder must be used in a vertical position during the welding process to:
   (a) prevent the escape of acetone
   (b) allow the acetylene gas to leave the cylinder
   (c) enable safe erection of equipment
   (d) conserve floor space.

19 Before connecting regulators to gas cylinders you should:
   (a) clean the threads with oily waste
   (b) open the valve on the regulator
   (c) lay the cylinder on its side
   (d) open the cylinder valve momentarily to dislodge any dirt.

20 One of the purposes of using a flux is to:
   (a) React with a base metal
   (b) Lower the solder's melting point
   (c) Increase the conductivity
   (d) Prevent oxidation.

8 Plumbing, qualifications and assessment

After reading this chapter you should be able to:

1 Understand the Plumbing Qualifications structure.
2 State the requirements for achieving qualification certification.
3 Describe the basic structure of the plumbing industry.
4 Understand the role of plumbing associated organizations.
5 Appreciate employment rights and responsibilities.

Qualifications

Plumbing qualifications have been developed to provide the basic training (in a college or training centre) for new entrants or existing unqualified workers in the plumbing industry. These qualifications are:

- City and Guilds Level 2 Certificate Basic Plumbing Studies
- City and Guilds Level 3 Certificate Plumbing Studies (Domestic).

These qualifications are often referred to as 'Technical Certificates' and are Related Vocational Qualifications (RVQ).

Plumbing NVQs have been developed by the Plumbing Industry organizations to produce qualifications that are a statement of the holder's occupational competence. These qualifications are:

- Level 2 NVQ in Mechanical Engineering Services (MES) Plumbing
- Level 3 NVQ in Domestic Plumbing.

The above qualifications comprise a number of units. Each unit contains one or more elements and the competence requirements covered by the elements is described by performance criteria and range statements (which detail the contexts in which the performance criteria have to be met).

The Technical Certificates have been designed to provide a basis for the training required to

support the NVQs in Plumbing and knowledge and practical assessments undertaken by Technical Certificate candidates can contribute to the evidence requirements of the NVQs.

It is important for candidates that their Technical Certificate programme provides them with training in current plumbing practices – materials, components and equipment of types that are in current use in the industry.

Assessment centres are required to provide installation facilities, which realistically simulate workplace conditions and to ensure that candidates carry out practical task assessments in a manner that reflects actual working procedures employed in a domestic plumbing workplace.

The Technical Certificate Scheme 6129 and Plumbing NVQ Scheme 6089

Qualification structure

The relationships between the Plumbing Technical Certificates and the Plumbing NVQs are as shown in Figure 8.1.

Candidates will be permitted to carry the results of Technical Certificate knowledge assessments and a number of practical assessments forward as evidence to be set against NVQ requirements. However, all NVQ Candidates must produce evidence from their workplace on a

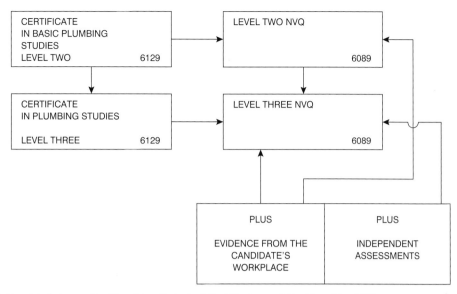

Figure 8.1 *Relationship between the Plumbing Technical Certificates and the Plumbing NVQs*

mandatory basis (laid down in NVQ Scheme Documentation) and Level 3 candidates undertake additional independent assessments (for Gas – ACS, Water Regulations and Unvented Domestic Hot Water Storage Systems).

The same knowledge assessment tests apply for the Technical Certificate and NVQs, with the exception of those competences for which independent assessments are undertaken.

Performance evidence from the Technical Certificate Practical Tasks can contribute to the performance evidence requirements of the NVQs (further detail is identified in the Plumbing NVQ Scheme Documentation).

Assessment procedures

Knowledge assessments – general requirements

The Technical Certificates at Levels Two and Three require candidates to undertake knowledge assessments, which include:

■ Multiple choice questions – centrally set and locally marked.

■ Multiple choice questions – centrally set and centrally marked.

■ Assignment on Job Planning requiring short written answers – paper centrally set and locally marked.

Candidates should undertake these assessments when they have completed the training in the appropriate subject areas to a satisfactory standard – the assessments are not intended to be 'end assessments' (with the exception, of the centrally set, centrally marked papers).

Centrally set and locally marked multiple-choice papers

The Awarding Body will provide for all Technical Certificate/NVQ Assessment Centres three series of multiple-choice question papers (Series A – Series B – Series C).

Each series includes a number of question papers which in total cover all the knowledge requirements of all the units and system ranges within the qualifications at Level Two and Three (with an exception at Level Three for certain knowledge competences which are covered by assignment style short written answer papers).

Level 2 candidates undertake a total of twelve papers.

Level 3 candidates undertake a total of seven papers.

In general all multiple-choice assessments are 'closed book' assessments; locally marked assessments have a basic pass mark of 80 per cent. The procedures for the conduct of these examinations and for candidates scoring less than the pass mark is explained to candidates prior to each examination and also set out in the Schemes Handbook.

Practical task assessment

General requirements

All Technical Certificate practical training activities are aimed at providing candidates with the practical competences to successfully complete the practical assessments required by the scheme. The awarding body has therefore developed a complete schedule of practical tasks, which candidates must complete on a mandatory basis to successfully complete the practical units.

All the assessment criteria in the practical tasks must be successfully met to complete the unit and achieve the award, additionally the assessment activity includes a maximum time duration within which the assessment activity must be conducted. Failure to complete in the time duration means that further training is required and the assessor will require the candidate to re-take the assessment activity.

The task requirements are laid down in:

■ City & Guilds Level 2 Certificate: Basic Plumbing Studies – Practical Task Manual – Level 2 Units

■ City & Guilds Level 3 Certificate: Domestic Plumbing Studies – Practical Task Manual – Level 3 Units.

The NVQ qualification structure

All units within the Level 2 and Level 3 Qualifications are mandatory units and all assessment requirements of all units must be successfully achieved for the award of the qualification. Scheme developments anticipate a number of additional or optional units may be included at NVQ Level 3 in the future.

Performance evidence

All candidates are required to produce evidence of performance required by the tasks detailed in the Candidate Workplace Evidence Record.

Recording evidence

The candidate's successful completion of workplace tasks will be recorded in the candidate's Workplace Evidence Record. Each task within the record has a number of assessment criteria all of which must be met for the task to be satisfactorily completed. In some cases the task must be completed on more than one occasion. In addition to direct observed assessment of performance, a variety of assessment tools may be used by the centre assessor to assess competence. These can include:

■ Oral questioning

■ Provision of supplementary evidence (including witness testimonials) – minimum supplementary evidence requirements are detailed for tasks in the Candidate Workplace Evidence Record

■ Provision of evidence of prior achievement.

The Candidate Evidence Workplace Record has been designed as the key document to compile the portfolio of evidence and to provide a record of assessment activities.

Employment: rights and responsibilities

A written contract of employment is useful to have because it sets out what you can expect from your job and from your employer. There are laws to protect workers, whether or not they have a written contract.

Every worker has the right to be paid and to enjoy a reasonably safe place of work. In the same way, you have duties to your employer, including a duty of honesty, loyalty, confidentiality and personal service.

You also have certain rights under Acts of Parliament, such as the right not to be discriminated against at work because of your sex, race or disability.

Also, the national Minimum Wage Act has imposed a minimum level of pay.

Working Time Regulations give you the right to four weeks' paid holiday (this includes public and statutory holidays such as Christmas and Bank Holidays). Other legislation gives you the right to statutory sick pay and maternity leave, and in some cases to statutory maternity pay.

Other rights may be included in your contract of employment. For example, if you have always received a Christmas bonus or extra holiday you may be able to prove that this is a legally-binding entitlement under your employment contract.

Every employee has a legal right to receive a written statement which sets out your terms of employment. You should get this within two months of starting your job. The terms include basic things like:

- The name of your employer
- Where you will be working
- When you started work
- How much you will earn and
- When you will be paid.

It will also include terms about:

- Your hours of work
- Your holiday entitlement (including public holidays)
- Holiday and sickness pay
- Pensions
- The length of notice you and your employer have to give if you leave or are dismissed
- How long your job is expected to continue, if it is temporary and
- Any disciplinary rules.

If you have to work outside the UK for more than a month at a time, it should also cover:

- The period of work outside the UK
- The currency you will be paid in
- Any extra pay and benefits you will get for working outside the UK and
- Any terms and conditions about your return to the UK.

In many cases, employers don't do all of this. They either give employees a short offer letter or nothing at all. This is partly because an employer can't be fined for not giving you a proper contract.

If you want a contract, but your employer won't give you one, all you can do is apply to an employment tribunal for a list of what should be included in your written statement.

Do I have the right to work in the UK?

If an employer employs someone who doesn't have the legal right to work in the UK, they will be breaking the Asylum and Immigration Act 1996. You may have to prove that you have the right to work in the UK. There are different documents that they may ask you to show to prove that you have the legal right. These include:

- A national insurance number
- A former P45
- A work permit or
- The right stamp in your passport.

If you do not have the legal right to work in the UK, but the employer still wants to employ you, they normally have to apply for a work permit for you from Work Permits UK at the Immigration & Nationality Directorate of the Home Office.

What is the minimum I should be paid?

We now have a national minimum wage in the UK. Your employer must not pay you less than this. Different rates apply to different types of people.

However, if you are between 18 and 22 years old, or you are over 22 but in the first six months of employment and taking part in training, the national minimum wage doesn't apply if you:

- Are under 18
- Are under 19 and employed as an apprentice or
- Are under 26 and in the first 12 months of an apprenticeship.

You cannot be forced or persuaded by your employer to sign away your right to the national minimum wage, or to agree to a lower amount. If you are not receiving the minimum wage (or you believe you are not); and you complain, you have legal protection from being unfairly dismissed or victimized by your employer.

If you are not being paid the national minimum wage, you can complain to the National Minimum Wage Helpline. The Inland Revenue is responsible for enforcing the law, and can make an employer pay you the national minimum wage, as well as back pay.

You or the Inland Revenue can take your employer to an employment tribunal or civil court. If that happens, it is up to your employer to prove that they were paying you the national minimum wage. Again, if the tribunal finds that your employer hasn't paid you what they should, you can claim backdated pay.

When can my employer change my salary?

Your salary, or the way it is worked out, will usually be written into your contract of employment. This contract is binding on both you and your employer, and it cannot be changed unless you both agree. However, the terms of your contract can say that there can be certain changes to your salary. But even where a change in salary is allowed under your contract, your employer can't avoid paying you the national minimum wage, or force you to agree to being paid less.

If your employer alters your salary in a way that breaks your employment contract without you agreeing to it, you may be able to resign and then claim that you have been 'constructively dismissed' (that is, you were forced to resign).

The only alternative for your employer would be to end your original contract, dismiss you and then immediately rehire you on new terms. But your employer would have to give you enough notice and meet any other obligations in your contract about terminating your employment. If you believe this is unfair, you may be able to make a complaint of unfair dismissal to an employment tribunal.

How many hours can my employer make me work?

Your employment contract or written statement of terms should tell you your normal working hours. However, as well as providing for annual holiday, the Working Time Regulations say that you should not work more than 48 hours a week on average, unless you have agreed in writing to work more.

If you have signed a form saying you will work more than 48 hours a week, you can change your mind, and say you no longer want to do this, by giving your employer up to three months' notice.

Does my employer have to recognize my trade union?

Under new laws your employer may have to deal with a union for the sake of 'collective bargaining' (negotiating workers' terms and conditions).

If your employer doesn't voluntarily agree to deal with your union, and you want to use the law to force them to, you will need to have a special vote by the workers, and there are other conditions which have to be met. The Central Arbitration Committee is the organization that deals with this.

What can I do if I've been unfairly dismissed?

Many people, if they are dismissed, talk about having been made redundant. In fact, redundancy is just one type of dismissal which generally happens when a business needs fewer people to carry on its operations. You can, of course, be dismissed for other reasons.

If you are dismissed (whether it is redundancy or otherwise), and you think you've been treated unfairly, there are two ways you may be able to claim compensation.

Contractual claim

If you are dismissed without your employer giving you the notice that is in your contract, and they have no good reason for dismissing you, you will be entitled to compensation. The amount of compensation will be the same as your net salary and fringe benefits (such as the use of a car or pension contributions) which you would have received during your notice period.

If you don't have a written contract of employment (which is not uncommon) you will be able to claim at least the statutory minimum period of notice. This depends how long you have worked for your employer:

- No notice during the first month
- One week during the first two years or
- One week for each year you have worked after that, up to a maximum of 12 weeks.

However, you may be able to argue for a longer period of notice based on what is 'reasonable' in the business or profession you work. For example, if colleagues employed at the same grade or level would normally get three months' notice, you would have a good reason for claiming the same.

How much compensation you can claim depends on how much notice you are entitled to. But if you are dismissed, you should try to reduce your losses by looking for (and taking) an acceptable and suitable new job.

Statutory claim

As well as your contractual rights, there are also those provided for by statute (by law). You must normally have had at least one year's continuous employment with your employer at the date you are dismissed to be eligible to bring a claim for unfair dismissal.

If that is the case, your employer must prove that the dismissal was for a 'fair' reason and that they used a fair procedure in reaching the decision to dismiss you. For example, if you were dismissed for gross misconduct, your employer usually has to show that they carried out a properly-conducted disciplinary hearing in which you had the chance to put across your case. You should be offered the right to appeal against any decision your employer reaches in a disciplinary hearing.

New laws now mean that if you go to a disciplinary or grievance hearing, you have the right to take someone with you (normally a colleague or a trade union official). If your employer won't let you take someone to a hearing, you can bring a claim for compensation through the employment tribunal for up to two weeks' pay.

If your employer needs to make some staff redundant, they should try to be as fair as possible. They should also speak to staff to see if they have any ideas about how the redundancies could be avoided.

If you've been dismissed along with a group of people

There are separate rules which provide employees with extra rights when an employer wants to make 20 or more employees redundant within a 90-day period. The employer must speak to trade unions representing the workforce (or, if there are no such trade unions, representatives chosen from the affected workforce). If the employer doesn't do this, the employees may be able to claim compensation.

How do I claim compensation if I've been unfairly dismissed?

You must make a claim for unfair dismissal to the employment tribunal within three months of being dismissed. For redundancy claims the time limit is six months, unless you have applied in writing to your employer for redundancy compensation within six months. In this case, you get an extra six months.

To bring any other claim for breach of contract in the employment tribunal the time limit is three months from when the event or matter you are complaining about happened. In the case of contractual claims, the employment tribunal can only deal with claims up to £25 000. If your claim is more than this, you must use the County or High Court.

You may also be able to bring a claim for breach of contract in the County or High Court.

The time limit for such claims is six years. You do not have to have worked for your employer for a certain length of time to bring such contractual claims.

The claims process

You bring a tribunal claim by filling out an application form, called an ET1, and sending it to the employment tribunal. The ET1 form is available from:

- Job centres
- Local law centres and
- Citizens Advice Bureaux.

The employment tribunal will send a copy to your employer. Your employer must then send back another form, giving their side of the story, within 21 days. When you send your claim, an officer from the Advisory, Conciliation and Arbitration Service (ACAS) will be assigned to your case. They look at the case impartially and will try and get you and your employer to agree a settlement.

If the ACAS officer cannot get you to agree, you and your employer will have to go to a hearing of the tribunal. You will both have the chance to give your case, and present evidence from witnesses. Employment tribunals consist of a chairman and two independent people who will consider your case. The tribunal is less formal than the court. You do not need a solicitor to represent you. You can present your own case or you can get someone else, such as a trade union official to present your case for you.

What can I do if I have been discriminated against?

Discrimination is still a common problem in the workplace, despite laws aimed at stamping it out. The main aim of the legislation is to prevent you being discriminated against because of your sex, race or a disability. But the law also protects part-time workers from unfair discrimination.

In England and Wales, there is no legislation to prevent discrimination because of religious beliefs (though there is in Northern Ireland), or to prevent age discrimination. But it is likely to happen in the coming years, following European Union rules which force European Union countries to develop such laws.

Also, there is no legislation preventing discrimination because of sexual orientation (if you are gay or lesbian), though the Employment Appeals Tribunal recently decided that doing so is unlawful. It is now more difficult for employers to discriminate against you because of this.

For more information about discrimination, at work and elsewhere, there are three separate Community Legal Service leaflets:

- 'Equal opportunities'
- 'Racial discrimination' and
- 'Rights for people with disabilities'.

What counts as discrimination?

There are two types of discrimination by an employer.

Direct discrimination

This is when you are treated less favourably because of your race or sex or a disability. For example, if you are refused a job or promotion because you are a woman or Asian. Sexual or racial harassment can also be direct discrimination.

Indirect discrimination

This is when a group of people (for example, a group of people of a particular race) cannot meet a condition or requirement of their work in the same way as the rest of the population.

An example of indirect racial discrimination would be an employer asking all employees (including one who is a Sikh) to wear safety helmets at work without good reason (for example, in an office). A Sikh who wears a turban for religious reasons could say he had been discriminated against if he had to wear the helmet to get the job, and that the need to wear the helmet was not justified.

Victimization can be similar to discrimination. For example, if you are singled out for unfair treatment because you were exercising rights or helping others to do so.

What action can I take?

If you think you have been discriminated against, you can take your case to the employment tribunal. If your claim succeeds, you can be awarded compensation.

What rights do I have if I work part-time?

New regulations, called the Part-Time Workers (Prevention of Less Favourable Treatment) Regulations, came into force in July 2000. The purpose of the regulations is to prevent part-time workers being treated less favourably than full-time workers. Part-time workers may only be treated differently from full-time workers if the treatment can be properly justified.

An important point is that the regulations apply to a wider category of workers than just employees. This means there may be people (such as contract workers) who are not covered by other employment legislation (such as the law on unfair dismissal) who will be protected under this legislation.

How do I know if I am 'part-time'?

The regulations describe a part-time worker only as someone who is not a full-time worker. It will usually be clear whether an employee works full or part-time by comparing the hours that people work in the same organization.

The new regulations mean that part-time workers on a pro rata basis (scaled down according to the hours they work) must receive the same:

- Rate of pay
- Promotion and training opportunities
- Sick pay and maternity benefits
- Access to company pension schemes

- Entitlement to annual holiday, health insurance, subsidised mortgages and discounts and
- Parental leave and career-break schemes.

An employer can ignore these regulations only if they can justify treating part-time workers less favourably.

However, when they first start work, part-time workers are not entitled to the same overtime rate as full-time workers. They have to have worked the same number of hours that a full-time worker has to work to get overtime rates.

Maternity

This is also covered under the 'New regulations' with the terms and entitlement listed.

Should any female worker require information on this matter they should consult *Employment: 'Your rights at Work'* information booklet.

Organizations within the plumbing industry

There are several organizations and institutions associated with the plumbing industry. They are:

- The Worshipful Company of Plumbers
- The Institute of Plumbing and Heating Enginers
- The Joint Industry Board for Plumbing Mechanical Engineering Services in England and Wales
- The Electrical, Electronic, Telecommunications and Plumbing Union
- The Council for Registered Gas Installers
- The National Association of Plumbing Teachers
- The Association of Plumbing and Heating Contractors
- The Scottish and Northern Ireland Plumbing Employers Federation
- British Plumbing Employers Council.

On the following pages, the history and functions of these various organizations are described.

The Worshipful Company of Plumbers

There are 94 Livery Companies. The oldest, the Mercers' Company, received its charter in 1394. The Plumbers' Company ranks 31st in the order of precedent having received its first Royal Charter in 1611. The charter was withdrawn by James II and subsequently restored in the reign of William and Mary. The ordinances governing how the company should operate are much older, the first being laid down in 1365 and subsequent ordinances in 1488.

The company was the instigator of the concept of a Register of Plumbers, a logical successor to the medieval role of the company. The Register of Plumbers was inaugurated in 1886 and efforts to get 'The Plumbers' Registration Act' through Parliament occupied the activities of the Company in the 1890s. In 1893 the company equipped a laboratory in King's College, London, during the Mayorality of Alderman Sir Stuart Knill, who was Master of the Company. Today the Institute of Plumbing carries on the tradition. The company maintains close links with the Institute of Plumbing, Copper and Lead Development Associations, and all branches of industry.

In addition, the company has sponsored and equipped a museum which is housed at the Weald and Downland Open Air Museum, Singleton, near Chichester, West Sussex. Every summer, plumbing students and staff from colleges and training centres demonstrate their skills in the Court Barn by working with lead and other traditional materials.

The company exists to:

- Foster, maintain and develop links with the plumbing craft and allied disciplines in the construction and other industries

Figure 8.2

- Promote as appropriate youth activities in the craft by financial and technical contributions to educational and vocational ventures
- Contribute to the City of London Charities
- Support the 'pursuit of excellence' so that through contact with other organizations the company is able to call on past experience for the benefit of future enterprises
- Provide a pleasant social ambience for those in the company.

The Institute of Plumbing and Heating Engineering

The Institute of Plumbing and Heating Engineering (IPHE) came into being on 4 June 2004 when the Institute of Plumbing amended their name. The institute is an independent organization embracing all interests in the plumbing industry.

The institute sees its prime role as the identification and promotion of competence, skill, good quality and standards for the public benefit.

Whereas most other professional institutions are somewhat élitist in outlook and cater only for those above craft level, the institute's requirements for entry are broad in scope, reflecting the largely practical nature of the UK plumbing industry.

The institute encourages young people to progress up the ladder of qualification it offers from apprentice to technician engineer.

The institute was founded as the Institute of Plumbers in 1906 by the National Association of Master Plumbers with the objective of developing both the industrial and technical aspects of the plumbing trade. Membership was limited to Master Plumber members of the Association.

In 1925 that Association became the National Federation of Plumbers and Domestic Engineers (now the National Association of Plumbing, Heating and Mechanical Services Contractors – NAPH & MSC). The institute withdrew from the industrial field and concentrated on devel-

oping the science and practice of plumbing. Membership was, however, still restricted to Master Plumbers.

The constitution of the Institute of Plumbers was revised in 1957 and the Institute of Plumbing then came into existence under the name which it still proudly enjoys. In that year, for the first time, membership extended to plumbers holding technical qualifications irrespective of their position in the industry.

In the late 1960s discussions began between the institute and the Registered Plumbers' Association with the object of effecting a merger of the two bodies. The main role of the RPA was to manage a voluntary register on behalf of the Worshipful Company of Plumbers.

After considerable negotiation, the IOP/RPA merger was completed in 1970, when it was realized there was a pressing need for a single organization to establish the technical authority of the plumbing industry in the wider field of building services and to manage the Register of Plumbers.

Today, the institute is an independent, non-political organization pursuing its major objectives of raising the science and practice of plumbing in the public interest and managing the Register of Plumbers. with compulsory registration in mind.

In 1979 the institute became a Registered Charity, thereby acknowledging that its aims and objectives are primarily in the public interest.

The institute's activities reflect its role as both a qualifying body and learned society. A wealth of information and expertise exists within the membership and this is used to great effect when the institute is called upon by government departments, technical and educational institutions to give its views on a wide range of topics.

At national level the institute is represented on a number of organizations and committees associated with its work. For example, the institute is represented on the DoE Standing Technical Committee on Water Regulations which prepares technical requirements for new water bye-laws.

The institute plays an active role in the drafting of standards and codes of practice affecting plumbing work, subsequently published by the British Standards Institution.

Figure 8.3

It supports the work of other organizations which promote high installation standards, namely the Council for Registered Gas Installers (CORGI), of which it is a founder constituent organization, and the National Inspection Council for Electrical Installation Contracting (NICEIC).

There is an excellent relationship between the institute and the manufacturing and distributing sectors of the plumbing industry. Communication is through the Industrial Associate category of membership and its Liaison Committee of representatives nominated by Industrial Associate members.

The institute's historical close links with the Worshipful Company of Plumbers continue today. There is contact through regular meetings and the institute is an active supporter of the Company's Plumbers' Workshop and Museum, established in the Weald and Downland Open Air Museum at Singleton, near Chichester, Sussex.

In the education and training field the institute gives help and guidance to the Plumbing Examinations Committee of the City and Guilds and the relevant committees of the Technician Education Council and the Scottish Technical Education Council. It is also represented on the Mechanical Engineering Services Committee of the Construction Industry Training Board, as well as several regional and local committees dealing with construction training. Many lecturers of plumbing at colleges are members of the institute.

At local level, the institute's District Councils arrange lectures, visits and other events to increase the knowledge and expertise of members and Registered Plumbers, updating their appreciation of new materials and methods, etc. and broadening their interests. They also arrange events that enable like-minded people to meet socially.

The institute regularly stages exhibitions in major towns and cities throughout the country and these often attract large attendances, not only of plumbers, but also of architects, engineers, merchants and others with an interest in plumbing. There is also an annual conference when significant papers are presented and important topics discussed.

AMICUS

The history of the Union Movement has been punctuated by mergers in which the smaller and more specialised Trade Unions come together to combine their resources and increase their bargaining power. The recent merger between AMICUS AEEU and AMICUS MSF is the latest in this chain of amalgamations.

The AEEU was formed in 1992 from the merger of the EETPU and the AEU.

The EETPU itself was the result of a merger of two unions the ETU (electricians) and the PTU (plumbers).

The PTU (Plumbing trade union) was formed in 1865, the modern union provides a number of financial benefits to its members and has its own legal department which provides aid to all members in respect of cases ranging from unfair dismissal to injury through accident at work.

Figure 8.4

The Joint industry Board for Plumbing Mechanical Engineering Services in England and Wales (JIB)

The JIB's main function is to establish national conditions of service for the Plumbing Mechanical Engineering Service Industry in England and Wales. After negotiations between the organizations representing the employers of plumbing and labour operative plumbers, the JIB lays down rates of pay, hours of work, holidays, etc. The JIB is also a registered body for the plumbing industry craft apprenticeship scheme and is involved in the design and operation of courses of further education and specialized training courses for the plumbing industry. The JIB's constituent bodies are:

The National Association of Plumbing, Heating and Mechanical Services Contractors

The Electrical, Electronic, Telecommunication and Plumbing Union

The Building Employers' Federation.

Plumbing is no longer concerned just with working in lead as the name would suggest, but has developed into an important component of both environmental and mechanical engineering, embracing a wide variety of systems, services and complex equipment. This is reflected in the scope of training and further education and has led to the change of title from Plumber Craftsman to Plumbing Mechanical Engineering Services Craftsperson.

The skills and related knowledge of such craftspersons are in demand for many industries beyond the construction industry and the limited requirements of domestic housing installations. The board's grading system provides sections of the industry with the opportunity of management status for those undertaking the available courses of further study.

The grading scheme

This scheme is designed to ensure that entrants to the industry have the opportunity to progress within the industry as their qualifications and experience increase. The industry pay structure is linked to the grading scheme so that progression through the grading scheme leads to increased rates of pay.

Pay allowances and benefits

1 All operatives and apprentices enrolled in the industry's Benefit Scheme are entitled to Sick Pay additional to that provided by the State Scheme. They are also covered for accidental loss of limbs and permanent disability.

2 A weekly Tool Allowance is paid by employers to all apprentices and operatives who possess the full range of tools as stipulated by the JIB.

3 Operative plumbers in possession of JIB Certificates of Competency in Welding are entitled to an hourly pay supplement.

4 Travelling time and expenses are paid by employers to apprentices and operatives when and where appropriate.

5 All operatives and apprentices employed in the industry are entitled to be enrolled in the Plumbing Industry Pension Scheme operated by Plumbing Pension (UK) Ltd.

The Council for Registered Gas Installers (CORGI)

During the 1960s, there was a major gas explosion in a tower block, which claimed several lives. Since then there has been a progressive tightening of the law relating to gas safety, allied to attempts to improve safety performance through voluntary trade bodies.

Despite the Gas Safety regulations of 1972 and 1984, which require gas work to be undertaken by 'competent' persons, continuing incidents in the 1980s led to public concern over gas safety.

In April 1990, the Gas Safety (Installation and Use) (amendment) Regulations 1990 were laid before Parliament. They require the compulsory registration of those engaged in the installation and servicing of gas equipment.

The Health and Safety Executive decided that CORGI should operate such a registration scheme. To comply with the law, which came into force on 30 March 1991, all gas installers must be registered with CORGI.

Who has to be registered?

All businesses – employers or self-employed people – who undertake work on gas fittings using piped gas must be registered.

'Piped gas' includes not only mains supply, but also the supply of liquefied petroleum gas (LPG) through pipes, e.g. metered supplies to an estate. It does not include LPG stored on the consumer's own premises in bulk tanks or bottled supplies, unless that gas is subsequently metered or 'sold'.

'Work on gas fittings' covers service, maintenance and repair as well as the actual installation of gas appliances, pipework, equipment and flues.

'Employers or self-employed people' includes not only those businesses which deal primarily with gas, but also, for example, the direct work departments of local authorities and health authorities: the operatives, kitchen installers or builders who may only work on gas fittings occasionally.

Those engaged in DIY do not have to register, though they may be liable to prosecution if they are not competent. Payrolled employees of installation and servicing businesses are also excluded from the need to register in their own right,

Figure 8.5

though their employer must be registered. Those whose work is restricted exclusively to factories or mines do not need to register.

What is CORGI?

CORGI was originally the 'Confederation for the Registration of Gas Installers', a voluntary body set up to improve gas safety standards, funded by British Gas. In December 1990 a new CORGI, the 'Council for Registered Gas Installers', was set up to operate the compulsory registration scheme. The new body is built on the experience and activities of the old CORGI, but is an independent, self-financing organization. It is controlled by a Board and Council, which are representative of the industry, and include gas suppliers, manufacturers, consumer groups, training organizations, trade associations, professional bodies and trade unions.

CORGI's costs are met from registration fees and British Gas is subject to the same registration standards as any other installer. Both the structure and the operational practices of CORGI have been set up to satisfy the Health and Safety Executive's demanding criteria for openness and fairness in the enforcement of gas safety. CORGI's objective is to help the industry towards better safety practices through practical and constructive advice.

CORGI is the only registration body approved under the regulations. Installers who remain outside CORGI and continue with piped gas work are not complying with the law.

How does the CORGI registration scheme work?

The rules for registration are set out in detail in a booklet that is supplied to each applicant.

First-time applicants must give some basic details about their business and its operational branches, sign a declaration undertaking to observe safety requirements, and pay a fee. Prior to enrolment on the CORGI register, new applicants will also need to demonstrate competence. Inspections are carried out to assess this.

If the inspections are satisfactory, registration will be processed and a certificate of registration will be issued to the business as a whole. This will expire at the end of the registration year, which is 31 March.

If the inspections are not satisfactory, remedial action and/or operative training will normally be requested, based on practical guidance. CORGI's aim is to bring about improved safety practices, not to catch people out over honest error. When CORGI is satisfied that competence has been achieved, a certificate will be issued.

If the application is not successful, there is a hearing and appeals procedure that can be invoked to ensure absolute fairness in the refusal of registration. Registration hearings and appeals are held at locations around the country as appropriate.

In subsequent years, CORGI will ask registered installers to make a renewal application. This renewal will normally be processed on the basis of the declaration and fee alone, but there will also be ongoing monitoring inspections and investigation of safety complaints, should they arise. Should any later inspection lead to a decision to deregister a business, a full hearing and appeals procedure is available to protect the installer's interest.

Administration operates from a national headquarters but each installer will be assigned to an inspector who will be part of a locally led team.

To further the main objective of practical and constructive support for better safety practice. CORGI also operates a technical 'hot line' and offers a publications service on gas safety matters.

The National Association of Plumbing Teachers

The National Association of Plumbing Teachers (NAPT) is a voluntary professional body and was founded by an eminent plumbing teacher, J. Wright Clarke. Its first meeting at The

Figure 8.6

Polytechnic, Regent Street, London, on 31 August 1907, makes it one of the oldest bodies in the plumbing industry and indeed one of the first associations for teachers in this country. Subsequently, other professional groups have followed this lead, and construction education owes a considerable debt of gratitude to Wright Clarke's vision and foresight.

The purpose of the National Association of Plumbing Teachers as defined in its constitution can be summarized as follows:

■ to consider educational and training aspects for the benefit of the industry

■ to preserve quality and standard of work

■ to keep its members informed of technical developments and educational trends.

To serve its members, spread widely across the United Kingdom, Ireland and overseas, the NAPT works through its regional councils which are linked to the national council with the president, elected annually, as its head. Although an independent association, the NAPT has established working relationships with other important bodies in the industry. It also has representation, at national level, on committees that formulate the implementation of educational and training policy.

The Scottish and Northern Ireland Plumbing Employers' Federation (SNIPEF)

Long before the federation was formed, there were active local associations of master plumbers in various parts of the country. A venture of major

importance in 1970 was the part the federation played in the constitution of the Joint Industry Board, and in the introduction of a grading scheme. In 1973, the federation took the initial steps to set up an industry-wide pension scheme for employees in the industry, which was then joined by the industry in England and Wales.

The 1980s was a period of significant development for the federation and saw the establishment of a managing agency for the operation of the Youth Training Scheme in the plumbing and mechanical engineering services industry in Scotland. The managing agency was awarded Approved Training Organisation status and the YTS operation was supplemented in 1989 by the introduction of an Adult Entrant Scheme using the Government's Employment Training Scheme as the first year of a three-year training programme. Also of major significance was the introduction in 1977 of the federation's Code of Fair Trading, which outlines general principles of good and fair trading between member firms and their customers and which provides a complaints procedure offering conciliation and arbitration in the event of disputes between customers and member firms. These provisions were strengthened in 1989 with the introduction of a Guarantee of Work Scheme.

In response to changes in technology, the federation has taken a lead in the provision of training courses for the industry. One particular area of success was the introduction of unvented hot water heating systems in the UK and the federation took the lead in arranging training and a registration system as agents of the British Board of Agrément. This was achieved through the establishment of a subsidiary association, the

Figure 8.7

Association of Installers of Unvented Hot Water Systems (Scotland and Northern Ireland). This was followed by the establishment of the Water Byelaws Certification Scheme under which suitably qualified firms become eligible to self-certify that their work complies with the Water Byelaws and other regulations.

British Plumbing Employers Council (BPEC)

The British Plumbing Employers' Council was reconstituted in April 1991 in response to the decision of the plumbing sector to withdraw from the scope of the statutory levy and grant system operated by the Construction Industry Training Board in October 1990. BPEC comprises representatives of the NAPH & MSC and SNIPEF.

BPEC has been recognized by the Employment Department as the Industry Training Organisation (ITO) and the Industry Lead Body for plumbing in Great Britain. Both founding organizations are Trade Associations, the NAPH & MSC covering England and Wales having a membership in the region of 3000 firms and SNIPEF covering Scotland and Northern Ireland with a membership of approximately 1100 firms. The scope of the membership of both associations varies usually depending upon the size of the company. In the main the companies will design, install, repair, maintain, commission and test a range of mechanical engineering services installations including plumbing, heating and gas services in domestic, industrial and commercial environments. The size of companies in membership again varies from sole traders up to companies employing in excess of 200 people.

The mechanical engineering services industry has seen major changes in skills and technology over the last 20 years. There has been a trend away from some of the traditional skills, such as lead joint wiping, and an increase in the use of increasingly sophisticated technological products, including combination and highly efficient

Figure 8.8

condensing boilers, 'intelligent' building control systems and a demand for the use of water purification systems and more efficient soil waste appliances. Many of these changes have resulted from an ever-increasing awareness of the need for energy conservation and the protection of the total environment. Dramatic changes in material technology have resulted in pipework systems being both easier to join and install. Consequently this has led to an increase, on the smaller domestic installations, in DIY, which has accounted for a reduction in the range of work available to skilled sole traders.

As a result of these changes in working practices and the continuing pace of technological advances, BPEC believes that the industry faces an enormous challenge in training terms in order that all firms operating in the plumbing industry are adequately geared to meet the demands imposed by these changes. This implies a consolidation of the considerable effort that has already been put into the training of new entrants to the in and the continuing development and promotion of training in new technology and skills at all levels of employment, including management.

Note

The National Association of Plumbing Heating and Mechanical Services Contractors (NAPH-MSC) have recently amended their association title to (APHC) Association of Plumbing and Heating Contractors.

Trade Associations and Professional Bodies

Association of Plumbing and Heating Contractors (APHC)
Ensign House
Westwood Way
Coventry CV4 BJA
Email enquiries@aphc.co.uk

Heating and Ventilating Contractors Association (HVCA)
Esca House
34 Palace Court
London W2 4JG
Tel: 020 7313 4900
Fax: 020 7727 9268
Email: contact@hvca.org.uk

Institute of Domestic Heating and Environmental Engineers (IDHE)
Dorchester House
Wimblestraw Road
Berinsfield
Wallingford
Oxon OX10 7LZ
Tel: 01865 343096
Email: info@idhe.org.uk
Web: www.idhe.org.uk

The Institute of Plumbing and Heating Engineering (IPHE)
64 Station Lane,
Hornchurch,
Essex
RM12 6NB
T: 01708 472791
F: 01708 448987
E: info@iphe.org.uk

Scottish and Northern Ireland Plumbing Employers Federation (SNIPEF) –
Scotland & Northern Ireland only
2 Walker Street
Edinburgh
EH3 7LB
Tel: 0131 225 2255
Email: info@snipef.org
Web: www.snipef.org

9 Answers to test questions

Question number	Safety	Tools etc.	Science	C. water	H. water	San. App.	Roofwork
1	d	a	d	b	d	b	b
2	b	a	d	b	a	d	d
3	c	a	d	b	b	b	a
4	b	d	a	d	b	d	d
5	d	d	b	b	a	a	a
6	a	d	d	d	c	b	d
7	d	b	a	a	a	d	c
8	c	b	a	d	d	b	c
9	d	c	c	a	d	b	a
10	a	a	b	c	d	c	b
11	c	a	b	c	c	c	b
12	b	d	b	b	b	c	a
13	d	a	a	b	c	a	c
14	c	a	b	c	a	a	c
15	c	a	d	b	c	a	b
16	a	b	d	a	d	d	a
17	c	a	c	d	a	c	d
18	c	a	a	c	d	b	a
19	b	c	b	b	b	d	d
20	b	a	d	c	c	b	d

10 Glossary

Absolute Zero   The coldest possible temperature (−273°C).

Absorption   Dispersal of molecules of one substance through the body of another. Absorption of a gas in a liquid may be selective and so used to separate gases, e.g. removal of carbon dioxide from air by an alkaline solution.

Abutment   An intersection between a roof surface and a wall rising above it.

Accelerator   Inorganic or organic substance added to speed up any chemical reaction.

Acetone   A liquid with the ability to dissolve 25 times its own volume of acetylene for each atmosphere of pressure applied to it.

Access pipe   A pipe provided with a removable cover for inspection and maintenance.

Accessory   Any device connected and used with other appliances/apparatus.

Acetylene   A gas produced from calcium carbide and water. It is combustible and is used with oxygen in welding processes.

Acid   A corrosive substance.

Adhere   Stick (hold) together.

Adhesion   The force of attraction between molecules of different substances.

Air   A mixture of gases with an average composition of nitrogen 78 per cent, oxygen 21 per cent, argon 0.9 per cent, carbon dioxide 0.03 per cent plus smaller quantities of other gases.

Air lock   The restriction or the stopping of the flow of a liquid by trapped air.

Air gap   The physical break between the water inlet or feed pipe to an appliance and the water in that appliance or cistern.

Alloy   A mixture of metals, or metal and non-metal substances.

Alnico   A carbon-free alloy used for permanent magnets.

Ambient   Surrounding.

Ambient temperature   The temperature of the surrounding element, i.e. air or water.

Ammeter   An instrument used for measuring direct electric current.

Anneal   To soften a metal by the application of heat.

Anode   The electrode to which negative ions move, the positive electrode.

Anti-flooding valve   A drain fitting incorporating a mechanical non-return device as a means of giving a measure of protection to a drain or sewer against surcharge.

Anti-splash device   A device fitted to the nozzle of a valve to ensure that the discharge will be non-splashing.

Anti-splash shoe   A rainwater fitting fixed at the lower end of a rainwater pipe and so shaped as to reduce splashing when rain water is discharging.

Apron flashing   A one-piece flashing, such as is used on the lower side of a chimney penetrating a sloping roof.

Anti-freeze   A substance added to cooling systems to reduce freezing point below probable ambient temperatures (often ethylene glycol).

Architect   One who designs and supervises the construction of buildings. His or her main duties are preparing designs, plans and specifications; inspecting sites; obtaining tenders for work and the legal negotiations needed before building commences.

Asbestos   A mineral substance consisting of thin, tough fibres which can withstand high temperatures. The Asbestos (Prohibition) Regulations 1985 prohibit the importation of raw asbestos fibres; or products containing them, into Britain, and also their use in the manufacture and repair of any other product. Asbestos spraying and the installation of new asbestos insulation are not allowed.

Asbestos cement   Cement mixed with asbestos fibre previously used in the manufacture of flue pipes, rainwater gutters and downpipes and cold water storage cisterns.

Atmospheric pressure   Pressure due to the atmosphere surrounding the earth.

Autogenous   A welded joint in which two parts of the same metal are welded together with or without a filler rod of the same metal.

Back fill   This term is used to indicate the return of excavated soil (spoil).

Back flow (cold water)   Flow in a direction contrary to the natural or intended direction of flow.

Back gutter   A gutter between the upper end of a chimney and a sloping roof.

Back siphonage   This is caused by negative pressure within the water system which could result in foul water being drawn back into the service pipe or even the main.

Backnut   A locking nut provided on the screwed shank of a valve or pipe fitting for securing it to some other object. A threaded nut, dished on one face to retain a grommet, used to form a watertight joint on a long threaded connector.

Baffles   Pieces of metal fitted inside a gas burner by its manufacturer to prevent more gas reaching one part of the burner than another.

Balanced flue   An arrangement of air intake and flue outlet commonly used for domestic gas fired appliances.

Bale tack   A lead tack used for securing sheet metal weatherings.

Ball-peen hammer   A hammer with an hemispherical peen (also pein or pane).

Bar   A unit of pressure, equivalent to the weight of a column of mercury of unit area and 760 mm in height. 1 bar = 1000 millibars.

Barff process   An anticorrosion treatment of iron or steel by the action of steam on the red-hot metal; the layer of black oxide formed gives protection.

Barometer   A device for measuring atmospheric pressure.

Base   Substance which in aqueous solution reacts with an acid to form a salt and water only.

Basement   A storey where the floor is at some point more than 1.2 m below the surface of the adjoining ground.

Base exchange   A water softening process in which water is passed through a bed of mineral reagent called zeolite, which absorbs those salts in the water which make it hard.

Bauxite   An ore of aluminium.

Bead   Formed on sheet metals for stiffening the edge or fixing it, usually bent round to a circular tube shape.

Bedding   The material on which an underground pipe is laid and which provides support for the pipe. Bedding for drainage pipelines can be concrete or granular material.

Bell-type joint   This is used on copper tube; one end of the pipe is opened to form a bell shaped socket which is then filled with bronze rod.

Benching   The sloping surfaces on either side of a channel at the base of an inspection chamber for the purpose of confining the flow of sewage.

Bend   A curved length of pipe, tubing or conduit. A 90° bend is called a quarter-bend, a 45° bend is called a one-eighth bend.

Bends   (A) *Set*   A single bend.
        (B) *Offset*   A double bend.
        (C) *Passover/* A pipe formed to
           Crank   pass over a pipe fixed at
                  right-angles to it.

Bib valve   A water valve which has an horizontal inlet connection which is male threaded.

Bidet   A sanitary appliance used for washing the excretory organs.

Birdsbeak   A method of terminating an overflow or warning pipe. The pipe end is cut across a centre line and this portion is opened and shaped to form a birds beak.

Blende   Zinc sulphide, the chief ore of zinc.

Blender   Also known as a mixer valve, is used on a hot water heating system to mix the hot flow and cooler return water to give a blended temperature flow.

Blinding   A material used to fill irregularities in an exposed trench bottom, which when compacted will create a firm uniform foundation on which to place the pipe bedding material. Hoggin, sand, gravel, all-in aggregate or lean concrete are commonly used.

Bobbin   An egg shaped boxwood tool which is drilled through its centre. It is threaded on to a strong cord and pulled through a bent section of pipework to remove distortions and return the pipe to its correct bore.

Boiler   A water heater in which, generally the water should not boil. It may be healed by gas, oil, electricity or solid fuel.

Bonding   (A)   This term used in soldering or welding and refers to the adhesion of the solder to the parent metal.

(B)   Where the electrical system is earthed through the cold water supply pipe, the term bonding refers to the earth connection.

Branch drain   A line of pipes installed to discharge into a junction on another line or at a point of access.

Brass   A large group of alloys based on copper and zinc.

Brazing   The joining of metals using a low melting point bronze rod as in capillary or silver soldering (hard soldering).

Borax   A white crystalline solid used as a flux.

Bore   The internal diameter of a pipe or fitting.

Boss   An attached fitting on a vessel or pipe which facilitates the connection of a pipe or pipe fitting.

Bronze welding   The joining of metals using a bronze filler rod with a high melting point. The deposited rod forms a built up joint pattern (also known as hard soldering).

Bush   A pipe fitting used for reducing the size of a threaded or spigot connection.

Butadiene   A colourless gas, which polymerizes to give a type of synthetic rubber.

Butane   A gas which is present in natural gas and is sold compressed in cylinders for industrial and domestic purposes.

Butt weld   A weld joining two pieces of metal both pieces being in the same plane.

Cable   One or more conductors provided with insulation. The insulated conductor(s) may be provided with an overall covering to give mechanical protection.

Calcium carbonate   The chemical name for chalk, limestone and marble.

Calorifier   A cylindrical vessel in which water is heated indirectly, by means of hot water or steam contained in pipe coils, a radiator or a cylinder within a cylinder.

Calorific value   The number of heat units which can be obtained from a measured quantity of fuel.

Candela   A unit of luminous intensity based on electric lamps.

Capacity   (A)   *Actual* is the volume of water that a vessel contains when filled up to its water line.

(B)   *Nominal* is the theoretical capacity of a vessel calculated using overall dimensions.

Cap   A cover usually with an internal thread or socket joint for sealing the end of a pipe.

Capacity   The quantity of liquid contained in a vessel (measured in litres).

Capillary attraction   A general term covering surface tension phenomena, e.g. the rise or fall of liquids in capillary tubes, fibres and materials.

Capillary joint   A fine clearance spigot and socket joint into which molten solder is caused to flow by capillary action.

Carbonizing flame   The flame produced at the blowpipe tip when there is an excess of acetylene being burned.

Cast iron   Iron containing 2.5–4 per cent carbon in the form of graphite. Cast iron is brittle but heat resisting.

Catalyst   A substance which increases the speed of a chemical reaction.

Cat ladder   A ladder or board with cross cleats fixed to it, laid over a roof slope to protect it and give access to workers for inspection or repairs.

Cathode   A negative electrode.

Cathodic protection   A sacrificial metal (anode), usually a block of magnesium, is fixed in the system.

Caulked joint   A spigot and socket joint in which the jointing material is compacted by means of a caulking tool.

Caulking compound   A cold jointing substance used on cast-iron drains.

Cavitation   A phenomenon in the flow of water consisting of the formation and collapse of cavities in the water bringing about pitting and wear on valve seatings and ball valve outlets.

Cement   A powder used in conjunction with sand to form a mortar used in jointing. It is produced by a breaking down process of a special rock.

Chamfer   The edge of a piece of metal or pipe shaped off to form an angle other than a right angle.

Change of state   Transformation of a substance from one of the physical states of matter (solid, liquid, gas) to another, i.e. melting. freezing. boiling or condensing.

Chain wrench   A steel pipe grip which holds the pipe by a chain linked to a bar which is grooved at the end touching the pipe.

Chalk line   A length of line well rubbed with chalk, held tight and plucked against a wall, floor or other surface to mark a straight line on it.

Channel pipe   An open pipe, semi-circular or three quarter section, used in drainage, particularly at manholes.

Chase wedge   A wooden wedge-shaped tool, used for setting in a fold on sheet metal work.

Cheek   The vertical side of a dormer.

Chemical change   A change in a substance involving a rearrangement of the atoms in the molecules to produce a different substance or substances.

Chloramine process   This involves adding ammonia to water to remove the taste of chlorine.

Chlorination   A method of treating water with chloride to sterilize it by destroying harmful bacteria.

Chromium   A hard white metal, widely used in the production of alloy (stainless) steels and as a plating metal.

Cistern   An open topped container for water in which the water is subject to atmospheric pressure only. The water usually enters the cistern via a float operated valve.

Circuit   An assembly of pipes and fittings, forming part of a hot water system through which water circulates.

Circuit-breaker   A mechanical device designed to open or close an electrical circuit.

Circulation   (A)   *Primary* is the circulation between the boiler and the hot storage vessel. (B)   *Secondary* is the circulation of the hot water to supply the appliances, and is taken from the vent pipe, returning into the top third of the cylinder. (C)   *Gravity* could also be called natural circulation because it is brought about naturally by the difference in weight between two columns of water (hot water is lighter than cold water and is therefore displaced. i.e. rises). (D)   *Forced* is the movement of water brought about by the introduction of an impellor (pump).

Clay   A material used in pottery, sanitary ware and drainage pipeline manufacture.

Cleaning eye   An access opening in a pipe or pipe fitting arranged to facilitate the clearing of obstructions and fitted with a cap, plug or cover plate.

Cladding   The non-loadbearing covering of the walls or roof of a building. The skin used to keep the weather out.

Close coupled   A term used to describe a toilet suite where the cistern bolts or clamps to a projection formed on the top back edge of the pan.

Clout nail   A galvanized nail usually between 12 mm and 50 mm long with a large round flat head.

Cohesion   The force of attraction between molecules of the same substance.

Collar   A pipe fitting in the form of a sleeve for joining the spigot ends of two pipes in the same alignment also called a socket.

Combined primary storage unit (CPSU)   A special category of storage combination boiler.

Combined system   A drainage system in which foul water and surface water are conveyed by the same pipes.

Combustion   The burning of a substance or fuel brought about by rapid oxidation.

Communication pipe   The pipe between the water authority's main and the consumer's stop valve on the boundary. It is part of the consumer's service pipe which belongs to the water authority.

Compression joint   A fitting used to joint copper, stainless steel and polythene tubes.

Compression   This term refers to the compressive stress applied to metal when jointing, or the throat of the pipe when bending.

Concave fillet   A fillet weld with an inward curve.

Conductor   The conducting part of a cable or functioning metalwork which carries current.

Connector   A means of connecting together pipes and cables.

Consumer unit   A combined fuseboard and main switch controlling and protecting a consumer's final sub-circuits.

Contraction   The reverse of expansion, i.e. the decrease in size of a solid or liquid substance.

Convex fillet   A fillet weld with an outward curve.

Cornice   A projection usually of concrete or stone located in a wall to throw rainwater clear of the wall face below, and improve the appearance of the building.

Corrosion   A chemical action which takes place on the surface of the metal, and usually started by atmospheric conditions, attack by acids, or electrolytic action. The 'eating away' of a surface.

Cover   (A)   Air-tight cover.
(B)   Material above pipe.

Cradle   A support shaped to fit the underside of a pipe, cylinder or appliance.

Cracking   Thermal splitting of substances, especially petroleum hydrocarbons into substances of lower molecular weight. Mineral oils can be converted into petrol using this method.

Cross vent   A short relief vent between a main discharge pipe and a main ventilating pipe.

Crown   The highest point of the inside surface of the pipe (also known as the soffit).

Curtilage   The area attached to a dwelling.

Cupro-solvency   The ability of some waters to dissolve copper.

Cylinders   (A)   *Ordinary*. Cylindrical closed containers in which hot water is stored under pressure from the feed cistern.
(B)   *Indirect*. Sometimes called calorifiers, they are used where domestic hot water and heating are fed from the same boiler or where the water is of a temporary hard nature. They come as cylinders within a cylinder or with annular rings designed to keep the primary and secondary waters separate (two cisterns are required).
(C)   *Patent indirect*. These do not require two cisterns, the filling, venting and expansion being taken care of in the unique patent cylinder.

Damp proof course   A layer of impervious material laid in a wall to prevent the passage of water.

Daywork   A method of payment for building work, involving agreement between the clerk of works and the contractor on the hours of work done by each worker and the materials used. Proof of this agreement in shown by the clerk of works' signature on the contractor's day work sheets. Payment to the contractor consists of expenses in labour and materials, plus an agreed percentage of overheads and profit.

Dead leg   (1) A section of hot water draw-off pipe in which the contained water does not circulate, except when it is being drawn off. The water cools down between draw-offs and the dead leg wastes both heat and water. Regulations specify the maximum permissible length of dead leg allowed.

Dead leg   (2) A section of distribution pipe which cannot be emptied via a valve fitted to an appliance. The section of pipework can only be emptied via a drain valve fitted to this section of pipework.

Dead soft temper   The softness of copper sheet required for roof weatherings.

Deep seal trap   A trap with a 75 mm water seal.

Density   A word used to indicate the mass or weight of a body of known or stated volume, thus giving a convenient method of comparison.

De-oxidized   This is the removal of oxygen from the material, i.e. the use of flux in soldering or welding.

Deoxidation   The process of separating oxygen from a substance. This process is also called reduction and is a necessary part of the treatment of many metallic oxides and ores.

Deposited metal   The actual metal deposited by the filler rod during a welding operation.

Detergent   A surface active agent which combines good wetting power with the ability to remove grease and dirt from surfaces.

Development   A geometrical method by which the whole of the surface area of a solid may be set out in one plane, as on sheet metal or drawing paper.

Dew Point   Dew point is defined as the temperature at which the water vapour present in the air is just sufficient to saturate it (thus forming condensation when this air is in contact with colder surfaces).

De-zincification   The term used when the zinc used in the manufacture of brass (copper-zinc) is partically destroyed (becomes a porous mass).

Die   An internally threaded metal block for cutting male threads on pipes or tubes. The die block is held within a stock.

Diffusion   The tendency of molecules of a gas to spread out and fill a container or space, or to intermingle with another gas; many liquids also mix by diffusion.

Discharge pipe   A pipe which conveys the discharge from sanitary appliances.

Distillation   The process of converting a substance into vapour, leading off the vapour, and converting it back into liquid by cooling. It is used to separate a liquid from substances that are soluble in it, or to separate liquids of differing boiling points.

Distortion   A term used in welding to indicate the deformation of the metal being welded, due to unequal expansion and contraction of the heated metal causing bending and buckling.

Distribution pipe   A pipe conveying cold water from the storage cistern to the appliances.

Dog ear   A corner or angle formed in sheet metal roofwork by folding the material. No cutting takes place.

Dormer   A vertical opening formed in a roof slope to give light or ventilation to rooms formed in the roof space.

Double branch   A branch fitting used to connect two branch pipes or channels from opposite sides to a main pipe or channel.

Double collar   A pipe fitting usually in the form of a sleeve for joining the spigot or plain ends of two pipes in the same alignment.

Drain   A line of pipes normally underground, including all fittings and equipment such as inspection chambers, traps and gullies.

Drain chute   A drain pipe tapered in its upper half at the point where a drain pipe enters or leaves a manhole. The pipe is shaped in such a way so that rodding shall be easy.

Drain plug   An expanding stopper, used to seal a drain pipe usually during a test.

Drain valve   A valve fitted at the lowest point of a water system or on any section of pipework which is not self-draining, to facilitate complete emptying of the system.

Draught   The pressure difference at the base of a chimney between the air outside and that inside the chimney. This pressure difference (caused by the air inside being hotter and lighter) draws air up through the burner or fuel bed into the chimney.

Dresser   A hand tool made from hard wood or a plastics material, used for flattening or dressing of sheet materials such as a lead, copper or aluminium.

Drip   A step formed in flat or low pitched roof usually at right-angles to the direction of the fall.

Drip edge   The free lower edge of a sheet material covered roof which drips into a gutter.

Duckfoot bend   A bend having a foot formed integrally in its base, also known as a rest bend.

Duct   A closed passage way formed in the structure or underground to receive pipes and cables.

Earth   Refers to the facility of a system being connected to a general mass of earth, i.e. by a wire and rod or through a metal pipe laid in the ground.

Eaves   The lowest overhanging part of a sloping roof.

Eaves fascia   A board or edge nailed along the foot of the rafters. It is used to carry the eaves gutter and may also act as a tilting fillet.

Eaves gutter   A rainwater gutter along the eaves.

Edge preparation   The shaping of the edges of the parent metal prior to welding.

Elasticity   The term referring to the elongation of a substance and its ability to return to its normal position when the load has been removed. It is a relationship of stress to strain within the elastic range of the material.

Elbow   A sharp corner of change of direction in a pipe, usually a manufactured fitting.

Electric immersion heater   This is manufactured as either a single or dual element heater to be fitted in the hot storage vessel; it can be used as the sole means of heating or as a booster to the existing method.

Electrolysis   This is a detrimental action between dissimilar metal, Increased action is brought about if there is an electrical contact, i.e. earth. One metal acts as a cathode, the other as an anode.

Element   A material composed entirely of atoms of one kind. An element cannot be split into anything simpler than itself.

Emitter   The name given to a radiator or appliance emitting heat for space heating.

Enamel Vitreous enamel is a glass like surface attached by firing to cast iron or pressed steel articles such as baths or sink unit tops.

Epoxide resin A synthetic resin used for glueing metal or concrete.

Equilibrium Refers to equal pressures.

Erosion The wearing away of a surface.

Eutectic A term used in metallurgy in connection with the solidifying or setting of alloys. When alloyed in varying proportions, one particular combination of metals will give the lowest solidifying point. This is known as the eutectic point.

Evaporation The loss of moisture in vapour form from a liquid.

Expansion An increase in the size of a material or substance usually brought about by an increase in temperature.

Expansion joint A joint fitted in pipework to allow for the linear expansion of the pipe material when the water temperature is raised.

Fatigue Failure of a metal sheet or component brought about by repeated stress. It is related to changes in the crystal structure of the metal.

Feed cistern Supplies cold water to the sanitary appliances and also cold water to the hot water system.

Feldspar A material used in the manufacture of ceramics and enamels.

Ferrous This is an iron-based material.

Filler metal The material that is added to the weld pool to assist in filling the joint space. It forms an integral part of the weld.

Flames (A) *Neutral flame* A clean flame burning equal amounts of oxygen and acetylene. (B) *Oxidizing flame* A flame with an excess of oxygen. (C) *Carburizing flame* A flame with an excess of acetylene gas.

Flange A projecting flat rim which may be cast, screwed or welded on to a pipe, fitting or vessel, and is used for making a joint.

Flash-back arrestor A safety device fitted to a cylinder regulator (outlet).

Flexible joint A joint designed and made to permit small angular deflection and small changes in the length of a drain without loss of watertightness.

Float A buoyant device floating on a water surface and actuating a mechanism or valve by its response to rise or fall of the water surface.

Fluorescein A dye used in water for drainage pipeline tracing purposes.

Flushing cistern A cistern provided with a device for rapidly discharging the contained water and used in connection with a sanitary appliance for the purpose of cleansing the appliance and carrying away its contents into a drain.

Flux A substance used in soldering and/or welding to prevent oxidation.

Foul water Any water contaminated by soil, waste or trade effluent.

Fresh air inlet A fitting, usually with a hinged flap, used to allow air to enter a system of drainage.

Furred A term to describe pipes, boilers or components which have become encrusted with hard water lime or other salts deposited from the water heated in them.

Fuse Usually a small piece of wire in an electric circuit which melts when the current exceeds a certain value. The fuse is protection against short circuiting.

Fusible link A metal link made from a material with a low melting point and usually incorporated into an anti-fire device.

Fusion welding Welding of metals which are in a molten state without the application of pressure.

Gable The triangular part of the end wall of a building with a sloping roof.

Galena The principal ore of lead.

Galvanize A process used for protecting metals, usually steel. The process involves dipping the metal to be protected into molten zinc.

Gas circulator A gas-fired water heating appliance connected to a hot water cylinder or hot water tank from which the hot water can be drawn. The connections are made by flow and return pipes in which gravity circulation takes place.

Gas rate The supply of gas at the correct pressure to an appliance burner.

Gas welding A process of joining metals, using the heat of combustion of an oxygen/fuel gas mixture to melt and fuse together the edges of the parts to be welded, generally with the addition of a filler metal.

Gauze   Perforated sheet metal often fitted below burner ports and used to 'even out' the internal gas pressure and reduce the danger of 'back lighting' in the burner port.

Governor   A device which automatically controls the gas pressure in a pipeline.

Grease trap   A chamber incorporated into a drainage pipeline used for preventing grease from passing into the system of drainage.

Ground water   Water occurring naturally at or below the water table.

Gully   A drainage fitting or assembly of fittings used to receive surface water and/or the liquid discharge from waste pipes. The top may have a grating or sealed access cover.

Gutter   A channel for collecting surface rainwater.

Hard water   (A)   Temporary hard water is water that can be softened by boiling. The hardness is due to carbonates of lime and magnesium.
(B)   Permanent hard water is water that cannot be softened by boiling. The hardness is due to sulphates of lime and magnesium.

Haunching   Additional concrete support at the sides of a drain pipe above the foundation bedding.

Head of water   This is the height to which the water will rise in a pipe under atmospheric pressure, i.e. in a domestic water system it is the water level in the cistern.

Heating   See Systems.

Heel of bend   This is the back of the bend which is under tension during the bending operation.

Hopper head   A flat or angle-backed rainwater fitting used to collect discharge of rainwater.

Hose union   A fitting consisting of a coupling nut for screwing to the external threaded outlet of a hose union valve and a serrated tail for insertion into a hose pipe.

Humidity   The measure of the water vapour present in a gas; it is measured by a hygrometer or wet and dry bulb thermometer.

Hydraulic mean depth   A factor used in calculating the rate of flow of a liquid in a pipe or channel. It is obtained by dividing the cross-sectional area of the liquid by the length of the wetted perimeter of the pipe or channel.

Hydraulic gradient   The 'loss of head' in liquid flowing in a pipe or channel, expressed per unit length of the pipe or channel.

Hydraulic pressure   Fluid pressure.

Hydrometer   An instrument used to find densities of liquids.

Immersion heater   An electrically heated rod type heater inserted in the hot water storage vessel, the heater is usually thermostatically controlled.

Infiltration   The unintended ingress of ground water into a drain or sewer.

Inspection chamber   A covered chamber constructed on a drain or sewer so as to provide access to the drainage pipeline.

Insulation   The opposite of conduction. Insulators are bad conductors of electricity or heat (e.g. rubber, PVC, glass, wood, cork).

Integral   Means part of, i.e. an overflow in a wash basin, or an 'O' ring in a drain pipe joint.

Interceptor   A trap fixed on a drain to prevent the passage of sewer gases or vermin into the drain. It is normally fixed on the outlet side of the inspection chamber.

Interceptor (reverse action)   A trap designed to be fitted on the inlet side of an inspection chamber with a reverse clearing arm giving access to the drain against the direction of water flow.

Interstitial   Occurring within the tissues of the fabric or structure of a building.

Invar   A nickel/steel alloy. Its coefficient of expansion is very low and for this reason it is used in bimetallic types of thermostat and steel measuring tapes.

Invert   (A)   The lowest point of the inside of the pipe or channel.
(B)   The lowest point of the internal surface of a drain, sewer or channel.

Iron   A grey, magnetic metal which forms the basis of a wide range of different types of steel.

Isolating valve   Any valve so positioned that a part of the system can be isolated.

Isometric   A drawing method based on the principle that all vertical lines are drawn vertical while all horizontal lines are drawn at an angle of 30°. With this method several surfaces of the object can be exposed to view.

Jig   An accessory usually purposely made to assist in the carrying out of an activity.

Joint box   A box forming part of an electrical installation in which the cables are joined.

Joule   A unit of work It is named after J. P. Joule who conducted research on the mechanical equivalence of heat.

Junction   A fitting on a drainage pipeline designed to receive discharges from a branch drain.

Key   (A)   A roughening or indentation made on a surface to provide better adherence of another surface, filler or jointing medium.
(B)   A tool or device provided to enable a person to operate a valve, or device or to lift an access cover.

Killed spirits   A term applied to zinc chloride which is used as a flux for soldering zinc, copper and brass. The chloride of zinc is made by dissolving small pieces of zinc in hydrochloric acid.

Kinetic   Due to movement or motion.

Kite Mark   This is the British Standards Mark; it is placed on all items manufactured up to an approved standard (ensures good quality).

Knuckle bend   This is a short radius bend.

Lagging   Material used for thermal or acoustic insulation.

Latent heat   The amount of heat required to change the state of a substance from a solid to a liquid or from a liquid to a gas. The heat applied does not bring about a temperature rise.

Lead tack   A lead casting or piece of sheet lead soldered or welded on to lead pipe or lead sheet and used to secure the lead to its supporting surface.

Lead welding   A process of fusing together pieces of lead sheet or pipe. The basic process involves forming a small molten pool of lead, and adding further lead from a rod in order to reinforce the joint.

Lever   A rigid bar pivoted about a fixed point of support called the fulcrum.

Lint   A substance which may choke up pre-aerated gas burners, and is derived from dust and fibres in the room of a dwelling.

Lime   (A)   Carbonates of lime cause temporary hard water. When water containing these carbonates of lime is heated to a high temperature (approximately 70°C or above) the lime is deposited as a scale or fur in the boiler or flow, pipe.
(B)   Sulphates of lime cause permanent hard water. These sulphates of lime are not removed when the water is heated, the water therefore remains hard after boiling.

Longscrew   A piece of low carbon steel tube threaded externally at each end, one end having the thread sufficiently long to accommodate a backnut and the full length of a socket. It is used to join two pieces of steel tube, neither of which can be rotated. (It is also called a connector.)

Loop vent   (A)   The continuation of a discharge pipe which is taken up as a vent to connect back into the stack vent or other ventilating pipe.
(B)   The branch ventilating pipe, which, after being carried up above the flood level of the appliance it serves, is connected back into the vertical section of the discharge pipe as near to the branch discharge pipe as is practicable.

Magnet   Mass of iron or other metal which attracts or repels other similar masses and which exerts a force on a current-carrying conductor placed near it.

Main   The pipe which carries the public water supply.

Main contractor   A contractor who is responsible for the bulk of the work on a site, including the work of the sub-contractors.

Make good   To repair as new.

Mallet   A tool like a hammer with a wooden, hide, rubber or plastics head.

Mandrel   (A)   A cylindrical piece of hardwood which is pushed through a lead pipe to remove distortion.
(B)   A wooden tool (elongated bobbin) which is driven through a lead pipe to ensure uniformity of bore.

Manhole   A term which is often used to describe an inspection chamber, i.e. a chamber constructed on a drain or sewer to provide access for inspection, testing or the clearance of an obstruction. The usual interpretation is that a shallow chamber is termed an inspection chamber, and chambers of such depth that an operative cannot work from ground level are referred to as manholes.

Manipulative joint   A compression joint in which the ends of the copper tubes are opened out.

Manometer   An instrument used to measure differences in pressure. The usual form consists of a U-shaped tube mounted vertically and filled with a liquid to a predetermined level.

Mansard roof   A roof which is often gabled, and has on each side a relatively flat top slope and a steeper lower slope, usually containing dormers.

Margin   The exposed surface of a slate or tile.

Masking   A form of tape, applied as a protection.

Mastic   A plastic permanently water-proof material, which hardens on its surface so that it can be painted. Used for sealing gaps in expansion joints, gutters and flashings.

Mechanical advantage   The mechanical advantage of a machine is defined as the ratio of the load to the effort used.

Meniscus   The curved surface of a liquid when it touches a solid object rising above the liquid level.

Metal coating   A thin film of copper, nickel, cadmium, chromium, aluminium or zinc applied to corrodible metal surfaces.

Milled lead   Lead rolled into sheets from cast slabs.

Mixing valve   A valve used to mix hot and cold water to give a temperature controlled outflow; it may be thermostatically controlled.

Module   A unit of length by which the planning of structures can to some extent be standardized.

Molecule   The smallest part of a substance which can exist independently and still retain the properties of that substance.

Monel metal   A copper-nickel alloy containing approximately 67 per cent nickel and 33 per cent copper.

Mortar   A mixture of Portland cement, sand and water.

Multimeter   An instrument used for measuring electrical current, voltage, etc.

Multi-point water heater   A water heater (usually gas) which supplies hot water to several taps.

Nail   A fixing device. Clout nails are large headed. Nails of brass, aluminium or alloy or copper are used to provide fixings in certain types of roofing.

Nail punch   A short blunt steel rod which tapers at one end. It is struck by a hammer to drive a nail head below its surrounding timber surface.

Neoprene   The trade name for an American synthetic rubber which has excellent properties of non-inflammability.

Nipple   A short section of pipe threaded at each end.

Nominal size   A numerical designation of the size of a pipe, fitting, or other component which is a convenient round number approximately equal to a manufactured dimension.

Non-manipulative joint   A compression joint in which the ends of the copper tube are cut square and the internal and external burrs removed. Jointing is usually achieved with the aid of a cone, ring or olive which is compressed into the tube wall by the action of the joint nut being screwed on to the joint body.

North-light roof   A sloping roof having one steep and one shallow slope. The steeper slope is usually glazed and faces north.

Notch   A groove in a timber to receive a pipe.

Nozzle   The open ended portion of a valve draw-off valve or swivel arm from which water is discharged.

Offset   A double bend in a pipeline, formed so that the pipe continues in its original direction.

Ordinary Portland cement   A hydraulic cement made by heating to clinker in a kiln a slurry of clay and limestone.

Orifice   A small opening intended for the passage of a fluid.

Overcloak   In sheet metal roofing, that part of an upper sheet which laps over a lower sheet or undercloak at a drip or roll joint.

Overflow   An overflow is sometimes confused with a warning pipe. Some appliances have an overflow as an integral part of the fitment, the overflow water discharging direct into the waste outlet. All cisterns must have either an overflow pipe or a warning pipe or both. A warning pipe is an overflow pipe which discharges in an obvious position to warn the householder of some malfunction of the ball-valve. An overflow pipe can discharge into a drainage system provided the cistern is still fitted with a warning pipe.

Oxidation   This is an action brought about by the element oxygen, it is generally of a destructive nature, i.e. iron oxide (rusting).

Oxygen   A gas ever present in the atmosphere; *it is a supporter of combustion.*

Pantile   A single-lap tile *shaped* like an 'S' laid horizontally.

Parallel thread   A thread screwed to a uniform diameter. Used on mechanical connections such as bolts, but not generally on low carbon steel pipe fittings except running nipples and connectors. Compression fittings and valves make use of parallel threads.

Parapet   A low wall guarding the edge of a roof.

Parent metal   A metal or component to be welded.

Partially separate system   see Systems.

Patina   A thin, protective film of, sulphate which forms on metals exposed to air, particularly the green coating on copper or its alloys.

Penetration   This refers to the position of the jointing material in relation to the parent metal in a joint.

Performance test   A test for the stability of the trap seals in above or below ground drainage systems.

Perspex   A transparent acrylic resin.

Pet-cock   A small drain valve.

Physical change   A change not involving any chemical change in a body or substance. Freezing, expansion and magnetization are examples of physical changes.

Pictorial views   (A)   Isometric projection. In this projection all horizontal lines are drawn at an angle of 30° to the horizontal, all vertical lines remain vertical.

(B)   Axonometric projection In this projection all vertical lines are drawn vertical while all horizontal lines are drawn at an angle of 45°.

(C)   Planometric projection In this the horizontal lines are drawn horizontal while all vertical lines are drawn at an angle of 60° to the horizontal.

Pillar valve   This type of valve is fixed to the top surface of the appliance with the outlet above the highest possible water level in the fitment.

Pilot hole   A guiding hole, drilled in a material to form a route for a larger drill or bit.

Pilot light   A small gas flame used to ignite the main burners on a gas appliance.

Pipe cutter   A tool for cutting copper iron or steel pipes. Cutting is achieved by hard steel discs or wheels which bite into the pipe walls as the tool is revolved around the pipe.

Pipe hanger   A support for a suspended pipe such as a drainage pipeline run beneath a ceiling. The hanger usually consists of a metal tube or bar with provision for adjusting the length, a flange at upper end to provide fixing and a pipe ring or purpose-made metal strap at the lower end to carry the pipe.

Pipe ring   A ring-shaped clamp or bracelet made in halves for screwing or bolting together, which forms part of an assembly for supporting a pipe.

Pipe wrench   A heavy wrench with serrated jaws for gripping, screwing or unscrewing low carbon steel pipes and fittings.

Pitch   The ratio of the height to the span of a roof or its angle of inclination to the horizontal.

Pitcher tee   A tee on which the branch is swept into the main pipe with a gently curved turn.

Pitting   A defect in welding, consisting of little hollows brought about by the use of an incorrectly adjusted flame. These hollows are also known as blowholes.

Plasticizers   Substances added to plastics to modify their elastic properties without altering their chemical properties.

Plastics   A name commonly used to describe a group of materials including poly vinyl chloride, polystyrene, polythene, perspex, etc. Plastics are either thermosetting (those which harden once and for all time when heated), or thermoplastics (those which soften whenever they are heated.

Plastics memory   Plastics pipes when heated and bent, and most plastics mouldings, have a residual strain in them. When this strain is released, i.e. by heating, the pipe will try to revert to its original position. This is its plastics memory.

Pliers   A holding or gripping tool pivoted like a pair of scissors. Some types have blades for cutting thin wire built into the jaws.

Plug   (A)   A fitting used to seal off a pipe or section of pipeline, usually fitting into the bore of the pipe.

(B)   An electrical device intended for connection to a flexible cord or flexible cable.

Plug   A small threaded fitting used to screw into an internally threaded connection to seal it off.

Plug cock   A simple value in which the liquid or gas passes through a hole in a tapered plug. The valve is opened or closed by turning the plug through 90°.

Poly tetra fluoro ethylene (PTFE)   A plastics material which is used as a thin tape or pate as a jointing medium in pipe threads.

Poly vinyl chloride (PVC)   A vinyl resin which is impervious to water, oils and petrol and is particularly incombustible. Used for making gutters, soil and waste pipes, storage cistern, drainage components.

Polythene   Abbreviated from polyethylene, a chemically inert synthetic rubber used for making pipes and cisterns for cold water services.

Polyurethane   A plastics which in foamed form is used as an insulating material in cavity walls.

Portable electric tool   A hand-held tool, driven by an electric motor, i.e. an electric drill.

Pouring rope   Also called a running rope or squirrel's tail, usually made from fibre rope and secured with a steel spring loaded clip. Used for containing poured molten lead when jointing cast iron pipes in a horizontal position.

Pressed steel   A sheet steel which is hot pressed into plumbing components such as baths, sink units and flushing cisterns. The steel is protected by a layer of vitreous enamel.

Pressure   Defined as the force acting normally per unit area (here the word normally means vertically).

Pressure (bar)   The reading of the air pressure is recorded as so many bars (one bar is equivalent to one atmosphere).

Primary flow and return   The pipes in which water circulates between a boiler and a hot water storage tank or cylinder.

Propane   A colourless gas which burns in air to carbon dioxide and water.

Pump   A mechanical device for causing a liquid to flow.

Purging   The cleaning out of the system.

PVC (polyvinyl chloride)   This is a tough lightweight plastic material produced from oil.

Quarter bend   A 90° bend in a pipe. Other bends are proportional to this, a one-eighth bend being 45°.

Radiator   A sealed container, usually for water with tappings for pipework connections if part of a central heating system.

Radius   A straight line running from the centre of a circle to any point on the circumference. Its length is half that of the diameter.

Rainwater head   The enlarged entrance at the head of a rainwater pipe, often used as a collection point for other rainwater pipes.

Rainwater pipe   A pipe for conveying rainwater from a roof or other parts of a building.

Ramp   A short length of pipeline or channel laid at a steeper gradient than the adjoining portions.

Raising piece   A fitting for extending the height of a gully or of a rainwater shoe. A gully raising piece may have branch inlets.

Reducer   A fitting which enables a pipe diameter or socket connection to be reduced in bore.

Refractory   A substance which is not damaged by high temperature.

Reinforcement bead   The built-up part of a weld above the parent metal.

Relief valve   An additional ventilating pipe connected to a discharge pipe at any point where excessive pressure fluctuation is likely to occur.

Rest bend   A bend with a foot or web formed integrally on its base. It is used to support a vertical pipe or line of pipes and is also known as a duckfoot bend.

Return pipe   A pipe in a hot water system which conveys water back to the boiler, or a pipe in a secondary hot water circuit through which the water flows back to the hot storage vessel.

Ridge   The apex or highest point of a roof.

Rodding   A method of clearing obstructions or blockages from drains or sewers by the use of flexible rods which can be connected together and pushed into the pipeline. Various attachments are available which fit on to the feed end of the flexible rods to assist with the clearing operation.

Rodding eye   A cover, plate, plug or *cap* which when removed provides access to the inside of a pipeline.

Rose   Head or outlet of a shower fitting.

Rose's metal   An alloy which consists of bismuth, lead and tin. It has a low melting point (94°C).

Rusting   The corrosion of iron by the combined action of oxygen, carbonic acid and water.

Saddle   A drainage fitting used to connect a branch drain to a larger drain or sewer pipe.

Saddle clip   A fixing which passes round the front of a pipe and is screwed to the surface behind the pipe via two lugs which are part of the saddle.

Saddle piece   A piece of sheet weathering, formed to cover a joint at a vulnerable position, e.g. where a ridge meets an abutment wall or another roof.

Sal-ammoniac   (NH_4Cl, ammonium chloride) A flux used in soldering.

Sanitary appliance   A fixed soil appliance or waste appliance.

Scale   see Lime.

Sarking felt   Bituminous flax felt laid over rafters prior to battening and slating or tiling.

Scribe   To cut a line on the surface of a material with a sharp pointed tool.

Seal   The depth of water contained in a trap to prevent foul air passing through. The seal depth is measured from the water level down to the crown of the U-shaped part of the trap.

Sealing compound   A material used to fill and seal the surface of an expansion joint. It can be applied like a mastic from a pressure cartridge or gun.

Sealing plate   A cover and frame which fit into the socket of a drain pipe, drain fitting or gully top and which finishes level with the ground or floor surface.

Seam   A joint, fold or welt formed in sheet metal weatherings. A seamed edge is often formed at the front of an apron flashing to stiffen the edge, provide a more attractive finish, resist the possibility of capillary attraction between the roof and slates or surface and the flashing, or to form a safety edge so that the material is easier to handle and fix.

Seaming pliers   A pair of pliers with jaws specially shaped and extended for forming seams or welts.

Secondary circulation   A pipework circuit which supplies hot water to appliances which are located some distance from the hot water storage vessel. The circuit may have gravity or pumped circulation.

Secret gutter   A nearly hidden gutter, in which the gutter is concealed by the roof covering.

Secret tack   A strip of lead, soldered or welded on to the back of a lead sheet. The strip is passed through a slot in the roof boarding and screwed to the inside of it, providing a hidden fixing. Often used for securing vertical lead bays or panels.

Se-duct   A flue which is also used as an air intake for gas appliances in multi-storey buildings, enabling many appliances to be supplied from the same flue.

Service pipe   A water or gas pipe between the main and the premises receiving the supply.

Self-aligning pipe   A pipe which by means of the shape of the socket and spigot is naturally held in a straight line.

Self-cleansing velocity   The velocity of the flow of the liquid in the pipe or channel necessary to prevent the deposition of solids.

Self-siphonage   The extraction of the water from a trap by siphonage set up by the momentum of the discharge from the sanitary appliance to which the trap is attached.

Separate system   A drainage system in which foul water and surface water are conveyed in separate pipes.

Sewer connection   The length of pipe between the last inspection chamber on a drain or private sewer and the public sewer.

Shave hook   A hand tool used for shaving or cleaning the surface of lead pipes or sheet before soldering or welding.

Sheet   A description of aluminum, copper, lead or zinc which is thicker than foil, thinner than 6.35 mm and more than 450 mm wide. Foil is thinner than the size quoted. Strip is narrower than the width quoted.

Sherardizing   The coating of small iron or steel components with zinc by heating them with zinc dust in a revolving drum at about 350°C.

Shoe A rainwater pipe fitting located at the base of a stack, used to direct water away from the structure.

Single stack system A form of one-pipe system which may have waste water and soil discharging into it, all or most of the trap ventilating pipes are omitted.

Sink A waste water fitting usually located in a kitchen or in an area where food preparation occurs.

Skylight A window or light incorporated into the slope of a roof.

Sleeve A length of pipe built into the fabric of a building to allow for the passing through of a smaller pipe, thus allowing for movement and giving protection to the pipe and building fabric.

Slop sink A hopper-shaped sink, with a flushing rim and outlet similar to those of a WC pan, for the reception and discharge of human excreta.

Small bore system A system of heating pipework in which the sizes of circuit involved do not exceed 19 mm bore. Circulation is pumped.

Smoke test A method of tracing leaks in sanitation or drainage pipework. The test used is an air (pneumatic) test, and smoke is introduced to find a suspected leak after the air test has failed. Smoke should not be used on PVC pipework systems.

Soakaway A pit dug into permeable ground and lined or filled with hard-core to form a covered perforated chamber to which surface water is conveyed and from which it may soak away into the ground.

Socket (A) The end of a pipe, or pipe fitting, with an enlarged bore for the reception of the plain or spigot end of another pipe, or pipe fitting, for the formation of a spigot and socket joint. (B) A pipe fitting in the form of a short cylindrical pipe, threaded on its inner surface, used for joining together two pipes with externally threaded ends.

Socket reducer A reducing fitting, for use with discharge or drainage pipes, which fits inside the socket of the larger pipe, thus adapting the socket to receive the spigot of a smaller pipe.

Soffit The highest point of the internal surface of a pipeline at any cross-section.

Soaker A small piece of flexible material bent to form a watertight joint at an abutment between a roof and a wall.

Socket outlet A device connected to the electrical installation to enable a flexible cord or cable to be connected by means of a plug.

Soil The discharge from water closets, urinals, slop hoppers and similar appliances, i.e. any water containing human or animal excrement.

Soil pipe A pipe which conveys to a drain the waste matter from WC, urinal, slop hopper, etc.

Solder An alloy used for joining other metals.

Soldered dot A method of securing sheet lead to flat, vertical or sloping surfaces.

Soldering (A) *Soft* The joining of metals using a lead-tin solder. (B) *Hard* The joining of metals using copper-zinc solder (bronze).

Solvent A liquid capable of dissolving solids.

Solvent cement (weld) A liquid used in the jointing of plastics. It has the property of eating into the plastics material so producing a homogeneous bond.

Space nipple A short section of threaded pipe with a *space* between the threads. Nipples with a formed grip surface are called hexagon nipples.

Sparge pipe A perforated pipe used for flushing urinal stalls.

Specific heat The specific heat of a substance is the heat required to raise a unit mass of that substance through 1°C.

Spigot The plain end of a pipe, inserted into a socket to make a spigot and socket joint.

Splash lap That part of an overcloak of a roll or drip which extends on to the fiat surface of the next sheet.

Split collar A collar cut lengthwise and secured together with a steel band or clip. Used on flue pipes to assist with disconnection or removal of the appliance.

Spreader A fitting which connects to the flush pipe and which spreads the flushing water around the surface of a bowl or slab urinal.

Spring A steel coil used for bending copper pipes.

Spun lead Lead prepared in long strands and twisted together like yarn and used for cold caulking, (also known as lead wool).

Stack A name used to describe a vertical soil, waste or rainwater discharge pipe.

Stainless steel A steel containing chromium and nickel. It is highly corrosion-resistant and is used for waste and sanitary appliances and small diameter plumbing pipework.

Standing seam   A raised seam in flexible metal roofing usually running from ridge to eaves.

Step turner   a hardwood tool used for forming the turned step (usually 25 mm wide through 90°) on sheet metal.

Step flashing   A sheet weathering built into the horizontal joints of brickwork to make a watertight joint between a wall or chimney and the sloping part of a roof. The flashing steps down the thickness of a brick and a bed joint with the slope of the roof.

Stop valve   A control valve used for regulating the supply of water in a service pipe or pressure pipeline.

Storage cistern   An open-topped vessel used for storing a quantity of cold water to supply cold water draw-off points at a lower level. The cistern should have an airtight lid or cover to keep its contents clean and uncontaminated.

Storage water heater   A gas or electric water heater which heats a quantity of water and stores it for use at a later time.

Straight edge   A long piece of seasoned timber or metal with parallel and straight edges often used in conjunction with a spirit level.

S trap   A trap in which the outlet is vertical and in line with the trap inlet.

Strap boss   A fitting designed to clip around and on to a pipe, enabling a branch connection to be made to that pipe.

Stopper   An inflatable bag used to seal a drainage pipeline when testing is being carried out.

Step iron   A step usually of malleable iron which may be either straight for building into corners or u-shaped for building into the walls of manholes or inspection chambers to facilitate access.

Stratification   The term given to the layers of hot water in the hot water vessel.

Subsoil water   Water occurring naturally in the subsoil.

Substitute natural gas (SNG)   A gas manufactured either as a direct substitute for natural gas or as a means of providing additional gas to meet peak demands.

Surcharge   Excess flow in a drain or sewer when the normal flow capacity is exceeded.

Surface tension   The elastic skin effect present in liquids; forces between the molecules cause a state of tension in the surface of the liquid.

Surface water   Water that is collected from the ground, paved areas and roofs, etc.

Surface water sewer   A sewer intended to convey surface water only.

Surround   The concrete completely encasing a pipe.

Switch   A means of connecting or disconnecting an electrical supply.

Swan neck   A name for an offset fitting, particularly that between an eaves gutter and a rainwater discharge pipe.

Sweat   To unite or bond metal surfaces together by allowing molten solder to flow between them and adhere to their surfaces.

Sweep tee   A pitcher tee usually for copper or low carbon steel pipes in which the branch connection gently curves into the main run.

System of drainage   An arrangement of drains and sewers.

(A)   *Combined system* A system of drainage or sewers in which foul water and surface water are conveyed in the same pipes.

(B)   *Separate system* A system of drainage or sewers in which foul water and surface water are conveyed in separate pipes.

(C)   *Partially separate system* A modification of the separate system in which some of the surface water is admitted to foul water drains and sewers.

Systems hot water   (A)   *Direct* In this system the water that is heated in the boiler is drawn off at the appliance. Water in this system is continually being changed (causing excessive furring and/or corrosion).

(B)   *Indirect* In this system the water in the boiler is not changed, the same water being heated again and again and used to indirectly heat the water that is drawn off at the appliances.

(C)   *Space heating* A heating system using heat emitters (radiators) to heat the air in the building.

(D)   *Single feed* An indirect hot water system using only one cold water feed cistern and a patented hot water storage vessel. They are also known as self-venting systems.

(E)   *Small-bore* A domestic space heating system involving the use of small diameter pipes (usually 15 mm) with pumped circulation.

(F)   *Mini-bore* As for small bore system only in this instance the pipes are even smaller, i.e. 6 mm diameter.

(G)   *Unvented* The unvented hot water system is a comparatively new system to the United Kingdom, it is also known as the 'pressurized system', the cold water supply to the hot water system and the appliances being fed direct from the authorities' water main.

Tack   A form of fixing or cleat, used mainly for securing sheet weatherings.

Tang   The pointed end of a steel tool such as a file, rasp or wood chisel, which is driven into the wooden handle.

Tank   A closed straight sided storage vessel generally used for storing water or oil.

Tan pin   A conical shaped steel tool used for opening out the end of a copper pipe.

Tap   A screwed plug accurately threaded, made of hard steel and used for cutting internal threads.

Taper pipe   A drainage fitting which is used as an increaser or reducer in a pipeline.

Taper thread   A standard screwed thread used on pipes and fittings to ensure a watertight or gas tight joint.

Tariff   A scale of charges.

Tee   A fitting, which is a short section of pipe with three openings, one of which is a branch which is usually set at a right angle and located midway between the other two openings.

Temper   To toughen steels and non-ferrous metals by the application of controlled heat and cooling.

Temperature   The degree of hotness or coldness of a body.

Template   A full size pattern of metal or wood used for forming shapes or testing the accuracy of a manufactured component.

Tension   This is the stretching of the material.

Terminal   A cowl or open hood used to finish the end of a gas flue, etc.

Test piece   A portion of a weld joint removed from a welded structure or component that has been welded together according to a specified welding procedure.

Test specimen   A portion of a test piece that has been removed and prepared for testing.

Tests for drainage systems

(A)   *Air test* A test for soundness, carried out by applying air pressure internally.

(B)   *Ball test* A test for obstruction in a drain, in which a steel ball, less in diameter than the bore of the drain, is rolled through the drain.

(C)   *Colour test* A test for tracing the flow in a drain or sewer by introducing colouring matter (fluorescein).

(D)   *Mirror test* A method of inspecting the interior of a pipeline by means of light reflected by a mirror.

(E)   *Smoke test* A test for soundness, in which smoke is introduced under pressure to locate leaks in the pipeline.

(F)   *Water test* A test for soundness, applied by filling the pipeline with water.

Thermometer   Any device which measures temperature.

Thermal movement   Movement caused by expansion or contraction due to temperature.

Thermoplastic   Description of a synthetic resin or other material which softens on heating and hardens again on cooling.

Thermostat   A device, usually electrical, for sensing temperature rise or fall. They are often used for maintaining a constant temperature, making or breaking a circuit, which turns off or on the gas or electrical supply.

Tilting fillet   A small timber of triangular cross-section, used in roofworks to tilt slates or tiles slightly less steeply than the rest of the roof.

Tingle   A tab or cleat, usually a strip of aluminium, copper, lead or zinc used for holding down the edge of a metal weathering.

Tinman's solder   A fine solder containing more tin than wiping solder (grade D), so that its melting point is lower.

Tinning   Coating copper, brass, lead or other metals with a film of tin or tin alloy (solder).

Tin snips   Strong scissors used for cutting sheet metals.

Trap   A U-shaped bend in a pipe or sanitary appliance used to retain a quantity of water to prevent foul air or gases passing through.

Throat of a bend   The surface of the pipe on the inside of the bend.

Transformer   A device used in electrical work to achieve change of voltage.

Toe board   A scaffold board set on its edge at the side of a scaffold to prevent materials or tools being kicked or knocked off the working platform.

Toggle bolt   A fixing device which enables a sound fixing to be made to thin board such as hardboard or plasterboard.

Tommy bar   A loose bar or rod inserted into a hole in a box spanner or capstan to provide the leverage for turning it.

Torus roll   A horizontal wooden roll usually located at the intersection between the two slopes of a mansard roof, weathered with a sheet material.

Tubular trap   A fitting formed from pipe, thus ensuring an unrestricted flow through.

Twist drill   A hardened steel bit with helical cutting edges, used in electric or hand drills for cutting circular holes in metal or wood.

Two-pipe system   (A)   A sanitation system consisting of a soil stack and a waste water stack, each with its own anti-siphonage or vent pipe.
(B)   A heating circuit with a flow and return connected to each radiator or heat emitter.

'U' gauge   A glass or plastics 'U' tube half filled with water, one end of the 'U' being connected by flexible tube to a system of pipes under an air pressure test (a manometer).

'U' value   A thermal transmittance value, determined by experiment for 1 m^2 of a certain floor, wall, roof in a particular situation.

Undercloak   The lower layer or cover of sheet weathering material at a drip or roll which is covered by the overcloak.

Undercut   The term used to indicate the thinning of the metal adjacent to a weld.

Underlay   A layer of fire resistant material laid beneath sheet metal weatherings to assist movement and provide a degree of noise and heat insulation.

Union   A screwed pipe fitting, usually of brass or low carbon steel. It enables pipes or appliances to be quickly connected or disconnected.

Upstand   That part of a sheet weathering or flashing which turns up against a vertical surface without being fitted into it. The upstand is usually covered with a flashing.

uPVC   This abbreviation means 'unplasticized polyvinyl chloride' and is a plastics material used in the manufacture of pipes and fittings.

Valley   The intersection between two sloping surfaces of a roof.

Valve   A device to open or close a flow of liquid, gas or air, or to regulate a flow.

Valves   (A)   *Check valve* This valve will allow the flow of liquid in one direction only. It is also known as a non-return valve or anti-gravity flow valve.
(B)   *Economy valve* As the name implies its function is to save fuel. It is used to divert the flow of water in the hot water system to heat only the top half of the storage.
(C)   *Reflux valve* An automatic valve for preventing reverse flow, being open when flow is normal and closing by gravity when flow ceases.

Vent   An outlet for air or gases.

Ventilating pipe   A pipe connected to the drainage system carried up above the ground terminating above the highest window. Its function is to facilitate the removal of air from the system and maintain equilibrium of pressure so protecting the water seals.

Ventilation   This is the movement of air in a system or work place by forced or natural means.

Venturi   A tube which tapers to a narrow 'throat' and gradually widens out again to its original diameter.

Verdigris   Green basic acetate of copper formed as a protective patina over copper exposed to the air.

Verge   The edge of a sloping roof which overhangs a gable.

Vermiculite   A mica which is used as a light insulating aggregate.

Vice   A screwed metal or timber clamp, usually fixed to a workbench or tripod and used for holding materials while they are being worked.

Viscosity   The internal friction or drag in a fluid which tends to prevent its easy movement.

Vitreous china   A ceramic material used for the manufacture of sanitary appliances.

Vitreous enamel   A hard, smooth, glass-like surface attached to cast iron or pressed steel.

Vitrified clay   A vitreous form of clay used for the manufacture of drainage pipes and components.

Volt   The unit of electrical pressure, related to the units of flow (amperes) and power (watts), watts = volts × amperes. Electromotive force.

Voltage   The term used to designate an electrical installation.

Warning pipe   An overflow pipe from a cistern, which discharges water in a conspicuous position.

Washer   A flat ring made of rubber, plastics or fibrous composition used to form or make a seal between two surfaces. Alternatively, a flat ring made of metal.

Waste discharge pipe   A pipe carrying waste water from a waste appliance.

Waste disposal unit   An electrically driven rubbish grinder in domestic properties located in the kitchen sink, used for cutting and shredding kitchen waste. The waste is disposed of via a trap and discharge pipe into the drainage system.

Waste pipe   A pipe which conveys the discharge from a sanitary appliance used for ablutionary, culinary or drinking purposes.

Waste water   This is the discharge from appliances that do not contain human or animal excrement, i.e. wash basins, baths, sinks and similar.

Water (types) (A)   *Soft water* is a water that lathers readily. This is because it contains only a small amount of lime.

(B)   *Hard water* is a water with which it is difficult to obtain a lather. This is because it contains a large amount of lime.

(C)   *Temporary hard water* is a water that contains carbonates of lime together with carbon dioxide gas which dissolves the lime. This type of hardness can be softened by boiling.

(D)   *Permanently hard water* is a water that contains sulphates of lime which are dissolved in the water without the assistance of gases. This type of hardness cannot be softened by boiling.

Water hammer   This is a hammering sound in a water pipe caused by violent surges and the sudden arresting of the flow of the water.

Water level   An instrument for setting out or transferring levels on a building site. It consists of a rubber tube connecting two vertical glass tubes containing water. The level of water in one tube is the same as that in the other if there are no kinks or air locks in the rubber tube.

Water line   A line marked inside a cistern to indicate the highest level of water at which the supply valve should be adjusted to shut off the supply.

Water seal   The water in a trap which acts as a barrier to the passage of air through the trap.

Water softener   A chemical plant for treating water. It removes from the water the calcium and manganese salts which cause hardness. The most used types are base exchange softeners, and the soda-lime process.

Water table   The level of water in the ground or subsoil.

Water test   A form of drain, testing, used generally for new installations. The drain is filled with water and subjected to hydrostatic pressure for a specified period of time.

Water waste preventer   A cistern for flushing a WC or a slop sink (abbreviation WWP).

Watt   A unit of power.

Weathering   A sheet covering to the roof or part of the roof of a building.

Wedge   A tapered piece of metal used for securing a flashing in between two bricks or similar.

Welt   A folded joint or seam used in sheet metal roofing or weatherings.

Wetted perimeter   This is the line of contact between the pipe or channel and the liquid flowing through it (used in calculating flow through pipes).

Wing nut   A thumb screw nut, which has wings, enabling it to be turned by hand without the use of a spanner.

Wiped joint   A joint made with plumber's solder (grade D) between suitable materials, e.g. lead to lead, or lead to brass. The solder is moulded around the joint with a wiping cloth.

Wire balloon   A bulbous-shaped wire guard fitted at the top of a ventilating pipe to prevent birds nesting there.

Wood roll   A piece of timber, round-topped and tapering towards its base, fixed on to a roof surface to enable flexible metal materials to be lapped over it and jointed.

Wrench   A form of gripping tool or spanner, usually adjustable.

Zeolites   Minerals which are used in the base exchange process of water softening.

Graphical symbols and abbreviations

Graphical symbols and the appropriate abbreviations are used on building drawings in order that components installed within the building may be clearly identified. Figure 10.1 outlines the main symbols and abbreviations used to show the plumbing and heating components. (For a complete list refer to BS 308.)

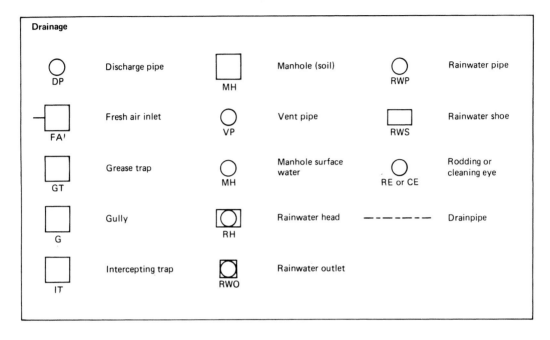

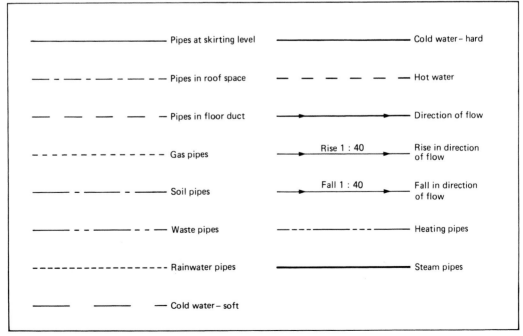

Figure 10.1 *Graphical symbols*

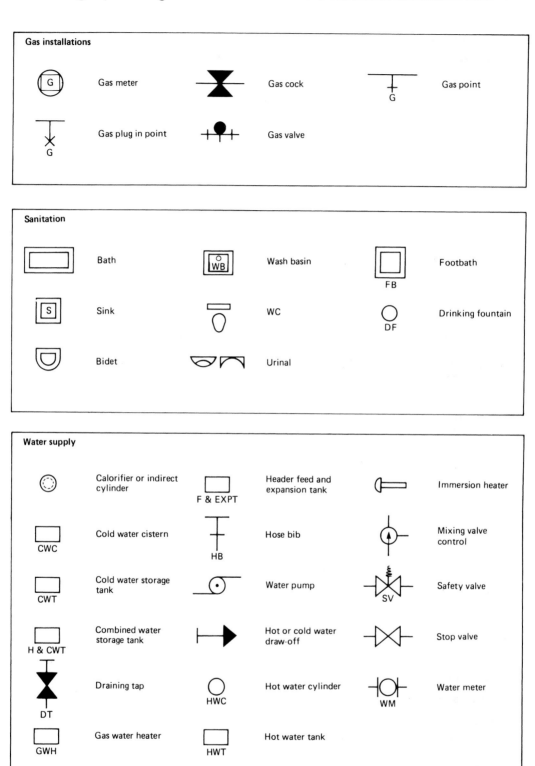

Figure 10.1 (continued)

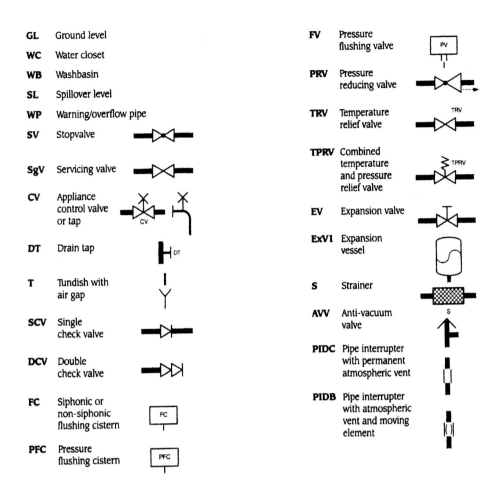

GL	Ground level
WC	Water closet
WB	Washbasin
SL	Spillover level
WP	Warning/overflow pipe
SV	Stopvalve
SgV	Servicing valve
CV	Appliance control valve or tap
DT	Drain tap
T	Tundish with air gap
SCV	Single check valve
DCV	Double check valve
FC	Siphonic or non-siphonic flushing cistern
PFC	Pressure flushing cistern
FV	Pressure flushing valve
PRV	Pressure reducing valve
TRV	Temperature relief valve
TPRV	Combined temperature and pressure relief valve
EV	Expansion valve
ExV1	Expansion vessel
S	Strainer
AVV	Anti-vacuum valve
PIDC	Pipe interrupter with permanent atmospheric vent
PIDB	Pipe interrupter with atmospheric vent and moving element

Table 10.1 *Units and symbols*

Item	Units	Symbol
Area	Square metres	m^2
Volume	Cubic metres	m^3
Capacity	Litres	I (better written as litres)
Mass	kg per cubic metre	kg/m^3
Flow rate	Litres per second	l/s
Volume flow	Cubic metres per second	m^3/s
Pressure of force	Newtons	N
Intensity of pressure	Newtons per square metre	N/m^2
Quantity of energy	Joules	J
Power	Watts	W

Appendix: British and European Standards and Codes of Practice

Some of the BS ENs listed below were still to be published at the time of weiting. On publication these BS ENs will often completely or partially replace a British Standard, which wil then by withdrawn or revised as appropriate. The exact title and possibly the number of parts may vary slightly when new BS ENs are actually published.

Users are advised to check the current status of these documents before using them, BSI have an excellent website where on-line checks can instantly be made. www.bsi-global.com

Hot and cold water supplies

Design

BS EN 805: Water supply – Requirements for systems and components outside buildings.

BS EN 806: Specification for installations inside buildings conveying water for human consumption.

BS EN 806 - 1: General.

BS EN 806 - 2: Design.

BS EN 806 - 3: Pipe sizing.

BS EN 806 - 4: Installation.

BS EN 806 - 5: Corrosion.

BS 6700: Specification for design, installation, testing and maintenance of services supplying water for domestic use within buildings and their curtilages.

Materials

BS 143 and 1256: Specification for malleable cast iron and cast copper alloy threaded pipe fittings.

BS EN 200: Sanitary tapware – General technical specification for single taps & mixer taps (nominal size 1/2˝) PN10 – Minimum flow pressure of 0.05MPa (0.5 bar).

BS EN 545: Ductile iron pipes, fittings, accessories and their joints for water pipelines – Requirements and test methods.

BS 699: Specification for copper direct cylinders for domestic purposes.

BS 853 - 1: Calorifiers and storage vessels for central heating and hot water supply.

BS 853 - 2: Tubular heat exchangers and storage vessels for building and industrial services.

BS EN 1057: Copper & copper alloys – Seamless round copper tubes for water and gas in sanitary & heating applications.

BS 1212: Float operated valves.

BS 1212 - 1: Specification for piston type float operated valves (copper alloy body) (excluding floats).

BS 1212 - 2: Specification for diaphragm type float operated valves (copper alloy body) (excluding floats).

BS 1212 - 3: Specification for diaphragm type float operated valves (plastic bodied) for cold water services only (excluding floats).

BS 1212 - 4: Specification for compact type float operated valves for WC flushing cisterns (including floats).

BS EN 1213: Building valves – Copper alloy stop valves for potable water in buildings – General technical specifications.

BS EN 1254: Copper and copper alloys – Plumbing fittings.

BS EN 1254 - 1: Fittings with ends for capillary solder.

BS EN 1254 - 2: Fittings with compression ends for copper tube.

BS EN 1254 - 3: Fittings with compression ends for plastic pipes.

BS EN 1254 - 4: Fittings combining other end connections with capillary or compression ends.

BS EN 1254 - 5: Fittings with short ends for capillary brazing to copper tubes.

BS EN 1452: Plastic piping systems for water supply – PVC-u

BS EN 1452 - 1: General.

BS EN 1452 - 2: Pipes.

BS EN 1452 - 3: Fittings.

BS EN 1452 - 4: Valves and ancillary equipment.

BS EN 1452 - 5: Fitness for purpose of the system.

BS EN 1452 - 6: Recommended practice for installation.

BS EN 1452 - 7: Assessment of conformity.

BS EN 1508: Water supply – Requirements for systems & components for the storage of water.

BS 1566: Copper cylinders for domestic purposes.

BS 1566 - 1: Specification for double feed indirect cylinders.

BS 1566 - 2: Specification for single feed indirect cylinders.

BS EN 1717: Protection against pollution of potable water in drinking water installations & general requirements of devices to prevent pollution by back flow.

BS 3198: Specification for copper hot water storage combination units for domestic hot water.

BS 4127: Specification for tight gauge stainless steel tubes, primarily for water applications.

BS 5154: Specification for copper alloy globe, globe stop and check, check and gate valves.

BS 6282: Devices with moving parts for the prevention of contamination of water by backflow.

BS 6282 - 1: Specification for check valves of nominal size up to and including DN54.

BS 6282 - 2: Specification for terminal anti-vacuum valves of nominal size up to and including DN54.

BS 6282 - 3: Specification for in-line anti-vacuum valves of nominal size up to and including DN42.

BS 6282 - 4: Specification for combined check and anti-vacuum valves of nominal size up to and including DN42.

BS 6283: Safety and control devices for use in hot water systems.

BS 6283 - 2: Specification for temperature relief valves for pressures from 1 bar to 10 bar.

BS 6283 - 4: Specification for drop-tight pressure reducing valves of nominal size up to and including DN 50 for supply pressures up to and including 12 bar.

BS 6572: Specification for blue polyethylene pipes up to nominal size 63 mm for below ground use for potable water.

Legionnaires' disease

BS 6068 - 4.12: Water quality – Microbiological methods. Detection and enumeration of legionella.

BS 7592: Methods for sampling for legionella organisms in water and related materials.

EN 13623: Chemical disinfectants and antiseptics – Bactericidal activities of products against legionella pneumophila – Test methods and requirements (phase2/step1).

Heating

BS BS3: Specification for vessels for use in heating systems.

BS EN 1264 - 1: Floor heating – Systems and components – Definitions and symbols.

BS EN 1264 - 2: Floor heating – Systems and components – Determination of the thermal output.

BS EN 1264 - 3: Floor heating – Systems and components – Dimensioning.

BS EN 1264 - 4: Floor heating – Systems and components. Installation.

BS 2767: Specification for manually operated copper alloy valves for radiators.

BS 2869: Specification for fuel oils for agricultural, domestic and industrial engines and boilers.

BS 4814: Specification for expansion vessels using an internal diaphragm, for sealed hot water heating systems.

B5 5250: Code of practice for control of condensation in buildings.

BS 5410 - 1: Code of practice for oil firing – Installations up to 45 kW output capacity for space heating and hot water supply purposes.

BS 5410 - 2: Code of practice for oil firing – Installations up to 45 kW and above output capacity for space heating, hot water and steam supply services.

BS 5440 - 1: Installation and maintenance of flues and ventilation of gas appliances of rated input not exceeding 70 kW net (1st, 2nd and 3rd family gases) – Specification for installation and maintenance of flues.

BS 54407 - 2: Installation and maintenance of flues and ventilation of gas appliances of rated input not exceeding 74 W net (1st, 2nd and 3rd family gases) – Specification for Installation and maintenance of ventilation of gas appliances.

BS5449: Specification for forced circulation hot water central heating systems for domestic premises.

BS 5970: Code of practice for thermal insulation of pipework and equipment in the temperature range of -100°C to +870°C.

BS 6351 - 1: Electric surface heating – Specification for electric surface heating devices.

BS 6351 - 3: Electric surface heating – Code of practice for the installation, testing and maintenance of electric surface heating systems.

BS 6798: Specification for installation of gas fired boilers of rated input not exceeding 70 kW net.

BS EN ISO 13370: Thermal performance of buildings – Heat transfer via the ground – Calculation methods.

Resource efficient design

BS 5918: Code of practice for solar heating systems for domestic hot water.

BS 8207: Code of practice for energy efficiency in buildings.

BS8211 - 1: Energy efficiency in housing – Code of practice for energy efficient refurbishment of housing.

BS EN 12975 - 1: Thermal solar systems and components Solar collectors – General requirements.

BS EN 12976 - 1: Thermal solar systems and components Factory made systems – Test methods.

Piped gas services

Design

BS 5482: Domestic butane and propane gas installations.

BS 5482 - 1: Specification for installations at permanent dwellings.

BS 5482 - 2: Installations in caravans and non-permanent dwellings.

Materials

BS EN 26: Gas fired instantaneous water heaters for the production of domestic hot water, fitted with atmospheric burners.

BS EN 88: Pressure governors for gas appliances for inlet pressures up to 20bar.

BS EN 89: Gas fired storage water heaters for the production of domestic hot water.

BS 143 and 1256: Threaded pipe fittings in malleable cast iron and cast alloy.

BS EN 297: Gas fired central heating boilers – Type B_{11} and B_{11BS} boilers fitted with atmospheric burners of nominal heat input not exceeding 70 kW.

BS EN 483: Gas fired central heating boilers – Type C boilers of nominal heat input not exceeding 70 kW.

BS EN 625: Gas fired central heating boilers – Specific requirements for the domestic hot water operation of combination boilers of nominal heat input not exceeding 70 kW.

BS 669: Flexible hoses, end fittings and sockets for gas burning appliances.

BS 669 - 1: Specification for strip-wound metallic flexible hoses, covers, end fittings and sockets for domestic appliances burning 1st and 2nd family gases.

BS 669 - 2: Specification for corrugated metallic flexible hoses, covers, end fittings and sockets for catering appliances burning 1st, 2nd and 3rd family gases.

BS EN 677: Gas fired central heating boilers – Specific requirements for condensing boilers of nominal heat input not exceeding 70 kW.

BS 746: Specification for gas meter unions and adaptors.

BS EN 1057: Copper & copper alloys – Seamless round copper tubes for water and gas in sanitary & heating applications.

BS EN 1254: Copper and copper alloys – Plumbing fittings.

BS EN 1254 - 1: Fittings with ends for capillary solder.

BS EN 1254 - 2: Fittings with compression ends for copper tube.

BS EN 1254 - 3: Fittings with compression ends for plastic pipes.

BS EN 1254 - 4: Fittings combining other end connections with capillary or compression ends.

BS 1387: Specification for screwed and socketed steel tubes and tubulars and for plain steel tubes suitable for welding or for screwing to BS 21 pipe threads.

BS 1552: Specification for open bottomed taper plug valves for 1st, 2nd and 3rd family gases up to 200 mbar.

BS 3016: Specification for pressure regulators and automatic change-over devices for liquefied petroleum gases.

BS 3212: Specification for flexible rubber tubing, rubber hose and rubber hose assemblies for use in LPG vapour phase and LPG/air installations.

BS 3554: Specification for gas governors.

BS 3554 - 1: Independent governors for inlet pressures up to 25 mbar.

BS3554 - 2: Independent governors for inlet pressures up to 350 mbar.

BS 3601: Specification for carbon steel pipes and tubes with specified room temperature properties for pressure purposes.

BS 3604: Steel pipes and tubes for pressure purposes – Ferritic alloy steel with specified elevated temperature properties.

BS 3604 - 1: Specification for seamless and electric resistance welded tubes.

BS 3604 - 2: Specification for longitudinally arc welded tubes.

BS 3605: Austenitic stainless steel pipes and tubes for pressure purposes.

BS 3605 - 1: Specification for seamless tubes.

BS 3605 - 2: Specification for longitudinally welded tube.

BS 4089: Specification for metallic hose assemblies for liquid petroleum gases and liquefied natural gas.

BS 4161: Gas meters.

BS 4161 - 8: Specification for electronic volume correctors.

BS 4250: Specification for commercial butane and commercial propane.

BS 7281: Specification for polyethylene pipes for the supply of gaseous fuels.

BS 7336: Specification for polyethylene fusion fittings with integral heating elements(s) for use with polyethylene pipes for the conveyance of gaseous fuels.

BS 7838: Specification for corrugated stainless steel semi-rigid pipe and associated fittings for low pressure gas pipework of up to 28mm.

Sanitary plumbing and drainage

Design

BS EN 752: Drain and sewer systems outside buildings.

BS EN 752 - 1: Generalities and definitions.

BS EN 752 - 2: Performance requirements.

BS EN 752 - 3: Planning.

BS EN 752 - 4: Hydraulic design and environmental considerations.

BS EN 752 - 5: Rehabilitation.

BS EN 752 - 6: Pumping installations.

BS EN 752 - 7: Maintenance and operations.

BS EN 858: Installations for separation of light liquids (oil & petrol).

BS EN 858 - 1: Principles of deign, performance, testing, marking and quality control.

BS EN 1085: Wastewater treatment – vocabulary.

BS EN 1091: Vacuum sewerage systems outside buildings.

BS EN 12109: Vacuum Drainage in Buildings.

BS EN 1295: Structural design of buried pipelines under various conditions of loadings – General requirements.

BS EN 1610: Construction and testing of drains and sewers.

BS EN 1671: Pressure sewer systems outside buildings.

BS EN 1825 Part 2: Installations for separation of grease.

BS 6297: Code of Practice for design and installation of small sewage treatment works.

BS 6465: Sanitary installations.

BS 6465 - 1: Code of practice for scale of provision, selection and installation of sanitary appliances.

BS 6465 - 2: Code of practice for space requirements for sanitary appliances.

BS EN 12056: - Gravity drainage systems inside buildings.

BS EN 12056 - 1: General performance requirements.

BS EN 12056 - 2: Sanitary pipework, layout and calculation.

BS EN 12056 - 3: Roof drainage, layout and calculation.

BS EN 12056 - 4: Wastewater lifting plants, layout and calculation.

BS EN 12056 - 5: Installation and testing, instructions for operation and use.

BS EN 12566: Small wastewater treatment plants less than 50 pt.

Materials

BS 65: Specification for vitrified clay pipes, fittings and ducts, also flexible mechanical joints for use solely with surface water pipes and fittings.

BS EN 124: Gully tops and manhole tops for vehicular & pedestrian areas – Design requirements, type testing, marking, quality control.

BS EN 274: Sanitary tapware – Waste fittings for basins, bidets & baths General technical specification.

BS EN 295: Specification for vitrified clay pipes and fittings and pipe joints for drains and sewers.

BS EN 295 - 1: Requirements.

BS EN 295 - 2: Quality control and sampling.

BS EN 295 - 3: Test methods.

BS EN 295 - 4: Requirements for special fittings, adaptors and compatible accessories.

BS EN 295 - 5: Requirements for perforated vitrified clay pipes and fittings.

BS EN 295 - 6: Requirements for vitrified clay manholes.

BS EN 295 - 7: Requirements for vitrified clay pipes and joints for pipe jacking.

BS EN 329: Sanitary tapware – Waste fittings for shower trays – General technical specifications.

BS EN 411: Sanitary tapware – Waste fittings far sinks – General technical specifications.

BS416 Part 1: Discharge and ventilating pipes and fittings, sand-cast or spun in cast iron – Specification for spigot and socket systems.

BS 437: Specification for cast iron spigot and socket pipes and fittings.

BS EN 476: General requirements for the components used in discharge pipes, drains and sewers for gravity systems.

BS EN 588: Fibre cement pipes for drains and sewers.

BS EN 588 - 1: Pipes, joints & fittings for gravity systems.

BS EN 588 - 2: Manholes & inspection chambers.

BS EN 598: Ductile iron pipes, fittings, accessories and their joints for sewerage applications – Requirements and test methods.

BS EN 607: Eaves gutters and fittings made from PVC-U – Definitions, requirements and testing.

BS EN 612: Eaves gutters and rainwater down-pipes made from metal sheet – Definitions, classification, requirements and testing.

BS EN 773: General requirements for components used in hydraulically pressurised discharge pipes, drains and sewers.

BS 882: Specification for aggregates from natural sources for concrete.

BS EN 877: Cast iron pipes and fittings, their joints and accessories for the evacuation of water from buildings – Requirements, test methods and quality assurance.

BS EN 1115: Plastic piping systems for underground drainage & sewerage under pressure – Glass reinforced thermosetting plastics (GRP) based on unsaturated polyester (UP).

BS EN 1115 - 1: General.

BS EN 1123: Pipes and fittings of longitudinally welded hot-dip galvanised steel pipes with spigot and socket for wastewater systems.

BS EN 1123 - 1: Requirements, testing, quality control.

BS EN 1123 - 2: Dimensions.

BS EN 1124: Pipes and fittings of longitudinally welded stainless steel pipes with spigot and socket for wastewater systems.

BS EN 1124 - 1: Requirements, testing, quality control.

BS EN 1124 - 2: System S – Dimensions.

BS EN 1124 - 3: System X – Dimensions.

BS 1247: Manhole steps.

BS 1247 Part 1: Specification for galvanised ferrous or stainless steel manhole steps.

BS 1247 Part 2: Specification for plastic encapsulated manhole steps.

BS 1247 Part 3: Specification for aluminium manhole steps.

BS EN 1253: Gullies for buildings.

BS EN 1253 - 1: Requirements.

BS EN 1253 - 2: Test methods.

BS EN 1253 - 3: Quality control.

BS EN 1253 - 4: Access covers.

BS EN 1254: Copper and copper alloys – Plumbing fittings.

BS EN 1254 - 1: Fittings with ends for capillary soldering or capillary brazing to copper tubes.

BS EN 1254 - 2: Fittings with compression ends for use with copper tubes.

BS EN 1254 - 4: Fittings combining other end connections with capillary or compression ends.

BS EN 1293: General requirements for components used in pneumatically pressurised discharge pipes, drains and sewers.

BS EN 1329: Plastics piping systems for soil and waste discharge (low and high temperature) within the building structure – Unplastisized polyvinyl chloride (PVC-U)

BS EN 1329 - 1: Specification for pipes, fittings and the system.

BS EN 1329 - 2: Guidance for assessment of conformity.

BS 1387: Specification for screwed and socketed steel tubes and tubulars and for plain steel tubes suitable for welding or for screwing to BS 21 pipe threads.

BS EN 1401: Plastic piping systems for non-pressure underground drainage & sewerage – Unplastisized polyvinyl chloride (PVC-U).

BS EN 1401 - 1: Specification for pipes, fittings and the system.

BS EN 1401 - 2: Guidance for assessment of conformity.

BS EN 1433: Drainage channels for vehicular and pedestrian areas – Classification, design & testing requirements, marking and quality control.

BS EN 1444: Fibre-cement pipelines – Guide for laying and on site work practices.

BS EN 1451: Plastic piping for soil and waste discharge (low and high temperature) within the building structure – Polypropylene (PP).

BS EN 1451 - 1: Specification for pipes, fittings and the system.

BS EN 1451 - 2: Guidance for the assessment of conformity.

BS EN 1453: Plastic piping systems with structured wall pipes for soil and waste discharge (low and high temperature) within the building structure – Unplasticized polyvinyl chloride (PVC-u).

BS EN 1453 - 1: Specifications for pipes and the system.

BS EN 1453 - 2: Guidance for the assessment of conformity.

BS EN 1455: Plastic piping for soil and waste discharge (low and high temperature) within the building structure Acrylonitrile–butadiene-styrene (ABS).

BS EN 1455 - 1: Specifications for pipes, fittings arid the system.

BS EN 1455 - 2: Guidance for the assessment of conformity.

BS EN 1456: Plastic piping systems for buried and above ground drainage and sewerage under pressure – Unplasticized polyvinyl chloride (PVC-u).

BS EN 1456 - 1: Specification for piping components and the system.

BS EN 1462: Brackets for eaves gutters – Requirements and testing.

BS EN 1519: Plastic piping for soil and waste discharge (low and high temperature) within the building structure Polyethylene (PE).

BS EN 1519 - 1: Specifications for pipes, fittings and the system.

BS EN 1519 - 2: Guidance for the assessment of conformity.

BS EN 1565: Plastic piping for soil and waste discharge (low and high temperature) within the building structure – Styrene copolymer blends (SAN + PVC).

BS EN 1565 - 1: Specifications for pipes, fittings and the system.

BS EN 1565 - 2: Guidance for the assessment of conformity.

BS EN 1566: Plastic piping for soil and waste discharge (low and high temperature) within the building structure Chlorinated polyvinyl chloride (PVC-c).

BS EN 1566 - 1: Specifications for pipes, fittings and the system.

BS EN 1566 - 2: Guidance for the assessment of conformity.

BS EN 1566: Plastics piping systems for non-pressure underground drainage and sewerage – Polypropylene (PP)

BS EN 1852 - 1: Specifications for pipes, fittings and the system.

BS EN 1852 - 2: Guidance for the assessment of conformity.

BS 2494: Specification for elastomeric seals for joints in pipework and pipelines.

BS 3868: Specification for prefabricated drainage stack units in galvanised steel.

BS 3943: Specification for plastic waste traps.

BS 4514: Unplasticized PVC soil and ventilating pipes of 82.4 mm minimum mean outside diameter, and fittings and accessories of 82.4 mm and of other sizes – Specification.

BS 4660: Thermoplastics ancillary fittings of nominal sizes 110 and 160 for below ground gravity drainage and sewerage.

BS 5911: Precast concrete pipes, fittings and ancillary products.

BS 5911 - 2: Specification for inspection chambers.

BS 5911 - 100: Specification for unreinforced and reinforced pipes and fittings with flexible joints.

BS 5911 - 103: Specification for prestressed non-pressure pipes and fittings with flexible joints.

BS 5911 - 110: Specification for ogee pipes and fittings (including perforated).

BS 5911 - 114: Specification for porous pipes.

BS 5911 - 120: Specification for reinforced jacking pipes with flexible joints.

BS 5911 - 200: Specification for unreinforced and reinforced manhole and soakaways of circular cross section.

BS 5911 - 230: Specification for road gullies and gully cover slabs.

Sanitaryware

BS EN 31: Pedestal wash basins – Connecting dimensions

BS EN 32: Wall hung wash basins – Connecting dimensions

BS EN 33: Pedestal WC pans with close coupled cistern – Connecting dimensions.

BS EN 34: Wall hung WC pans with close coupled cistern – Connecting dimensions.

BS EN 35: Pedestal bidets over rim supply only – Connecting dimensions.

BS EN 36: Wall hung bidets over rim supply only – Connecting dimensions.

BS EN 37: Pedestal WC pans with independent water supply – Connecting dimensions.

BS EN 38: Wall hung WC pans with independent water supply – Connecting dimensions.

BS EN 80: Wall hung urinals – Connecting dimensions.

BS EN 111: Specification for wall hung hand rinse basins – Connecting dimensions.

BS EN 232: Baths – Connecting dimensions.

BS EN 251: Shower trays – Connecting dimensions.

BS EN 695: Kitchen sinks – Connecting dimensions.

BS EN 997: WC pans with integral trap.

Pumps and pumping

Materials

BS 5257: Specification for horizontal and end suction centrifugal pumps (16 bar).

BS EN 12050: Wastewater lifting plants for buildings and sites – Principles of construction and testing.

BS EN 12050 1: Lifting plants for wastewater containing faecal matter.

BS EN 12050 2: Lifting plants for faecal-free wastewater.

BS EN 1205043: Lifting plants for wastewater containing faecal matter for limited applications.

BS EN 12050 A 4: Non-return valves for faecal-free wastewate and wastewater containing faecal matter.

Fire protection services

Design

BS 5306: Fire extinguishing installations and equipment on premises.

BS 5306 - 0: Guide for the selection of installed systems and other fire equipment.

BS 5306 1: Hydrant systems, hose reels and foam inlets.

BS 5306 - 2: Specification for sprinkler systems.

BS 5306 - 3: Maintenance of portable fire extinguishers Code of practice.

BS 5306 - 4: Specification for carbon dioxide systems.

BS 5306 - 5.1: Halon systems – Specification for halon 1301 total flooding systems.

BS 5306 - 5.2: Halon systems – Halon 1211 total flooding systems.

BS 5306 - 6.1: Foam systems – Specification for low expansion foam systems.

BS 5306 - 6.2: Foam systems – Specification for medium and high expansion foam systems.

BS 5306 - 8: Selection and installation of portable fire extinguishers – Code of practice.

BS 5588: Fire precautions in the design, construction and use of buildings.

BS 5588 - 0: Guide to fire safety codes of practice for particular premises/applications.

BS 5588 - 1: Code of practice for residential buildings.

BS 5588 - 4: Code of practice for smoke control using pressure differentials.

BS 5588 - 5: Code of practice for firefighting stairs and lifts.

BS 5588 - 6: Code of practice for places of assembly.

BS 5588 - 7: Code of practice for the incorporation of atria in buildings.

BS 5588 - 8: Code of practice for means of escape for disabled people.

BS 5588 - 9: Code of practice for ventilation and air conditioning of buildings.

BS 5588 - 10: Code of practice for shopping complexes.

BS 5588 - 11: Code of practice for shops offices, industrial, storage and other similar buildings.

Materials

BS 336: Specification for fire hose couplings and ancillary equipment.

BS EN 1254: Copper and copper alloys – Plumbing fittings.

BS EN 1254 - 1: Fittings with ends for capillary solder.

BS EN 1254 - 2: Fittings with compression ends for copper tube.

BS EN 1254 - 4: Fittings combining other end connections with capillary or compression ends.

BS 1387: Specification for screwed and socketed steel tubes and tubulars and for plain steel tubes suitable for welding or for screwing to 5521 pipe threads.

BS 3601: Specification for carbon steel pipes and tubes with specified room temperature properties for pressure purposes.

BS 3604: Steel pipes and tubes for pressure purposes Ferritic alloy steel with specified elevated temperature properties.

BS 3604 - 1: Specification for seamless and electric resistance welded tubes.

BS 3604 - 2: Specification for longitudinally arc welded tubes.

BS 3605: Austenitic stainless steel pipes and tubes for pressure purposes.

BS 3605 - 1: Specification for seamless tubes.

BS 3605 - 2: Specification for longitudinally welded tube.

BS EN 671: Fixed fire fighting systems – Hose systems.

BS EN 671 - 1: Hose reels with semi-rigid hose.

BS EN 671 - 2: Hose systems with lay-flat hose.

BS EN 671 - 3: Maintenance of hose reels with semi-rigid hose and hose systems with-lay flat hose

BS 750: Specification for underground fire hydrants and surface box frames and covers.

BS 1635: Graphic symbols and abbreviations for fire protection drawings.

BS 3169: Specification for first aid reel hoses for fire brigade purposes.

BS 3251: Specification – Indicator plates for fire hydrants and emergency water supplies.

BS 5041: Fire hydrant systems equipment.

BS 5041 - 1: Specification for landing valves for wet risers.

BS 5041 - 2: Specification for landing valves for dry risers.

BS 5041 - 3: Specification for inlet breechings for dry riser inlets.

BS 5041 - 4: Specification for boxes for landing valves for dry risers.

BS 5041 - 5: Specification for boxes for foam inlets and dry riser inlets.

Steam and condensate

BS 845 - 1: Methods for assessing thermal performance of boilers for steam, hot water and high temperature heat transfer fluids – Concise procedure.

BS 845 - 2: Methods for assessing thermal performance of boilers for steam, hot water and high temperature heat transfer fluids – Comprehensive procedure.

BS 2486: Recommendations for treatment of water for steam boilers and water heaters.

BS 5122: Specification for rubber hoses for low pressure and medium pressure saturated steam.

BS 5292: Specification for jointing materials for installations using water, low pressure steam or 1st, 2nd and 3rd family gases.

BS 5342: Specification for rubber hoses for high pressure saturated steam.

BS 54107/2: Code of practice for oil firing – Installation of 40 kW and above output capacity for space heating, hot water and steam supply services.

BS 6023: Glossary of technical terms for automatic steam traps.

BS 6068 - 6.7: Water quality – Sampling – Guidance on sampling of water and steam in boiler plants.

BS 6759 - 1: Safety valves – Specification for safety valves for steam and hot water.

BS EN 26553: Specification for marking of automatic steam traps.

BS EN 56554: Specification for face to face dimensions for flanged automatic steam traps.

BS EN 26704: Classification of automatic steam traps.

BS EN 27841: Methods for determination of steam loss of automatic steam traps.

BS EN 27842: Methods for determination of discharge capacity of automatic steam traps.

Pipework Expansion

BS 4618 - 3.1: Recommendations for the presentation of plastic design data – Thermal properties - Linear thermal expansion.

BS 6129 - 1: Code of practice for the selection and application of bellows expansion joints for use in pressure systems Metallic bellows expansion joints.

BS EN 26801: Rubber or plastic hoses – Determination of volumetric expansion.

Mechanical Ventilation

BS 5720: Code of practice for mechanical ventilation and air conditioning in buildings.

BS EN 13141 - 1 7: Ventilation for buildings – Performance testing of components/products for residential ventilation Part 7: Performance testing of a mechanical supply and exhaust ventilation units (including heat recovery) for mechanical ventilation systems intended for single family dwellings.

Designing for the disabled

BS 5588 - 8: Fire precautions in the design, construction and use of buildings – Code of practice for means of escape for disabled people.

PD 6523: Information on access to and movement within and around buildings and on certain facilities for disabled people.

BS 8300: Design of buildings and their approaches to meet the needs of disabled people – Code of practice.

Swimming pools

BS 6785: Code of practice for solar heating systems for swimming pools.

BS 8007: Code of practice for design of concrete structures for retaining aqueous liquids.

BS 8110 1: Structural use of concrete – Code of practice for design and construction.

Electrical earthing and bonding

BS 951: Electrical earthing – Clamps for earthing and bonding. Specification.

BS 7671: Requirements for electrical installations – IEE Wiring Regulations – Sixteenth edition.

Miscellaneous

BS 1387: Specification for screwed and socketed steel tube and tubulars and for plain end steel tubes suitable for welding or for screwing to BS 21 pipe threads.

BS 1710: Specification for identification of pipelines and services.

BS 4800: Schedule of paint colours for building purposes.

Index